Catalytic and Biocatalytic Methods for the Efficient Synthesis
of Biologically Relevant Non-Proteinogenic Amino Acids

Catalytic and Biocatalytic Methods for the Efficient Synthesis of Biologically Relevant Non-Proteinogenic Amino Acids

Dissertation

Zur Erlangung des mathemathisch-naturwissenschaftlichen Doktorgrades
"Doctor rerum naturalium"
der Georg-August-Universität Göttingen

Im Promotionsprogramm
Catalysis for Sustainable Synthesis (CaSuS)

vorgelegt von
Anke Lemke
aus Flensburg

Göttingen, 2013

Bibliografische Information der Deutschen Nationalbibliothek

Die Deutsche Nationalbibliothek verzeichnet diese Publikation in der Deutschen Nationalbibliografie; detaillierte bibliografische Daten sind im Internet über http://dnb.d-nb.de abrufbar.
1. Aufl. - Göttingen : Cuvillier, 2014
  Zugl.: Göttingen, Univ., Diss., 2013

Die vorliegende Arbeit wurde in der Zeit von Oktober 2009 bis November 2011 am Institut für Organische und Biomolekulare Chemie der Georg-August-Universität Göttingen und in der Zeit von Dezember 2011 bis Juni 2013 am Department Chemie der Universität Paderborn unter der Leitung von Prof. Dr. Christian Ducho angefertigt.

<u>Betreuungsausschuss:</u>

Prof. Dr. C. Ducho, Universität Paderborn, Department Chemie

Prof. Dr. U. Diederichsen, Institut für Organische und Biomolekulare Chemie

Prof. Dr. D. B. Werz, Technische Universität Braunschweig, Institut für Organische Chemie

<u>Mitglieder der Prüfungskommission:</u>

Referent: Prof. Dr. C. Ducho, Universität Paderborn, Department Chemie

Korreferent: Prof. Dr. U. Diederichsen, Institut für Organische und Biomolekulare Chemie

<u>Weitere Mitglieder der Prüfungskommission:</u>

Prof. Dr. D. Stalke, Institut für Anorganische Chemie

Prof. Dr. C. Steinem, Institut für Organische und Biomolekulare Chemie

Dr. I. Siewert, Institut für Anorganische Chemie

Prof. Dr. D. B. Werz, Technische Universität Braunschweig, Institut für Organische Chemie

Tag der mündlichen Prüfung: 19. November 2013

"In the fields of observation, chance favors only those minds which are prepared." [1]

— LOUIS PASTEUR

Table of Contents

1 Introduction

1.1 Antibiotics

Penicillin G[®], *Vancomycin*[®], and *Amoxicillin*[®] are drug names that probably everybody has read on a prescription once, or at least heard of. These trade names of antibiotics illustrate only 3 of the 80 different therapeutically established antibiotics in Germany.[2] Thus, it is not surprising that antibiotics still belong to the most prescribed drugs. The estimated total consumption of antibiotics in human medicine in Germany lies between 250-300 t per year. A quantum of 85% were prescribed in the outpatient care (GERMAP 2008).[3] In 2002, the global market was estimated at US $25 billion, and 6 antibiotics were topping US $1 billion each.[4] The mainstay of antibiotic scaffolds, including the 6 bestsellers, was represented by 3 structural classes for decades: the β-lactams (e.g. penicillin, amoxicillin, ceftriaxone), the macrolides (e.g. azithromycin, clarithromycin) and the quinolones (e.g. ciprofloxacin, levofloxacin). Additional structural classes are described by sulfonamides, polyketides (e.g. tetracyclin), glycopeptides (e.g. vancomycin), streptogramines, oxazolidines and lipopetides (e.g. daptomycin).[5-6]

The incredibly fast rise of antibiotics started in the early 1940s when the demand for a cure against wound infections increased during the Second World War. The first clinically used antimicrobial drug was the prominent penicillin, which was isolated from the mold *Penicillium notatum* by *Florey* and *Chain* in 1940.[7] It had already been discovered by *Fleming* in 1928 though.[8] Although the new hyped wonder drug was the beginning of the golden age of microbiology and led to the discovery of numerous new antimicrobially active substances,[9] the known curative effect of molds goes way back to Chinese medicine in 1000 B.C., when mold-cultured soybean-curd was used to cure skin infections.[10] Moreover, the Middle American Indians used to treat purulent inflammations with wild mushrooms,[10] and in the Hashemite Kingdom of Jordan red soil is still used today as an inexpensive alternative to antibiotics.[11] But despite these and other numerous anecdotes about the occurrences of antibiotic-like effects from all over the world,[12] the first scientific report of antimicrobial activity did not appear until 1877 when *Pasteur* observed an antagonism between bacteria in the same culture medium.[13]

But when do we call a compound antibiotic, and how do antibiotics work? The word 'antibiosis' was first used by *Vuillemin* in 1889 to describe the concept of one active organism destroying the life of the passive one to maintain its own life.[14] The word 'antibiotic' was primarily defined by *Waksman* with 'antibiosis' meaning the inhibition of growth of one organism by another.[15] While in *Waksman's* definition of 'antibiotic', only secondary metabolites of bacteria and certain mushroom species were included, nowadays, the term also covers entirely synthetic compounds without a natural lead structure, which are used in the treatment of bacterial infectious diseases.

Antibacterially active substances can mainly act in two different ways: bacteriostatic (limiting the growth of bacteria) or bactericidal (killing the bacteria). In this process, they can target different essential functions in bacteria, like the bacterial cell wall synthesis, DNA- and RNA-replication, bacterial protein synthesis and folic acid metabolism (see chapter 2.1).[4] The most successful antibiotics of our time hit these four classical targets only, and they only offer a few different modes of action. In contrast, there are approximately 200 conserved essential proteins in bacteria. The number of the currently exploited targets is very small though, and it still bears potential for the discovery of new antibiotic lead structures and modes of action.[16]

1.2 Antibiotic Resistance and the Critical Need for New Antibiotics

Due to the expansive discovery of new antibiotics in the 20th century, life threatening diseases or epidemics like cholera, diphtheria, pneumonia, or tuberculosis seemed more or less under control. But if the major infectious diseases of the 20th century are defeated, where is the need for the laborious and expensive discovery of new antibiotics? While the pharmaceutical industry found their answer in a decreasing antibiotic research, which resulted in an innovation gap between 1960 and 2000 (Figure 1.1),[4,12] the threat of antibiotic resistance aroused. The first report of antibacterial resistance, represented by the penicillinase, appeared simultaneously with the introduction of penicillin into the clinical market.[17]

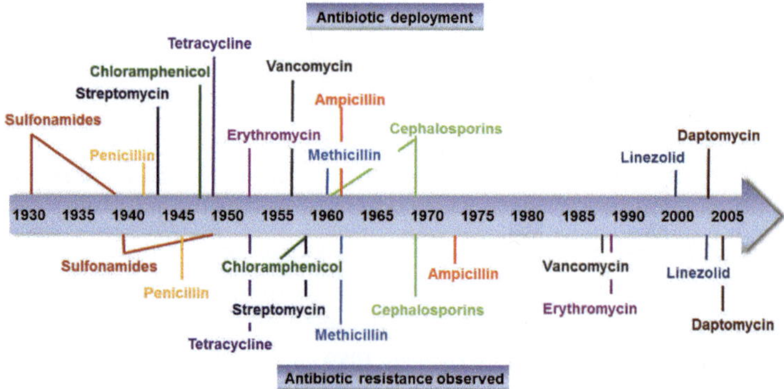

Figure 1.1: Timeline of the antibiotic deployment and the emerged resistance (adapted from: A. E. Clatworthy et al., *Nat. Chem. Biol.* **2007**, *3*, 541).[18]

Penicillinase is an enzyme of the β-lactamase family and hydrolyzes benzylpenicillin. This enzymatic transformation of penicillin describes only one of the six major mechanisms of antibiotic resistance (see also Figure 2.1). The other five mechanisms are a modification of the molecular target,[19] an active efflux from the cell interior,[20] a reduced entry of the compound due to alterations of penetration barriers,[21] finding a bypass for the inhibited sequence, or an increase of the production of the target metabolite.[16] The reason for the development of such mechanisms is evolutionary pressure, which leads to the selection of the resistant mutated organisms. Not only was the introduction of penicillin directly followed by its observed resistance. Nearly every antibiotic that had been clinically introduced entailed a significant resistance only a few years later (Figure 1.1).[18] Approximately 70% of the hospital-acquired infections are resistant to one or more antibiotics.[2,18] In this context, the occurrence of more and more multidrug-resistant bacteria is another alarming fact. The methicillin-resistant *Staphylococcus aureus* (MRSA)[22] and the vancomycin-resistant *Enterococcus faecium* (VRE) are only the most prominent examples. Multidrug resistance results from the ability of bacteria to transfer genetic resistance traits not only among their own, but also among different species. This horizontal gene transfer is mainly accomplished through transduction (via bacteriophages), conjugation (via plasmids and conjugative transposons), and transformation (via incorporation into the chromosome of DNA or plasmids).[12,23]

The very rapidly spreading multidrug resistance presents a new challenge to modern antibiotic research, in particular because in the last decades only a few new antibiotics were introduced into the clinical market.[24] These therapeutic agents are mainly based on the established scaffolds and are therefore missing new modes of action and the potential for clinical use. This is particularly critical if we think of extensively drug resistant (XDR) bacteria like *Mycobacterium tuberculosis*, which can be resistant to most antibiotics with classical targets.[25] These so called 'superbugs' are sparking fear for public health issues, especially since the latest news headlines announce: "*Superbugs: A ticking time bomb*" (CBS News),[26] or "*Europe 'losing' superbugs battle*" (BBC News Health).[27] Titles from fanatic blogs, such as "*How medicine is killing us all: Antibiotics, superbugs and the next global pandemics*" (NaturalNews.com),[28] even illustrate an apocalyptic scenario. But are we really on a critical edge of a post-antibiotic era, and the pharmaceutical industry is leading us there as some newspapers and maniacs want to make us believe?

Of course, due to a saturated antibiotic market in the 1960s and a high financial risk within antibacterial drug discovery, most of the large pharmaceutical companies and many biotechnology companies have left the area, leaving a gap in innovative strategies for today. But with the achievement of the first completely sequenced bacterial genome in 1995, several companies moved back into the antibacterials' area, hoping to unveil a whole treasure trove of new targets by a genomic-derived, target-based screening approach. Despite the promising identification of a whole new bunch of essential genes, the desired breakthrough could not be achieved, missing optimized lead structures suitable for clinical trials.[29] This example shows that there is not only an urgent need for new antibiotics, but also for new or improved approaches in antibacterial discovery research.

The limited scientific resources are not the only big problem in the fight against antibiotic resistance though. Antibiotic misuse, poor hygienic standards in hospitals, and non-controllable release into the environment by households and animal husbandry are further problems to solve. Starting points here are standardized regulations in antibiotic usage and hygiene and awareness of educational responsibility.[30-31] Reports like the GERMAP 2008 about the antibiotic usage in Germany[3] and the 'Deutsche Antibiotika-Resistenzstrategie' (DART) are first steps into the right direction.[32] From 2008 to 2014, the DART initiative has invested €80 million in different projects for antibiotic research. Such investigations are hopefully enhancing the attractiveness for companies to reenter the antibiotic market or to increase their efforts in the antimicrobial research field. In the context of the combat against biological terrorism, the US Health and Human Services Department announced an agreement with *GlaxoSmithKline* this summer (2013), with the potential of as much as $94 million funding under the so called '10 x '20 Initiative'. The aim is to create a "sustainable global drug research and development enterprise with the power in the short term to develop 10 new, safe, and efficacious systemically administered antibiotics by 2020".[33] Having the effect of the antibiotic research progress of the Second World War in mind, such governmental support might enormously push the antibiotic innovation field.

Looking at the FDA approvals of antimicrobial drugs since 1998 (Table 1.1),[33] the scenario of a post-antibiotic era with antibiotic research as a lost cause seems unlikely. However, only a few of the approved drugs display new mechanisms, such as linezolide (oxazolidines) with binding to the ribosomal 50S subunit, daptomycin (lipopeptides) by membrane depolarization, followed by a disturbed bacterial ionic management, tigecyclin (glyciclines) with the ability to subvert common tetracycline resistance, and telavancin (glycopeptides) exerting an additional mode of action by depolarization and permeabilization of the bacterial membrane.[34]

Antibacterial	Year of FDA approval	Novel mechanism?
Rifapentine	1998	No
Quinupristin/Dalfopristin	1999	No
Moxifloxacin	1999	No
Gatifloxacin	1999	No
Linezolid	2000	Yes
Cefditoren pivoxil	2001	No
Ertapenem	2001	No
Gemifloxacin	2003	No
Daptomycin	2003	Yes
Telithromycin	2004	No
Tigecyclin	2005	Yes
Doripenem	2007	No
Telavancin	2009	Yes
Ceftaroline fosamil	2010	No

Table 1.1: Systemic antibacterial drug approvals since 1998.[33]

Since 2007, only two systemic drugs have been approved. There are still potential candidates waiting for approval, but the pipeline is nearly dried nowadays. But where do new antibiotic lead structures come from, and what are promising strategies in antimicrobial research? Developing new antibiotics on established scaffolds with highly specifically improved properties is basically a good strategy. Nevertheless, these derivatives are more likely to enter a resistance stage that is critical for clinical use and are therefore still reflecting the urgent need for innovative strategies.

There are multiple strategies away from traditional antibiotic pathways, which are currently discussed, such as the design of antibacterial peptides,[35] modulating immunity, targeting virulence factors, the use of bacteriophages, prodrug concepts, and new delivering methods, to name only a few.[12,34] New creative strategies always bear the danger of failing though, as the 'genomic disaster' has proven.[29] Especially with new targets at hand, a large screening library with chemical diversity is essential. In this context, natural product leads still bear the highest potential of furnishing an active antimicrobial because they offer cellular permeability and defined structures for specific interactions with proteins. These in-nature-established pharmacodynamic properties are hard to design de novo. 'Back to the roots' is another promising guiding principle right now.[36] Searching for new natural lead structures in underexploited new areas, such as marine sediments or old areas like soil from all over the world, can deliver a whole bunch of structurally diverse compounds. In addition, quite old microbiological methods, like whole-cell assays, are rediscovered in modern research. Even the screening and improvement of fermentation conditions could deliver secondary metabolites that are only produced under a certain pH value or by the addition of special nutrients.[37] In this context, the activation of biosynthetic gene clusters, which are silent under standard laboratory conditions, is another interesting strategy.[38] This approach, also known as genome mining, bears the potential to discover numerous novel secondary metabolites.

Overall, there are still several promising antibacterial drugs with novel mechanisms of action in development.[37] But as new types of targets are emerging, it is more important than ever to take advantage of the large portfolio of biotechnological techniques, better knowledge of bacterial genetic function and the chemistry of natural lead structures, and to create an effective interdisciplinary field of research. This should lay the foundations for the development of effective antibiotics addressing new or clinically non-established targets. Promising key features of these antibacterially active compounds are often non-proteinogenic amino acids offering an extended chemical diversity and promising prospects in drug development.

2 Literature Review

2.1 Clinically Established Antibiotic Classes and Their Targets

In order to understand the mode of action of certain antimicrobial substances and antibiotic classes, the classical antibiotic targets will be explained: interference with bacterial cell wall synthesis (a), DNA- and RNA-replication (b), bacterial protein synthesis (c) and folic acid metabolism (d) (Figure 2.1).[4]

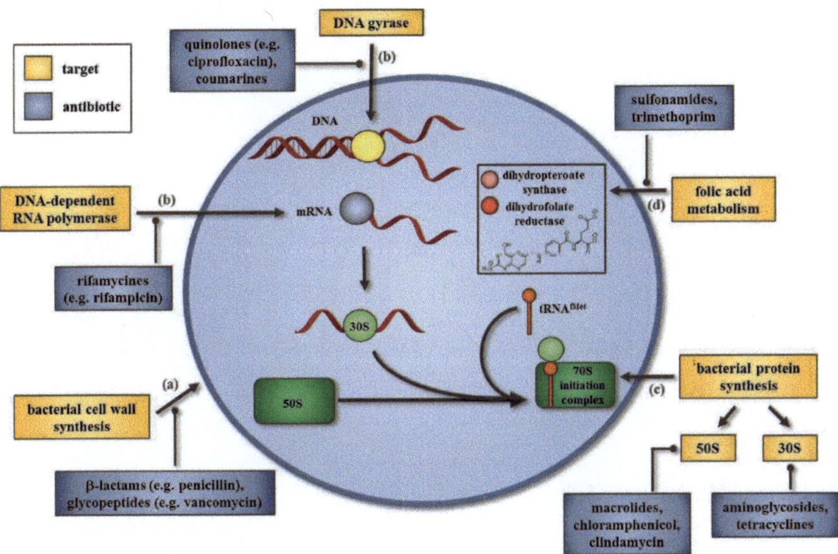

Figure 2.1: The four classical targets of established antibiotics (adapted from: K. Lewis et al., *Nat. Rev. Drug Discov.* **2013**, *12*, 371).[16]

In contrast to mammalian cells, folic acid biosynthesis is essential for bacterial survival. By acting as alternative substrates, structural analogues like sulfonamides inhibit the key enzyme dihydropteroate synthase. Another antibiotic targeting folate metabolism is trimethoprim, a 2,4-diaminopyrimidine, with the ability of selective inhibition of dihydrofolate reductase, which catalyzes the reduction of dihydrofolate to the crucial cofactor tetrahydrofolate.[39] As distinct binding sides of the ribosomal RNA subunits 50S and 30S provide the potential to block multiple steps in protein biosynthesis, numerous antimicrobial compounds, such as aminoglycosides, macrolides, tetracyclines, chloramphenicol, and clindamycin, were developed using this mode of action.[39-42] Quinolones, a well-established class of antibiotics (e.g. ciprofloxacin), inhibit DNA gyrase, which controls the topology of DNA. While promising DNA supercoiling inhibitors with new modes of action (e.g. coumarines) are still in the pipeline,[43] the DNA-dependent RNA polymerase, a major enzyme in the regulation of prokaryotic gene expression, remains quite underexploited in contrast, as it is only targeted by one class of clinically used antibiotics, the rifamycines (e.g. rifampicin).[39] The most established target for antibiotics in clinical use still remains the formation of the bacterial cell wall. In this context, the β-lactam antibiotics (e.g. penicillin) and glycopetides (e.g. vancomycin), two of the 'early-stage' antibiotic classes, are two of the first known inhibitors. Peptidoglycan, the essential cell wall building block, is a three-dimensional meshwork of peptide-cross-linked sugar polymers.[44] The interference with its biosynthesis or structure results in the loss of cell shape and integrity, followed by an inevitable bacterial death.[45] Peptidoglycan biosynthesis can be divided into three distinctive stages (**i**)-(**iii**) (Figure 2.2).

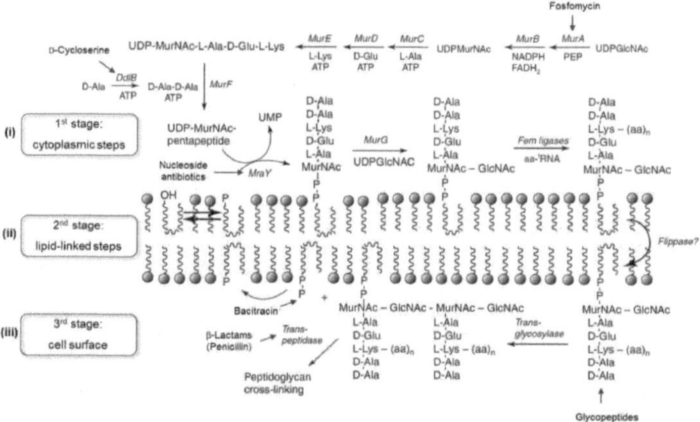

Figure 2.2: The peptidoglycan biosynthetic pathway showing sites of action of natural product inhibitors (adapted from: T. D. H. Bugg et al., *Trends Biotechnol.* **2011**, *29*, 168).[46]

The cytoplasmic steps which represent the first stage of peptidoglycan biosynthesis (**i**), lead from UDP-*N*-acetyl-glucosamine (Glc*N*Ac) to the peptidoglycan monomer UDP-*N*-acetyl-muramic acid (Mur*N*Ac) pentapeptide.[47-49] The second stage (**ii**) can be described as lipid-linked steps which involve the lipid carrier undecaprenyl phosphate.[50] After linking the monomer to the membrane, it is transferred to the cell surface, where, in a third stage (**iii**), it is polymerized and cross-linked to the pre-existing cell wall.[47,51] While well-established antibiotics like glycopeptides and β-lactams inhibit the late extracellular steps of the third stage, there are only few known antibacterial drugs in clinical use (D-cycloserine, fosfomycin, bacitracin) that target biosynthetic steps in stage one and two.[39] Due to this fact, bacterial cell wall assembly still remains an attractive target for the development of new antimicrobial compounds.[46]

2.2 MraY as Target for Nucleoside Antibiotics

One target in the peptidoglycan biosynthetic pathway for which no clinically established inhibitor exists is the formation of lipid I from UDP-Mur*N*Ac pentapeptide (Figure 2.3). This reaction is catalyzed by the enzyme translocase I (MraY), which is integrated into the membrane.

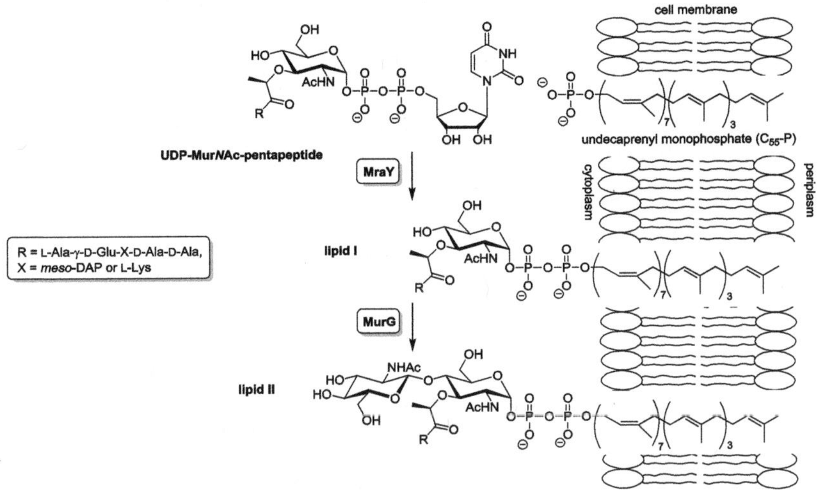

Figure 2.3: Biosynthesis of the peptidoglycan precursors lipid I and lipid II (adapted from: A. Matsuda et al., *J. Med. Chem.* **2011**, *54*, 8421).[52]

Early mechanistic studies showed that MraY utilizes the two substrates undecaprenyl phosphate and UDP-Mur*N*Ac pentapeptide. The transferase activity is fully reversible, and it also catalyzes an exchange between UMP and UDP-Mur*N*Ac pentapeptide, suggesting a two-step mechanism: (1) the formation of an enzyme-substrate complex under the release of UMP and (2) further reaction to lipid I.[53-55] In vivo, this reversible two-step reaction is coupled to the subsequent lipid II formation, which is catalyzed by the transferase MurG.[56] Although the encoding gene *mraY* had already been identified in 1991[57] and a first topology model suggesting MraY as an integral trans-membrane protein had been postulated in 1999,[58] it was not until 2004 that MraY was placed in the focus of scientists as an attractive target for antibiotic research. Due to the significant overexpression, purification, and characterization of MraY by *Mengin-Lecreulx* and coworkers,[56] a first model for the active site could be developed.[59] In 2011, *Bernhard* and coworkers were able to express MraY by cell-free methods.[60] Recently, the crystal structure of MraY was reported by Chung et al.[61] Although the binding and inhibition mechanism of MraY has still not been completely understood, this achievement will pave the way for the discovery of potent inhibitors. As MraY is essential for bacterial viability and only present in bacteria, compounds inhibiting this enzyme are quite attractive for the development of antibiotics. As the inhibition of lipid I formation represents a whole new target, there are no MraY inhibitors on the clinical market yet.

There are different natural products known which have the ability to inhibit MraY. A quite interesting group is represented by nucleoside antibiotics.[51,62] This class of structurally complex compounds shares a uridine-based motif. The nucleoside building block is connected to structurally different scaffolds via the 5'-C, depending on the compound set (Figure 2.4).

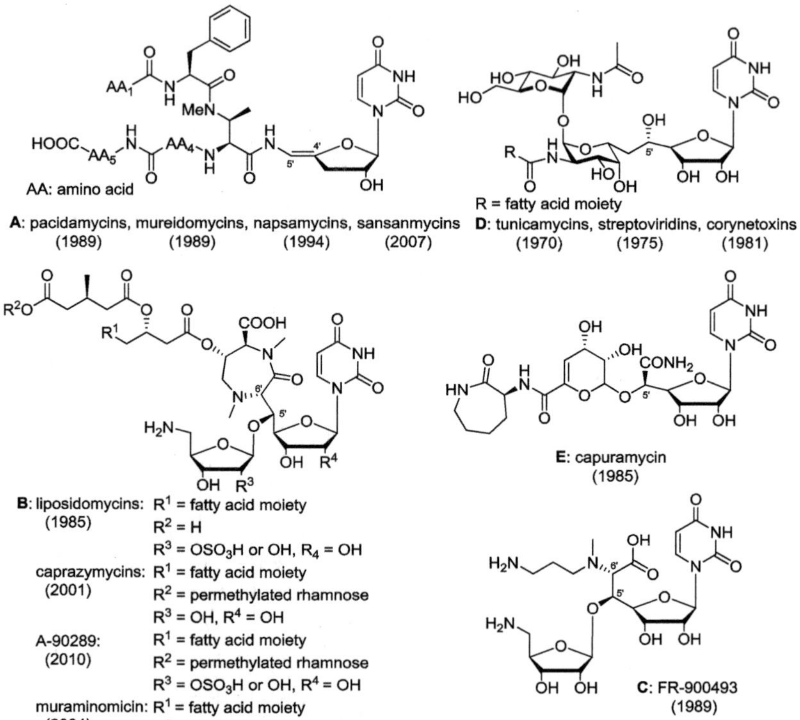

Figure 2.4: Structures of the nucleoside antibiotic groups **A-E** inhibiting MraY. The year of the first isolation is given in parentheses.

The pacidamycins,[63] mureidomycins,[64] napsamycins,[65] and sansanmycins[66] (**A**) show a 3'-deoxyuridine core structure attached to an *N*-methyl 2,3-diaminobutyric acid (DABA) residue via a unique 4',5'-enamide linkage. Via the α- and β-nitrogen atom of DABA, different peptide chains are connected to the core, depending on the substance class. The liposidomycins,[67] caprazamycins,[68] muraminomicins,[69] and A-90289[70] (**B**) stand out as they display a diazepanone ring. They show a glycyl-uridine core structure, are additionally glycosylated in the 5'-position to an aminoribose unit, and they comprise a fatty acid side chain. The muraminomicin nucleosides are 2'-deoxygenated. The structurally related muraymycins[71] (**C**) comprise the same glycyl-uridine motif, 5'-glycosylated amino ribose, and a fatty acid side chain. Another feature is their aminopropyl-linked urea peptide moiety. What is particularly noticeable is the non-proteinogenic amino acid epicapreomycidine, a cyclic arginine derivative. In contrast, the FR-900493[72] nucleoside comprises no peptide chain at all, and the lipopetidyl unit is also missing. The tunicamycins,[73] streptoviridins,[74] and corynetoxins[75] (**D**) are a structurally different group of nucleoside antibiotics due to an additional *N*-acetylglucosamine (Glc*N*Ac) unit and a unique 11-carbon dialdose sugar (tunicamin). The last class of compounds is represented by capuramycin (**E**).[76] This uracil nucleoside owes its name to its caprolactam substituent, which was proven to be essential for biological activity. Polyoxins[77] and nikkomycins[78] are nucleoside antifungals, and they show no inhibition of MraY, but of chitin synthase. They are only named in this context due to their structural relationship to groups **B** and **C**, which is based on the interesting glycyl-uridine moiety.

The tunicamycin group does not only inhibit MraY, but also a wide variety of membrane-associated glycosyl transferases, not only in bacteria but in eukaryotic glycoprotein maturation as well.[79] With respect to this fact, especially the non-toxic high-carbon nucleosides **B** and **C** have the potential to act as antibiotic lead structures. Therefore, they are quite attractive for structure-activity relationship (SAR) studies. In this context, the unique 5'-modifications, a glycine unit linked via a 5'/6'-C-C bond and the aminoribofuranoside attached via an *O*-glycosidic bond, might be quite interesting structural features. The most active member of the muraymycin family, muraymycin A1, shows staphylococcal (MIC 2-16 µg/mL), enterococcal (MIC 16 to >64 µg/mL), and Gram-negative (MIC 8 to >64 µg/mL) activity. First SAR studies with semi-synthetic derivatives showed the importance of a free amino group of the aminoribose for muraymycin activity.[80] The amino group might be involved in ion or hydrogen bonding to the putative Asp-Asp active site in the cytoplasmic loop (CL) 2 of the enzyme. Further studies with truncated muraymycin derivatives were also conducted.[81] A recent study revealed that a drastic simplification of the muraymycin structure does not result in an activity loss and is, therefore, feasible. The impact of the lipophilic side chain on antibacterial activity is very large. Derivatives with a fatty acid moiety exhibited good activity against MRSA and VRE (Gram-positive bacteria). This means that the lipid residue most likely contributes to membrane permeability. The SAR of the urea-peptide moiety indicates an interaction with the carbohydrate recognition (CR) domain in CL 5 of MraY (Figure 2.5).[52] Such a postulated binding model would leave the uridine unit free to compete with the MraY substrate UDP-Mur*N*Ac pentapeptide and result in inhibition. This proposed model represents only a hypothetical scheme. However, it could give new impulses for the development of effective inhibitors as potential new generation antibiotics.

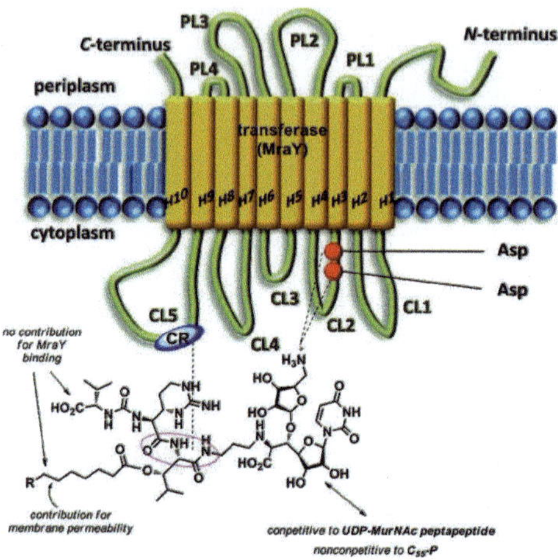

Figure 2.5: Schematic hypothetical model for the inhibition of MraY by muraymycins (from: A. Matsuda et al., *J. Med. Chem.* **2011**, *54*, 8431).[52]

2.3   Pathways in the Biosynthesis of Nucleoside Antibiotics

In 1996, the nikkomycin gene cluster was identified and sequenced.[82] This achievement gave first insights into the putative biosynthetic pathways of biologically active nucleosides, but it was not until 2009 that *Gust* and coworkers identified and analyzed the caprazamycin gene cluster.[83] Through homology studies, putative functions could be assigned to the identified genes and enzymes. Further mutant disruption studies made a more detailed investigation possible and revealed essential genes for the biosynthesis, which led to a first hypothetical pathway for the biosynthesis of nucleoside antibiotics.[83] Until now, the biosynthetic gene clusters of capuramycins (2009),[84] pacidamycins (2010),[85-86] liposidomycins (2010),[87] A-90289 (2010),[88] tunicamycins (2010),[89] napsamycins (2011),[90] muraymycins (2011),[91] sansanmycins (2012, draft genome sequence),[92] and most recently muraminomicins (2013)[93] were identified and analyzed. Structurally related compounds are most likely to share biosynthetic pathways, and the comparison of the gene clusters can help to find the missing piece in the often complex biosynthetic jigsaw puzzle.

2.3.1 Biosynthetic Formation of the Nucleoside Moiety

The origin of the glycyl-uridine has been a key question in the biosynthesis of high-carbon nucleoside antibiotics for a long time (see **B** and **C** in Figure 2.4). As metabolic labeling studies with tunicamycin-producing strains showed that uridine **1** is incorporated directly,[94] it was proposed that the biosynthesis of caprazamycins and related compounds also starts with an oxidation of uridine **1** to 5'-aldehyde **2** with a subsequent aldol addition of glycine. In the biosynthesis of pacidamycins in *Streptomyces coeruleorubidus* it could be demonstrated that the formation of the nucleoside building block proceeds through three steps from uridine **1** to build intermediate **5** (Figure 2.6).[95]

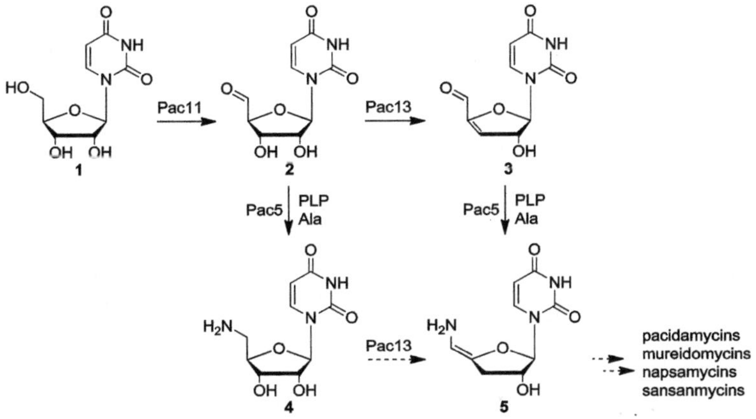

Figure 2.6: The biosynthesis of the pacidamycin nucleoside moiety with **5** as the potential common precursor in the biosynthesis of pacidamycins, mureidomycins, napsamycins and sansanmycins.[95]

Detailed precursor-based experiments combined with a knock-out mutant construction clearly showed that uridine **1** is oxidized to its aldehyde **2** by the flavin-dependent dehydrogenase Pac11. The subsequent dehydration is catalyzed by the Cupin family enzyme Pac13 and the transamination is mediated by Pac5. The last two steps can follow a randomized order due to the relatively flexible substrate recognition of Pac13 and Pac5. With UMP **6** as the substrate instead of uridine **1**, no conversion could be observed. In contrast, UMP **6** was found to be the essential precursor for the formation of the nucleoside moiety in the biosynthesis of nikkomycins.[96-97] As nikkomycins share the glycyl-uridine motif with the lipouridyl antibiotics, it is not surprising that UMP **6** also proved to be the fundamental precursor in the biosynthetic formation of A-90289. This finding also led to a revised postulated pathway in the biosynthesis of caprazamycins (Figure 2.7).[98-99]

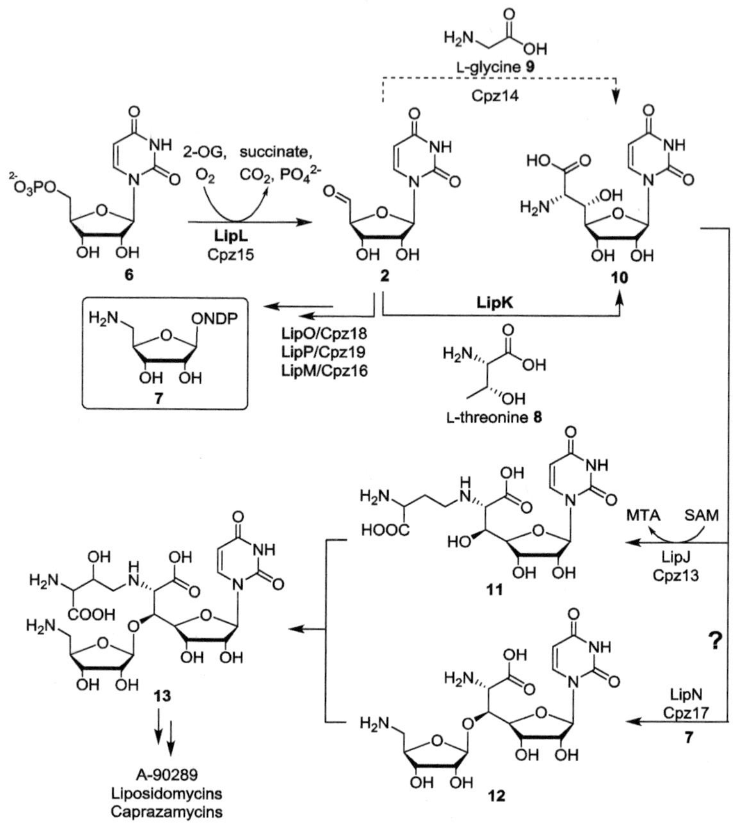

Figure 2.7: Hypothetical pathway for the formation of the nucleoside moiety of A-90289, liposidomycins, and caprazamycins. The enzymes written in bold letters were assigned using experimental data as well. The dashed arrow shows an old proposal. Abbreviations: SAM: *S*-adenosylmethionine, MTA: methylthioadenosine, 2-OG: 2-oxogluterate, NDP: nucleoside diphosphate.

The oxidation of UMP **6** to aldehyde **2** was found to be catalyzed by LipL, a non-heme, Fe(II)-dependent 2-oxoglutarate:UMP dioxygenase. Subsequently, aldehyde **2** undergoes an aldol-type reaction. This reaction introduces the glycyl-motif. One open reading frame (orf) was revealed to encode a serine hydroxylmethyltransferase (SHMT). Such enzymes catalyze the reversible conversion of glycine **9** to L-serine, using N^5,N^{10}-methylene tetrahydrofolate as the C1-donor and pyridoxal-5'-phosphate (PLP) as a cofactor. It was shown that tetrahydrofolate-independent reactions underwent an aldol-type reaction, in which aldehydes and glycine produced β-hydroxyl-α-amino acids. Thus, it was proposed that glycine **9** most likely acts as a substrate for LipK/Cpz14. However, recent studies revealed that LipK acts as a PLP-dependent L-threonine transaldolase, generating acetaldehyde and (5'*S*,6'*S*)-**10** with L-threonine as the amino acid substrate.[100] In this context, a gene-targeting approach aimed to identify potential L-threonine:uridine-

5'-transaldolases led to the discovery of sphaerimicin, a new MraY inihibitor containing a glycyl-uridine moiety.[101] However, the order of events after the formation of glycyl-uridine **10** has not been elucidated yet. Two hypothetical pathways might be possible for the formation of intermediate **13**: (1) the transfer of a 3-amino-3-carboxypropyl group from *S*-adenosylmethionine (SAM) to the 5'-amino group, resulting in **11** (LipJ/Cpz13) with a subsequent hydroxylation (LipG/Cpz10) and a glycosylation (LipN/Cpz17), or (2) an early-stage glycosylation to yield **12**, followed by side chain attachment and derivatization. Only the putative function of LipN as 5'-amino-5'-deoxy-α-D-ribosyl-transferase could be assigned by evidence from experimental data yet.[98] Thereby, the sugar-donor **7** for glycosylation is also generated from the UMP **6** precursor pool via 5'-aldehyde **2** (Figure 2.8).

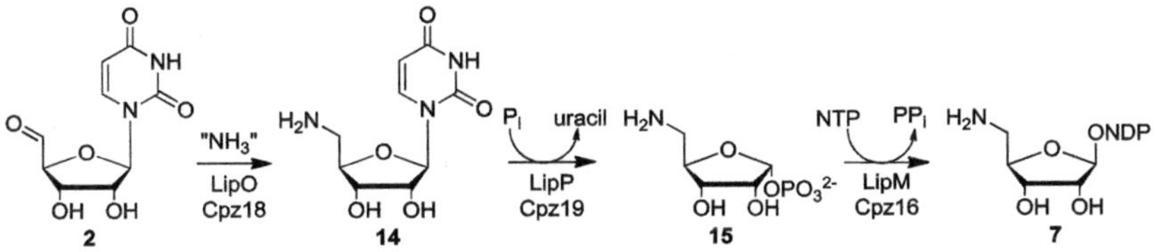

Figure 2.8: Biosynthetic formation of the aminoribosyl sugar **7** based on experimental data from studies in A-90289 biosynthesis by *Van Lanen* and coworkers.[98] Abbreviations: NTP. nucleoside triphosphate, NDP: nucleoside diphosphate, P_i: phosphate, PP_i: pyrophosphate.

The 5'-amino-5'-deoxyuridine **14** is generated by the aminotransferase LipO, which belongs to the PLP-dependent aspartate superfamily. Incubation with L-aspartate and L-glutamate resulted in poor conversion only. Because L-methionine gave the highest activity compared to other tested amino acids, it is most likely used as the amine donor. Phosphorolysis by LipP generates amino-substituted α-D-ribose-1-phosphate **15**, which is subsequently activated for the ribosyltransfer by LipM.[98-99]

In summary, there are most likely two different fundamental precursors for the biosynthetic formation of the uridyl scaffold of nucleoside antibiotics, uridine **1**, and UMP **6**. The compounds with 4',5'-enamine linkage, such as pacidamycins, mureidomycins, napsamycins, and sansanmycins (**A**), are postulated to originate from uridine **1**. Building block **5** might act as an interesting common intermediate here. Uridine **1** proved to be the source of the nucleoside moiety of tunicamycins as well. Considering this, the same is proposed for streptoviridins and corynetoxin (**D**). In the biosynthetic pathway towards the high carbon nucleosides, such as caprazamycins, liposidomycins, A-90289, muraymycins, and FR-900493 (**B**, **C**), the unique (5'*S*,6'*S*)-glycyl-uridine motif is generated from UMP **6**. Other shared intermediates in the biosynthesis might be the glycosylated glycyl-uridine **12**, or the diaminopropyl-derivative **11**. Considering **11** as a building block in the context of the only speculative biosynthesis of muraymycins, one could think of a decarboxylation reaction to yield a 1-uracil-6-diaminopropane uronic acid. Capuramycins (**D**) might be generated by the carboxylation of glycyl-uridine **10**. The 2'-deoxygenated muraminomicin is postulated to originate from 2'-deoxy-UMP or TMP. Its biosynthesis shares no common intermediate with **B**, but it follows an analogous pathway.[93] For a reliable complete picture of nucleoside formation in the biosynthesis of uridyl antibiotics, further experiments with potentially shared building blocks, for example, are necessary. For the elucidation of arylsulfate sulfotransferase activity, which is involved in the biosynthesis of caprazamycins, different synthetic glycosylyl nucleoside precursors already proved to be helpful tools.[102]

2.3.2 NRPS Assembly Line in Peptide-Containing Nucleoside Antibiotics

NRPS stands for nonribosomal peptide-synthetase. In contrast to ribosomes, they are independent from *m*RNA and highly specialized for the production of the desired peptide. Nonribosomally generated peptides often contain unique non-proteinogenic amino acids. NRPS are organized in highly functional assembly lines or modules. Each module consists of different domains with defined functions and is responsible for specific steps:[103]

- Initiation: activation with ATP by the adenylation domain (A), and loading onto the protein by the thiolation domain (T) via a serine-attached 4'-phospho-pantetheine (Ppant) side chain (PCP domain).
- Elongation: Loading a specific amino acid onto the PCP domain, and the condensation of the amino acids by the generation of an amide bond. Additional domains for cyclization or epimerization, for example, can occur as well.
- Termination: Hydrolysis of the polypeptide chain (thioesterase, TE), or the reduction of the thioester, and the release of the peptide by an R-domain.

Typically, multiple modules are encoded within a single protein, like in the biosynthesis of penicillin.[104] The analysis of the clusters of pacidamycins and muraymycins revealed genes, which encode highly fragmented NRPS modules. The pacidamycins, which comprise a tetrapeptidyl scaffold, show a well examined ten protein assembly line, leading to a quite complete picture of the biosynthetic NRPS steps for the formation of the peptide chain.[79,105] Thereby, the non-proteinogenic amino acid 2,3-diaminobutyric acid (DABA) acts as a trifunctional central building block to which the rest of the peptide scaffold is connected. In the biosynthesis of pentapeptidyl pacidamycins, another unique feature was discovered, a *t*RNA-dependent aminoacyl transferase linking ribosomal and nonribosomal peptide synthesis.[106]

The NRPS assembly line of muraymycins consists of six proteins and has not been elucidated in detail yet. However, by bioinformatic analyses (e.g. Toolbox-NRPSpredictor)[107] combined with homology studies, a hypothetical pathway for peptide chain formation could be proposed by *Chen* and coworkers.[91] In Figure 2.9 and

Figure 2.10, this pathway is depicted, and similar enzymes from homology studies are shown with their identity accordances, which explain the deduced functions. The formation of the non-proteinogenic amino acid epicapreomycidine is highly speculative. Nevertheless, attempts were made to identify the responsible enzymes. It is proposed that Mur16 mediates the β-hydroxylation of L-arginine **16** to form (3*S*)-3-hydroxy-L-arginine (*S*)-**17**. The homologous enzymes Cpz15 and LpmM are putative dioxygenases, catalyzing the β-hydroxylation of the 3-amino-3-carboxylgroup in the biosynthesis of caprazamycins and liposidomycins. As such a hydroxyl-substituent does not exist in the muraymycin structure, *mur16* supposably plays another role in the biosynthesis of muraymycins. The gene *mur15* shows homology to a multifunctional cupin 4 type protein and is co-transcripted with *mur16*. While a closer look at the formation of the 3-epimer capreomycidine during the biosynthesis of viomycin and a comparison with the viomycin-encoding gene cluster in *Streptomyces vinaeceus*[108-110] (see also chapter 2.4.1) furnished no similar gene, Mur15 is assigned to catalyze the cyclization of (*S*)-**17** to form epicapreomycidine **18** in a similar manner. Subsequently, the formed epicapreomycidine **18** is activated and loaded onto Mur12 (C-A-T), and it is condensed with L-valine intermediate **21** covalently tethered to Mur14, thereby generating an unusual ureido linkage (see also Figure 2.9).

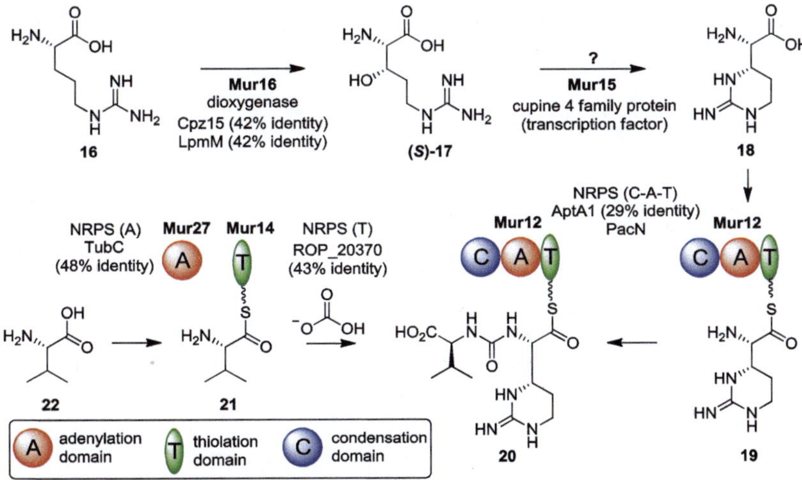

Figure 2.9: First putative steps in the NRPS assembly line of the biosynthesis of muraymycins and the highly speculative biosynthetic formation of epicapreomycidine **18**. Similar enzymes from homology studies are given with their identities in parentheses.[91]

An analogous linkage can be found in the structures of pacidamycins. The biosynthetic pathway towards the ureido motif during the formation of pacidamycins has not been experimentally proven yet. However, the peptide scaffold of syringolins, a family of proteasome inhibitors, comprises the same unusual ureido bond as in pacidamycins and muraymycins. Feeding studies with [^{13}C]bicarbonate, [1-^{14}C]-L-valine, and [^{18}O]bicarbonate/[^{18}O]water revealed that the NRPS enzyme SylC (A,T,C domains) iteratively activates two amino acid monomers, and that it forms the ureido linkage by the integration of bicarbonate/CO$_2$. This incorporation most likely occurs by the cyclization of an initial *N*-carboxy-aminoacyl-*S*-Ppant enzyme intermediate.[111] Based on these findings, a similar reaction pathway could be proposed for the formation of the ureido bond in the biosyntheses of muraymycins and pacidamycins. This would result in thioester intermediate **20** covalently tethered to Mur12, which can then be coupled with the activated and hydroxylated L-leucine intermediate **24**.

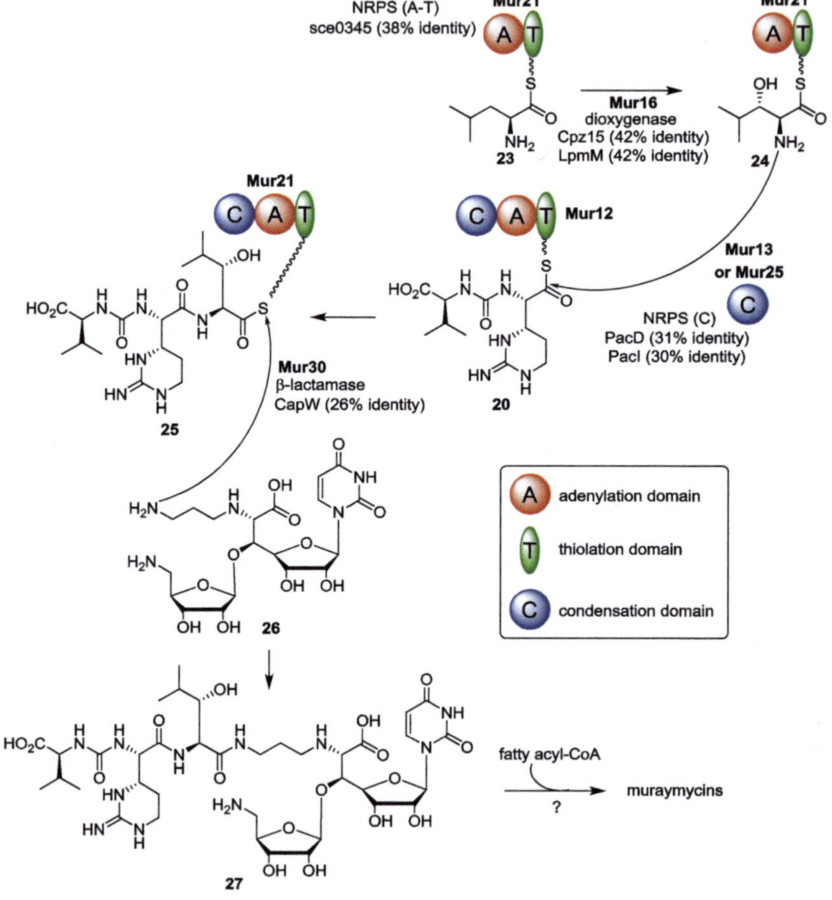

Figure 2.10: Putative late steps in the NRPS assembly line of the biosynthesis of muraymycins and the subsequent tailoring steps. Similar enzymes from homology studies are given with their identities in parentheses.[91]

This reaction is proposed to be mediated by Mur13 or Mur25 (Figure 2.10). Subsequently, the tripeptide building block can be attacked by the previously formed nucleoside intermediate **26** (see also chapter 2.3.1) via the catalysis of Mur30. This would terminate the NRPS-mediated steps and release the muraymycin precursor **27**. Further derivatization steps, such as the methylation of the ribofuranoside, and the attachment of the fatty acid side chain, are still unclear. In the biosynthetic gene clusters of caprazamycins and liposidomycins, the lipase-encoding genes *cpz23* and *lpmU* were found. These findings suggest an attachment of the lipophilic side chain using fatty acyl-CoA as a cosubstrate.

2.4 Non-Proteinogenic Amino Acids as Essential Building Blocks in Natural Products

Non-proteinogenic amino acids are often characteristic structural motifs of natural products with antimicrobial activity. They do not only contribute to the chemical diversity of the peptide backbone, but they can also transform the compound into their bioactive form. Nonribosomal peptides, for example, have recently gained rising attention in drug discovery.[103,112] 'Standard' peptides, however, often are not applicable as clinically used drugs because of their undesirable physiochemical and pharmacological properties. The presence of a non-proteinogenic moiety in a peptide can often improve bioavailability, stability, membrane permeability, and conformational rigidity.[113-114] This indicates that non-proteinogenic amino acids as key features in the peptide backbone of complex natural products might have a similar function. Although structure-activity relationship (SAR) studies of potential lead structures might suggest that structurally simpler motifs are sufficient for good biological activity, it cannot be ruled out that, in vivo, the non-proteinogenic amino acid is essential for activity. It has been estimated that about 500 naturally occurring amino acids have been identified to date,[115] thereby displaying diverse structures and functionalities. In the following, only the non-proteinogenic amino acids, which are relevant for this work, will be described closer.

Looking at the peptide-derived nucleoside antibiotics, especially the unusual non-proteinogenic amino acids 2,3-diaminobutyric acid (DABA) in pacidamycin-like compounds and epicapreomycidine in muraymycins attract attention. For the use of such amino acid motifs in semi-synthetic approaches or metabolic gene engineering, a detailed understanding of the biosynthetic pathway is essential. As described before, both non-proteinogenic amino acids mentioned are integrated over highly dissociated NRPS assembly lines into the peptide scaffold (see also chapter 2.3.2). While the NRPS assembly lines are in general quite well understood, the biosynthetic origin of non-proteinogenic amino acids often remains intriguing. The biosynthesis of DABA has already been exploited, revealing a β-replacement reaction of hydroxyl with ammonia from L-threonine. This reaction set is catalyzed by PacS, PacQ, and PacT based on a PLP-dependent mechanism.[85,116] In contrast, the biosynthetic formation of epicapreomycidine **18** is still unclear. Homology studies only lead to putative encoding genes and a highly speculative biosynthetic pathway (see also chapter 2.3.2). This makes the formation of epicapreomycidine **18** in muraymycin-producing *Streptomyces* sp. an appealing object for precursor-based studies.

2.4.1 Epicapreomycidine and Capreomycidine

Epicapreomycidine **18**, or more precisely (2*S*,3*S*)-capreomycidine, was first found as a component of the protease inhibitors chymostatin[117] and elastatinal.[118] Despite its occurrence in muraymycins, little is known about this unusual cyclic arginine derivative. In contrast, its epimer (2*S*,3*R*)-capreomycidine ('capreomycidine') is well studied. It is a constituent of natural products like the tuberactinomycin peptide antibiotics.[119-120] The capreomycins[121] and viomycin,[122] for example, are both used clinically as tuberculosis drugs.[123] These cyclic peptides bind to the bacterial 70S ribosome, and they inhibit the self-splicing reaction of group I introns.[124-125] The guanidine group of the capreomycidine moiety is therefore believed to offer a potential guanosine-like binding side.[126] However, early studies proved the essentiality of capreomycidine for the antibacterial activity of viomycin.[127] The biosynthesis of capreomycidine as part of the biosynthesis of viomycin in *Streptomyces vinaceus* was elucidated (Figure 2.11).

Initial feeding studies with deuterium labeled L-arginine and elucidation of the biosynthetic gene cluster already indicated that α-β-didehydroarginine might act as an important intermediate during the formation of capreomycidine.[128] Later, PLP was proposed to act as a stabilizing cofactor.[129] In 2004, *Zabriskie* and coworkers confirmed this proposal (Figure 2.11).[108-110] L-arginine **16** is first stereoselectively hydroxylated at the C3-position, catalyzed by the non-heme 2-oxogluterate (2-OG) dependent Fe(II) oxygenase VioC, to give (3*S*)-3-hydroxy-L-arginine (*S*)-**17**. Recently, crystallization trials proved the assigned stereospecificity of this VioC-catalyzed β-hydroxylation reaction of L-arginine as well. This is the first example for a crystal structure of a clavaminic acid synthase-like (CSL) oxygenase[130] that catalyzes the formation of an *erythro* diastereomer.[131] Other members of the CSL superfamily produce *threo* diastereomers.

The β-hydroxy amino acid (*S*)-**17** then undergoes a ring-closure to yield **29** with a formal inversion of the configuration at C3. This intramolecular *Michael*-type addition is PLP-dependent. The elimination of water generates α,β-unsaturated intermediate **28**, followed by an intramolecular attack in the β-position to yield **31**. By *M. Büschleb*, this biosynthetic ring-closure of didehydroarginine **28** towards capreomycidine **29** could be synthetically mimicked by a domino-guanidinylation-aza-*Michael*-addition sequence to yield capreomycidine **29** and epicapreomycidine **18** in racemic form.[132]

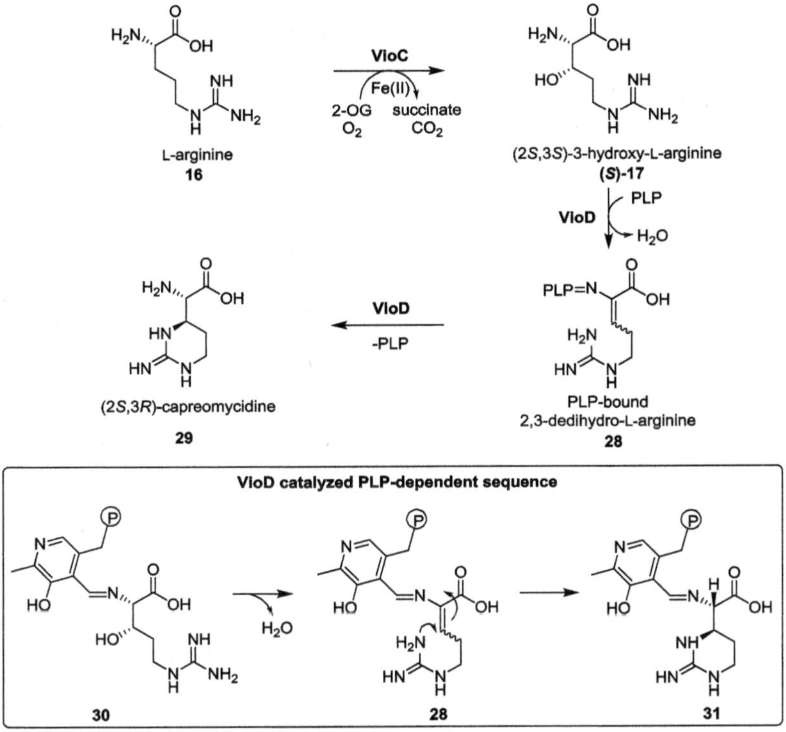

Figure 2.11: The biosynthesis of capreomycidine **29** in *S. vinaceus* via a two-step mechanism. The second step, mediated by VioD, is PLP-dependent.[108-110]

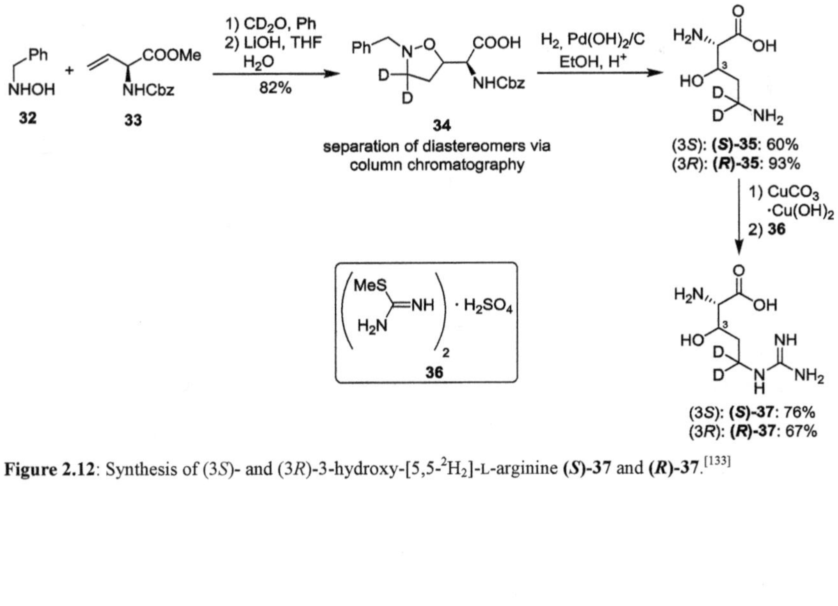

Figure 2.12: Synthesis of (3*S*)- and (3*R*)-3-hydroxy-[5,5-^{2}H$_2$]-L-arginine **(S)-37** and **(R)-37**.[133]

Only enzyme assays with isotope-labeled L-arginine and deuterium-labeled (3*S*)- and (3*R*)-3-hydroxy-[5,5-^{2}H$_2$]-L-arginine (*S*)-37 and (*R*)-37 could provide a solid identification of the VioC-produced compounds. The required deuterium-labeled compounds (*S*)-37 and (*R*)-37 were synthesized via a reaction route developed by *Gould* and coworkers (Figure 2.12).[133-134] The deuterium label at C5 was introduced via a 1,3-dipolar cycloaddition. The required nitrone was generated in situ from *N*-benzylhydroxylamine 32 and formaldehyde-d$_2$. The 1,3-dipolar cycloaddition of the nitrone with Cbz-protected L-vinylglycine 33 led to a diastereomeric mixture, which could be converted into oxazolidine 34 by hydrolysis with lithium hydroxide. At this stage of the synthetic route, a separation by column chromatography was feasible. From the pure diastereomers (*S*)-34 and (*R*)-34, (3*S*)- and (3*R*)-3-hydroxy-[5,5-^{2}H$_2$]-L-arginine (*S*)-37 and (*R*)-37 were accessible over two steps. This reaction sequence allowed the introduction of deuterium labels only at the C5-position.

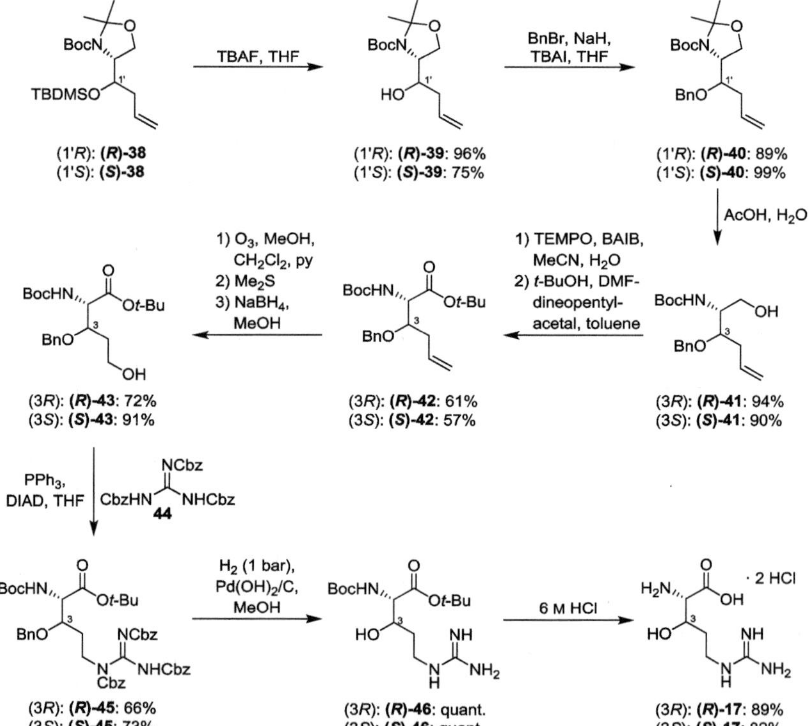

Figure 2.13: The synthesis of (3*S*)- and (3*R*)-3-hydroxy-L-arginine (*S*)-17 and (*R*)-17 from the diastereomerically pure silylated homoallylic alcohols (*S*)-38 and (*R*)-38.[135]

During my diploma thesis, a concise synthesis of both diastereomers (3*S*)- and (3*R*)-3-hydroxy-L-arginine (*S*)-**17** and (*R*)-**17** was developed, with the potential to prepare not only the C5-deuterated, but also the C3-deuterated compounds (Figure 2.13).[135] Starting from (*R*)-*Garner*'s aldehyde, the established silylated homoallylic alcohols (*S*)-**38** and (*R*)-**38** could be synthesized and separated by column chromatography.[136] Further transformation required a change of the protecting group strategy, as the TBDMS protecting group of the secondary alcohol was essential for the chromatographic separation of the diastereomers, but it was not stable under the acidic conditions of the subsequent acetonide deprotection. The oxidation and the subsequent esterification of (*S*)-**41** and (*R*)-**41** generated compounds (*S*)-**42** and (*R*)-**42**, which could be used in an ozonolysis reaction, followed by reduction to yield the alcohols (*S*)-**43** and (*R*)-**43**. The guanidine moiety was introduced by a *Mitsunobu* reaction. A final deprotection of (*S*)-**45** and (*R*)-**45** delivered both of the diastereomers (*S*)-**17** and (*R*)-**17** in overall yields of 21% each over nine steps from (*S*)-**38** and (*R*)-**38**, respectively. Both of the prepared 3-hydroxy-L-arginine derivatives (*R*)-**17** and (*S*)-**17** were used as valuable standards by the group of *Schofield* for the elucidation of an oxygenase-catalyzed ribosome hydroxylation.[137]

2.4.2 Enduracididine

Another interesting cyclic non-proteinogenic amino acid is enduracididine **47**. It represents the five-ring analogue of epicapreomycidine **18** and was found in seeds of *Leguminosea*,[138] marine ascidian *Leptoclinides dubius*,[138] and as a component of antimicrobial substances like enduracidin from *Streptomyces fungicidicus*[139] and minosaminomycin from *Lonchocarpus sericeus*.[140] In addition, the hydroxylated derivatives of enduracididine **49** and **50** (Figure 2.14) are components of mannopeptimycins from *Streptomyces hygroscopicus*.[141]

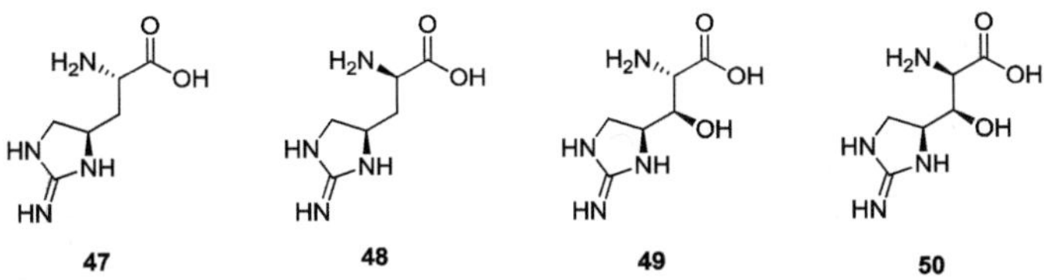

Figure 2.14: Naturally occurring enduracididine derivatives 47-50.

Enduracidin and mannopeptimycin are cyclic peptide antibiotics. They both show excellent activity against Gram-positive bacteria by binding to lipid II, and thereby, they are inhibiting the biosynthesis of peptidoglycan.[142] Against MRSA, enduracidin shows a minimal inhibitory concentration (MIC) of 0.063-0.125 µg/mL. Remarkably, no transmittable mechanism of resistance to enduracidin or mannopeptimycin has been reported yet. This fact makes these cyclic polypeptides quite attractive for clinical application.[142-143] Enduracididine **47** occurs in the peptide scaffold of enduracidin in L- and D-configurations. This makes the biosynthetic pathway towards both of these stereoisomers intriguing. Initial feeding experiments identified L-arginine as the precursor for enduracididine **47** and alloenduracididine **48**. Most likely, the epimerization occurs after the cyclization, as the NRPSs of enduracidin contain rare, dual-function epimerization domains.[143] Based on the gene cluster, which had been identified and analyzed for the biosynthesis of enduracidin,[144] and a comparison with the biosynthetic gene cluster of mannopeptimycin,[145] a possible route could be proposed (Figure 2.15).

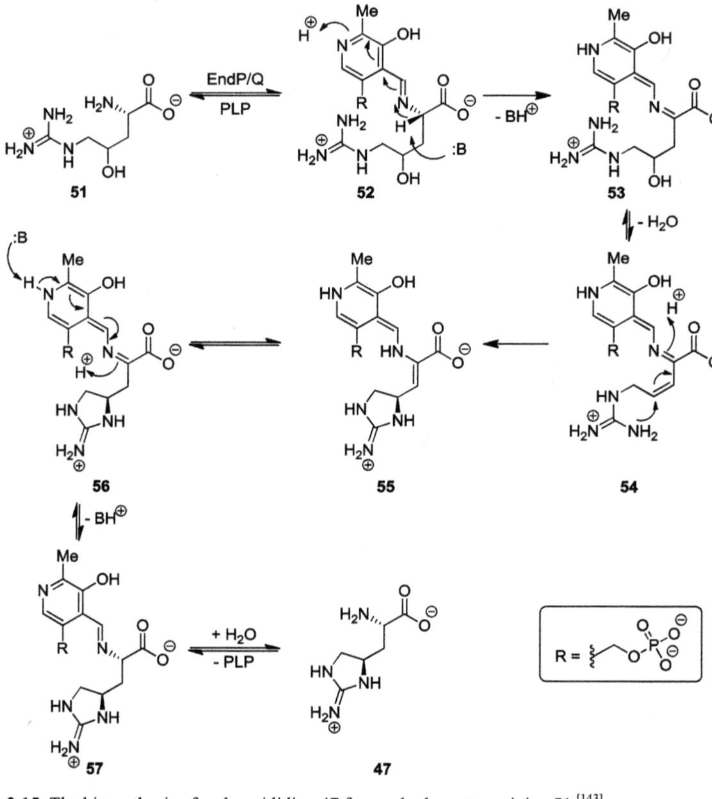

Figure 2.15: The biosynthesis of enduracididine **47** from γ-hydroxy-L-arginine **51**.[143]

Feeding experiments revealed γ-hydroxy-L-arginine **51** as an important intermediate.[143] This reaction sequence is catalyzed by EndP/Q and depends on PLP as a cofactor. From γ-hydroxyarginine **53**, which is covalently linked to PLP, water is eliminated to yield the unsaturated intermediate **54**, which can undergo a ring-closure reaction to yield the desired product **47**.

The first synthesis of enduracididine **47** was described by *Shiba* and coworkers in 1975 (Figure 2.16).[146]

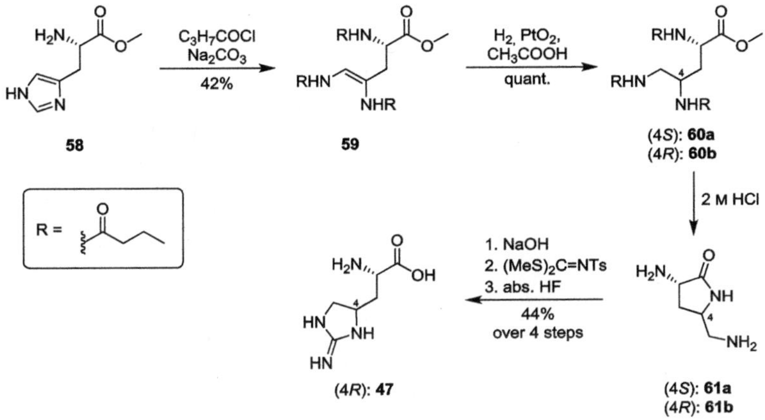

Figure 2.16: The synthesis of enduracididine **47** from L-histidine methyl ester **58** by *Shiba* and coworkers.[146]

The Bamberger cleavage of L-histidine methyl ester **58** delivered dienamine **59**, which could be hydrogenated under catalysis by platinum oxide in acetic acid to yield a diastereomeric mixture of the triamines **60a** and **60b**. Acidic hydrolysis resulted in a 5-exo-trig cyclization to give **61a** and **61b**. Alkaline hydrolysis and guanidinylation with S,S'-dimethyl N-tosyliminodithiocarboximidate delivered the analogous enduracididine derivatives. After acidic hydrolysis, chromatographic separation of the diastereomers was possible. Thus, enduracididine **47** was synthesized from L-histidine methyl ester over 6 steps with an overall yield of 6%.

In 2004, *Dauban* and coworkers reported a synthesis of protected enduracididine derivative **70**, which is suitable as a building block for peptide synthesis (Figure 2.17). [140] An enzymatic chiral resolution of Boc-protected, racemic allylglycine ethyl ester **62** led to unconverted (R)-configured ester **(R)-62** and the desired (S)-allylglycine **63** with an optical purity of 98%. As copper-catalyzed aziridinations with N-Boc-protected allylglycine only gave very poor yields, the phenyl-fluorenyl protecting group (PhF) was introduced. After esterification, the fully protected allylglycine **65** could be aziridinated using 2-(trimethylsilyl)ethane sulfonylamide (SesNH$_2$) with a yield of 28%. The diastereomeric mixture **66** was obtained in a ratio of 7:3. Ring opening with sodium azide generated diastereomers **67a** and **67b**, which could be separated. After *Staudinger* reduction of **67b** and subsequent guanidinylation, deprotection of the Ses group with caesium fluoride resulted in ring-closure. The protected enduracididine derivative **70** could thus be synthesized over 10 steps with an overall yield of 1.6%. A synthesis of D- and L-configured β-hydroxyenduracididine derivatives starting from diacetone-D-glucose was published by *Oberthür* and coworkers in 2009.[147]

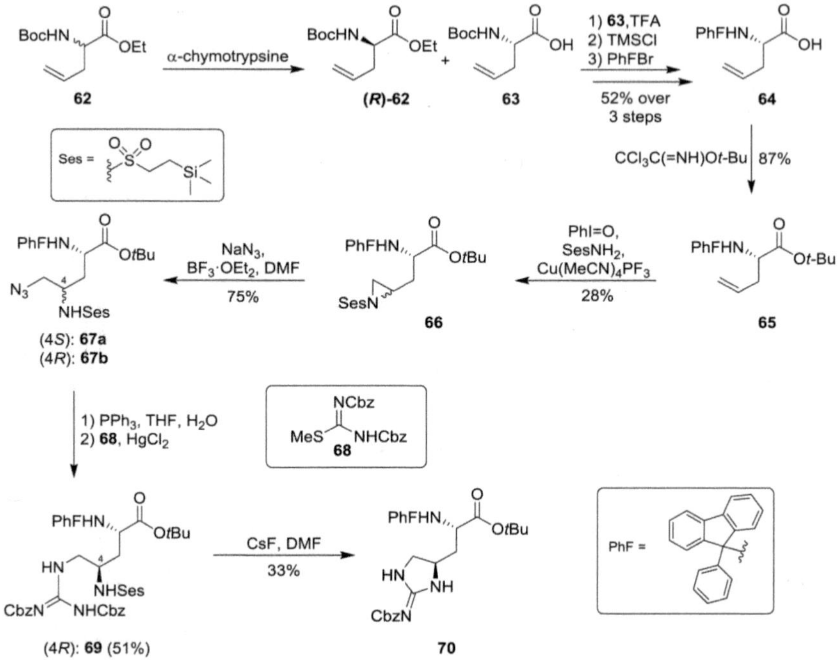

Figure 2.17: The synthesis of the protected enduracididine derivative **70** reported by *Dauban* and coworkers.[140]

2.5 Established Synthetic Work on Nucleosyl Amino Acid Building Blocks

Attempts on the total synthesis of muraymycin derivatives and related nucleoside antibiotics created several established synthetic routes towards the nucleosyl amino acid scaffold. In this context, the generation of the glycyl-uridine motif with two stereogenic centers in C5'-and C6'-position is intriguing. *Yamashita* and coworkers used an aldol reaction between a protected 5'-uridine-aldehyde derivative and the enolate of a protected glycine derivative as a key step towards this unusual motif.[81] A strategy for the stereoselective synthesis of *trans*-epoxides from 5'-uridine-aldehyde and sulfur ylides was developed by *Sarabia* and coworkers.[148] Another interesting strategy is the application of an asymmetric *Sharpless* aminohydroxylation. *Matsuda* and *Ichikawa* used this method to achieve the formation of glycyl-uridine for the total synthesis of (+)-caprazol, caprazamycin analogues, and muraymycin D2 (Figure 2.18).[149-151]

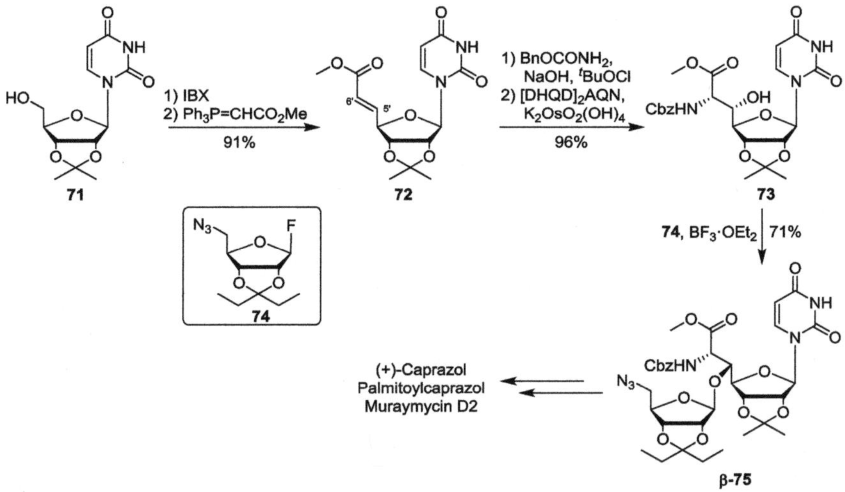

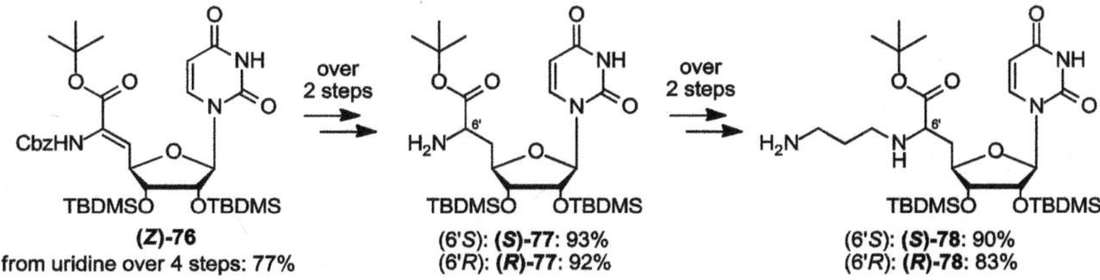

Figure 2.18: Synthesis of nucleoside building block β-75 as part of total syntheses of nucleoside antibiotics by *Matsuda* and *Ichikawa*.[149-151]

After oxidation, protected uridine **71** was converted into olefin **72** in a *Wittig* reaction with a phosphorous ylide. From this (*E*)-configured olefin, the conversion into the corresponding glycyl-uridine derivative **73** was possible via an asymmetric *Sharpless* aminohydroxylation. This was achieved by using *tert*-butylhypochlorite as the oxidizing agent, potassium osmate(VI) dihydrate as the catalyst, hydroquinidine (anthraquinone-1,4-diyl) diether ([DHQD]₂AQN) as the chiral ligand, and benzylcarbamate as the nitrogen source. Glycosylation with the fluoride donor **74** delivered the protected nucleoside building block β-75. In our laboratory, several synthetic strategies towards uridine-derived building blocks were developed as well.[152]

For the synthesis of 5'-deoxy derivatives, the route depicted in Figure 2.19 was used.[152-154]

Figure 2.19: Synthesis of the 5'-deoxy nucleoside building blocks (*R*)-77, (*S*)-77, (*R*)-78, and (*S*)-78.[152-154]

The α,β-didehydroamino acid ester (*Z*)-76 was generated from uridine using a *Wittig-Horner* reaction as the key step. The highly stereoselective homogeneous asymmetric hydrogenation of (*Z*)-76 with the chiral rhodium(I) catalysts (*S,S*)-Me-DUPHOS-Rh and (*R,R*)-Me-DUPHOS-Rh yielded the diastereomerically pure compounds (*S*)-77 and (*R*)-77, respectively, depending on the configuration of the catalyst. The aminopropyl linker could be introduced by reductive amination. To obtain the 5'-hydroxyl-substituted derivatives, three different routes were applied, depending on the desired configuration of the target compound (Figure 2.20).[152,155-157] The epoxides **79** and **85** were synthesized based on diastereoselective sulfur ylide chemistry, while **82** was prepared via *Sharpless* epoxidation of a uridine-derived allylic alcohol. For the formation of the desired building blocks, these epoxides were opened in a S_N2-like reaction with excellent diastereoselectivities. Using azide ions for the opening of the oxirane ring, compounds **80** and **83** were generated. With a fitting amine, the aminopropyl-substituted compounds **81** and **84** could be synthesized. To achieve the naturally occurring configuration of compound **86**, a strategy with a double inversion of the configuration in C6'-position was required. Therefore, the epoxide **85** was first opened by a bromide ion, which was followed by substitution with an azide salt. Particularly nucleoside building blocks like **86**, **(*S*)-78**, **(*R*)-78**, or **75** are attractive precursors for the synthesis of potential biosynthetic intermediates, which might be used in biological studies.

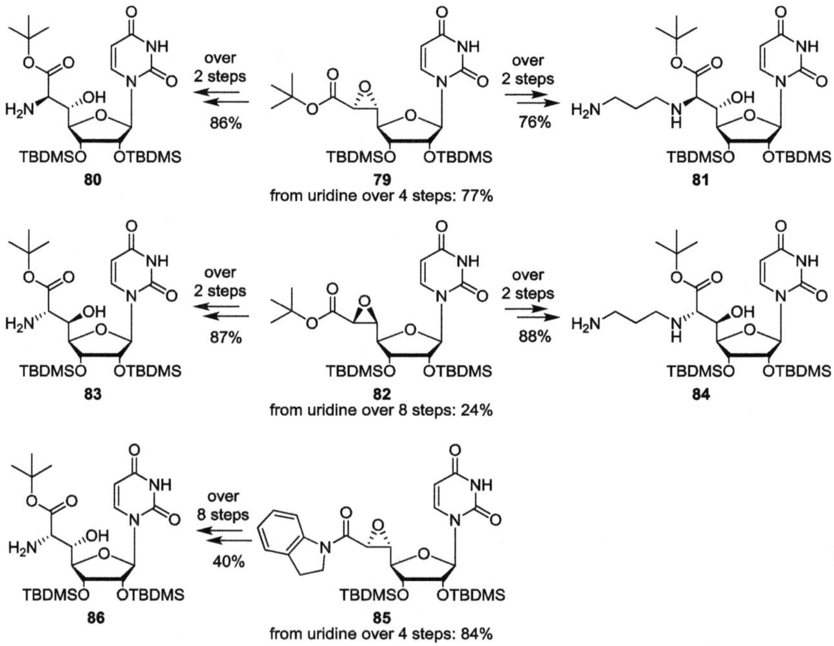

Figure 2.20: Syntheses of different nucleoside building blocks by an epoxide opening strategy.[152,155-157]

3   Aim of This Work and Retrosynthetic Considerations

3.1   Aim of This Work

Catalytic and Biocatalytic Methods for the Efficient Synthesis of Biologically Relevant Non-Proteinogenic Amino Acids

Non-proteinogenic amino acids with unusual structures are often important constituents of biologically potent natural products. The synthesis of these amino acid structures is often time-consuming, non-trivial and not sustainable. Thus, on the one hand it was planned to elucidate the biosynthetic pathway towards selected amino acids with the long-term goal of an enzymatic transformation or a biocatalytic fermentation-based production. On the other hand, it was planned to develop concise syntheses including catalytically performed key steps to create unusual amino acid structures. In this context, the aim of this work can be divided into three major parts **A-C**, as depicted in Figure 3.1.

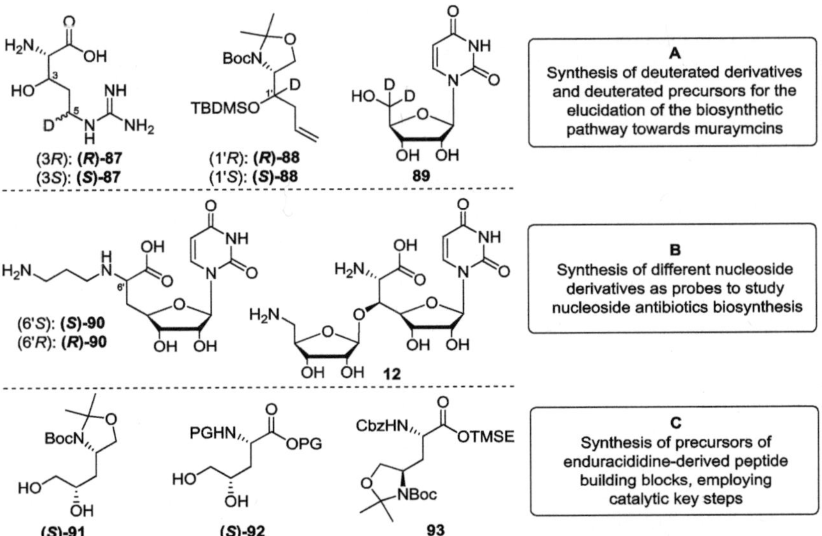

Figure 3.1: Aim of this work divided into three parts and the envisioned target compounds (PG = protecting group).

These three parts include the syntheses of **A)** deuterium-labeled derivatives for the elucidation of the formation of muraymycins in *Streptomyces* sp., namely the stereochemically pure 3-hydroxyarginine derivates **(R)-87** and **(S)-87**, the deuterium-labeled precursors **(R)-88** and **(S)-88** as well as the deuterated uridine derivative **89**;

B) different nucleoside derivatives as probes to study nucleoside antibiotics biosynthesis, namely the nucleoside building blocks **(*S*)-90**, its isomer **(*R*)-90** and the glycosylated nucleoside **12**, and **C)** peptide building block precursors towards the non-proteinogenic amino acid enduracididine, namely **(*S*)-91**, **(*S*)-92** and **93**. Especially for the preparation of the nucleoside building block **12** and the enduracididine precursors **(*S*)-91**, **(*S*)-92**, and **93**, catalytically performed key steps should be employed. Parts **A** and **B** combined were envisioned to furnish a chemical 'toolbox' for studies on nucleoside antibiotics biosynthesis.

The non-proteinogenic amino acid epicapreomycidine **18** is one of the unusual structural building blocks of muraymycins. The biosynthesis of its 3-epimer **29**, capreomycidine, was elucidated before and utilizes (3*S*)-3-hydroxy-L-arginine **(*S*)-17** as an intermediate.[108-110] Thus, it is proposed that 3-hydroxy-L-arginine also serves as an intermediate in the biosynthesis of **18**, but the stereochemical course of the formation of epicapreomycidine from L-arginine **16** is still unclear. In Figure 3.2, different possible biosynthetic pathways for the formation of epicapreomicidine **18** are given. It cannot be ruled out that epicapreomycidine **18** is formed via enzymatic epimerization of capreomycidine **29** at the C3-position, as there is evidence for unusual epimerization reactions in other biosynthetic pathways in bacteria.[130]

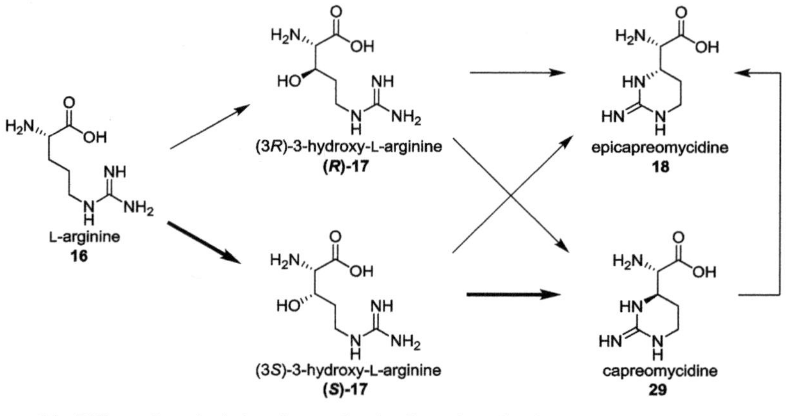

Figure 3.2: Different hypothetical pathways for the formation of epicapreomycidine **18**, based on the elucidated biosynthesis of **29** (bold arrows).

Therefore, selectively isotope-labeled derivatives of both **(*S*)-17** and **(*R*)-17** are essential for a detailed investigation. The synthetic route towards 3-hydroxy-L-arginine, which had been established by *Gould* and coworkers,[134] only allows the selective introduction of a deuterium label at the C5-position. However, a derivative selectively deuterated at the C3-position would be essential for the elucidation of a potential epimerization reaction. After a concise synthesis had already been developed for both potential intermediates

(*S*)-**17** and (*R*)-**17**, it was envisioned to generate the C5- and C3-deuterated analogues of (*S*)-**17** and (*R*)-**17**. By the use of these compounds for feeding studies or in vitro assays, it should be possible to identify the biologically relevant diastereomer used during the biosynthetic formation of epicapreomycidine **18**, and it should be possible to predict the detailed stereochemical course of the reaction.

In the context of the biosynthesis of muraymycins, the formation of the glycyl-uridine scaffold is intriguing as well. Potential nucleoside intermediates would be of particular interest for the determination of the biosynthetic order of glycosylation and side chain attachment, as this has not been elucidated in the biosynthetic pathways of structurally related nucleoside antibiotics yet, such as liposidomycins, caprazamycins, and A-90289. Considering this fact, suitable nucleoside building blocks, which are unknown in literature, should be synthesized (see also Figure 3.1). As compound (*S*)-**90** is deoxygenated in the 5'-position, this compound would be suitable for inhibition experiments, or it might serve as a substrate analogue for the biosynthetic generation of novel derivatives. Biological studies with (*S*)-**90**, (*R*)-**90**, and **12** might also provide crucial insights into the pathways of structurally related compounds.

Having potential feeding studies in mind, another essential aim was to establish fermentation methodologies for muraymycin-producing *Streptomyces* sp. in our laboratory. The strains of NRRL30471 should not only be cultivated according to the patent issued by the *Wyeth Holdings Corporation*,[158] but the production of muraymycin derivatives should also be analytically verified. In addition, the application of tandem mass spectrometry (MS/MS) should show the possibility to get information by the fragmentation pattern of muraymycin derivatives. This method would be another important tool for a detailed interpretation of feeding experiments, and it would therefore complete the chemical 'toolbox' resulting from the first two parts of the synthetic work. For initial feeding trials, structurally simple, isotopically labeled potential intermediates should be sufficient. These compounds should be easily accessible, or they should be obtained commercially.

Aside from the biocatalytic approach for the synthesis of non-proteinogenic epicapreomycidine **18**, the development of a concise synthetic route towards enduracididine **47** was envisioned, deploying a catalytic key step. As enduracididine **47** represents the 5-ring analogue of epicapreomycidine **18**, the developed route should bear the potential to generate suitable peptide building blocks with the long term-goal of incorporation into the peptide backbone of biologically active natural products such as the muraymycins for structure-activity relationship (SAR) studies. To date, the established synthetic routes towards enduracididine **47** consist of multiple steps with moderate to poor yields, which would make the provision of synthetically relevant amounts difficult. Therefore, a synthetic strategy with a catalytically mediated key step should set the stage towards a sustainable process for the production of enduracididine-derived peptide building blocks.

3.2 Retrosynthetic Considerations

Part A

The established synthetic route for both diastereomers of 3-hydroxy-L-arginine (R)-**17** and (S)-**17**, which was in principle already developed in my diploma thesis,[135] offers two possibilities for the introduction of deuterium labels by using deuterated reducing agents (Figure 3.3): (a) reduction of the methyl ester (R)-**101** to yield dideuterated alcohol **100**, which is a precursor of C3-deuterated 3-hydroxy-L-arginine derivatives (R)-**97** and (S)-**97**; (b) reduction of aldehydes (R)-**96** and (S)-**96** to yield C5-deuterated diols (R)-**95** and (S)-**95**.

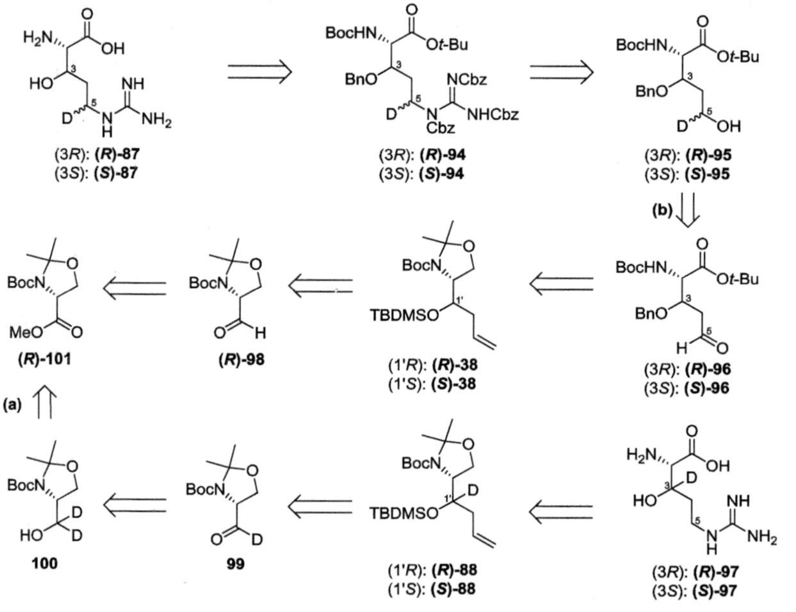

Figure 3.3: Possible ways to introduce deuterium labels in the established synthetic routes towards 3-hydroxy-L-arginine.

The synthesis of *Garner*'s aldehyde (R)-**98** is well established, and it employs D-serine as the starting material.[159] The subsequent protection of D-serine leads to methyl ester (R)-**101** in three steps. Using compound (R)-**101**, the known synthesis can be modified for an introduction of a deuterium label. In this manner, the deuterated *Garner*'s aldehyde **99**, which is not known in literature, should be accessible via reduction with lithium aluminium deuteride. By using a *Grignard* addition and subsequent silylation, both diastereomers (R)-**88** and (S)-**88** can be prepared. At this stage of the synthetic route, a chromatographic separation of the diastereomers should be possible. This separation delivers both diastereomeric precursors **(S)-88** and **(R)-88** for the synthesis of (3S)-[3-^{2}H]-3-hydroxy-L-arginine **(S)-97** and (3R)-[3-^{2}H]-3-hydroxy-L-arginine **(R)-97**. The non-deuterated, TBDMS-protected homoallylic alcohols **(R)-38** and **(S)-38** were already synthesized by *Zabriskie* and coworkers.[136] As the TBDMS protecting group was proven to be unsuitable for further transformations,[160] the change to the benzyl group was established by our group. Via protecting group manipulations and an ozonolysis reaction with subsequent oxidative workup, the crude aldehydes **(R)-96** and **(S)-96** are accessible. Reduction with sodium borodeuteride should lead to a single deuterium label at the C5-position. Guanidinylation followed by deprotection should give access to the C5-deuterated 3-hydroxy-L-arginine derivatives **(R)-87** and **(S)-87**.

For the analysis of feeding studies, a second deuterium label at the C5-position might be advantageous, as a mass difference of $\Delta m/z = 2$ might be easier to identify. The introduction of two deuterium labels should be feasible by an oxidation-reduction sequence (Figure 3.4). Activation of carboxylic acids **(R)-104** and **(S)-104** followed by reduction with a suitable deuterated reducing agent should give access to compounds **(R)-103** and **(S)-103**, and therefore, it should lead to a second deuterium label at the C5-position in the target compounds **(R)-102** and **(S)-102**.

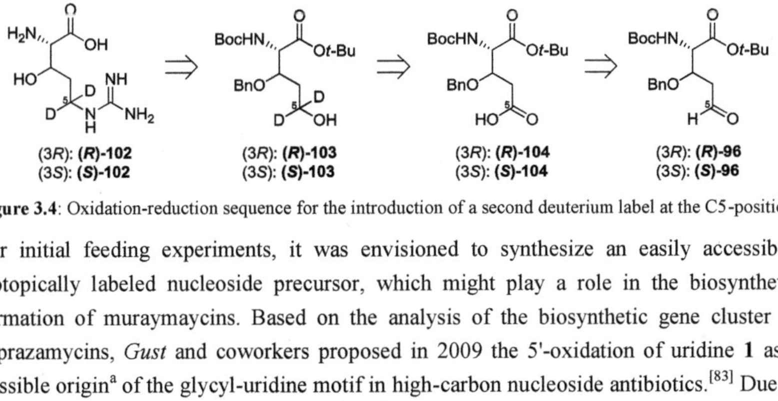

Figure 3.4: Oxidation-reduction sequence for the introduction of a second deuterium label at the C5-position.

For initial feeding experiments, it was envisioned to synthesize an easily accessible, isotopically labeled nucleoside precursor, which might play a role in the biosynthetic formation of muraymaycins. Based on the analysis of the biosynthetic gene cluster of caprazamycins, *Gust* and coworkers proposed in 2009 the 5'-oxidation of uridine **1** as a possible origin[a] of the glycyl-uridine motif in high-carbon nucleoside antibiotics.[83] Due to this fact, it was envisioned to obtain [5',5'-^{2}H$_2$]uridine **89** by a short synthetic sequence using an oxidation-reduction strategy (Figure 3.5). Activation of the protected uridyl carboxylic acid **105**, followed by reduction with a suitable deuterated reducing agent and acidic hydrolysis should deliver the desired deuterated uridine **89**.

---

[a] This hypothesis was revised by *Gust* and coworkers in 2013,[99] based on detailed investigations on the biosynthetic origin of the nucleoside moiety in the biosynthesis of A-90289 antibiotics by the *Van Lanen* Group in 2011.[98]

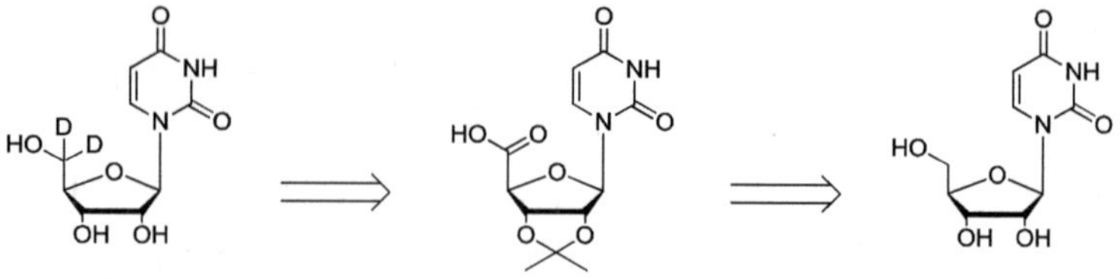

Figure 3.5: Oxidation-reduction strategy for the synthesis of $[5',5'-^2H_2]$uridine **89**.

Part B

Based on the synthetic work on nucleoside building blocks for the preparation of (+)-caprazol by *Matsuda* and *Ichikawa* (see also Figure 2.18),[149] a route towards the glycosylated nucleoside building block **12** is conceivable (Figure 3.6).

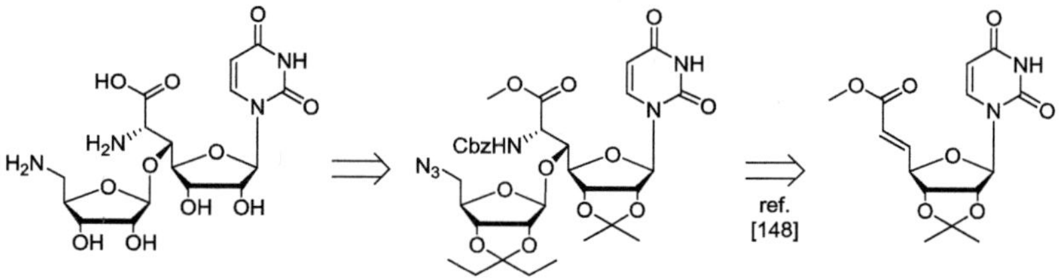

Figure 3.6: Retrosynthetic route towards the glycosylated nucleoside building block **12**, based on established strategies by *Matsuda* et al.[149]

A suitable deprotection strategy of **β-75** should give access to target compound **12**. The reduction of the azide to the corresponding amine should be possible in one step together with the hydrogenolytic removal of the Cbz protecting group. The methyl ester should be removable by alkaline hydrolysis, and the isopropylidene- and isopentylidene protecting groups should be cleavable by hydrolysis under acidic conditions. The synthesis of the glycosylated nucleoside derivative **β-75** has already been accomplished by *Matsuda* and coworkers, using an aminohydroxylation of **72** as a key step. The olefin **72** was synthesized by a *Wittig* reaction of the corresponding protected 5'-uridine aldehyde with a suitable phosphorous ylide (see also chapter 2.5, figure 2.18). The challenge in the deprotection sequence might be, on the one hand, the 'right' order of deprotection steps and on the other hand, to handle purification problems due to the polarity of the obtained compounds. Even by HPLC purification, it might thus be difficult to remove possibly formed isomers.

For the synthesis of the nucleoside building block **(S)-90** and its (6'R)-epimer **(R)-90**, suitable precursors were provided by *A. Spork*.[152-154] The synthetic strategy towards the precursors **(S)-78** and **(R)-78** is summarized in chapter 2.5. After acidic hydrolysis of **(S)-78** and **(R)-78**, the corresponding deprotected nucleosides **(S)-90** and **(R)-90** should be accessible (see Figure 3.7). The non-naturally occurring (6'R)-epimer **(R)-90** might be useful as a reference compound, or for studies regarding enzyme stereospecificity.

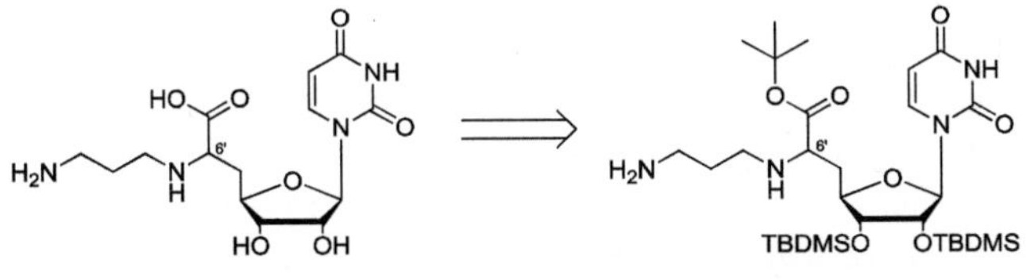

(6'S): **(S)-90**
(6'R): **(R)-90**

(6'S): **(S)-78**
(6'R): **(R)-78**

Figure 3.7: Retrosynthesis of the deprotected nucleosides **(S)-90** and **(R)-90** from the provided precursors **(S)-78** and **(R)-78**.

Part C

Regarding retrosynthetic considerations for the synthesis of the non-proteinogenic amino acid enduracididine **47**, an orthogonal protecting group strategy is essential since the generated enduracididine derivative should be suitable as a building block for peptide synthesis. Selecting a Cbz protecting group for the α-amino group and 2-trimethylsilyl ethanol (TMSE-OH) for the carboxy group, acidic conditions can be used for the deprotection of other protecting groups, such as Boc or isopropylidine, during the synthesis. Building block **106** should be accessible by guanidinylation of **107** under *Mitsunobu* conditions (Figure 3.8).

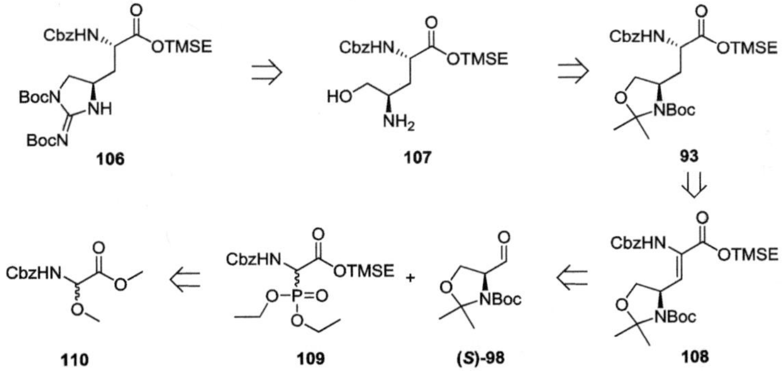

106

107

93

110

109

(S)-98

108

Figure 3.8: Retrosynthesis of the enduracididine derivative **106** with an asymmetric hydrogenation as the catalytic key step.

Amino alcohol **107** can be obtained by acidic hydrolysis of the compound **93**. Using the chiral rhodium(I) catalysts (S,S)-Me-DUPHOS or (R,R)-Me-DUPHOS in asymmetric hydrogenation reactions of (Z)-didehydroamino acids is a well-established procedure for the selective preparation of diastereomerically pure amino acids in our laboratory.[152-154,161-165] This procedure should also be applicable for the generation of the stereogenic center in **93**. The required (*Z*)-olefin **108** should be accessible by a *Wittig-Horner* reaction of phosphonate **109** with (*S*)-*Garner*'s aldehyde (*S*)-**98**. The (*S*)-*Garner*'s aldehyde (*S*)-**98** can be prepared by using the same established reaction sequence as for the (*R*)-*Garner*'s aldehyde (*R*)-**98**.[159] *Schmidt* and coworkers as well as *Zoller* and coworkers described a synthesis of the corresponding methyl ester of **109** via **110** with glyoxylic acid as the starting material.[166-167] Following a protocol by *Toone* and coworkers,[168] transesterification with TMSE-OH should lead to the TMSE-esterified phosphonate **109** (Figure 3.8). As the *Wittig-Horner* reaction as well as the asymmetric hydrogenation of these amino acid precursors represents a new, unusual synthetic pathway, an alternative route for the synthesis of enduracididine peptide building blocks was envisioned as well (Figure 3.9).

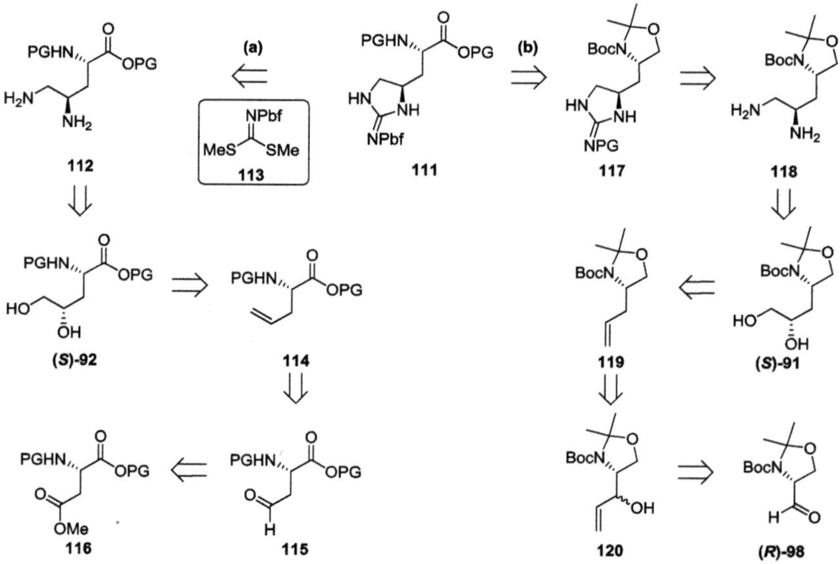

Figure 3.9: Alternative retrosynthetic strategies (**a**) and (**b**) for the synthesis of the enduracididine derivative **111** with an asymmetric dihydroxylation as the catalytic key step.

The second strategy is based on an asymmetric *Sharpless* dihydroxylation as the catalytic key step. In principle, the synthesis of peptide building block **111** should be possible via two alternative routes. Following route (**a**), building block **111** may be obtained by a cyclization of **112** with a suitable guanidinylation reagent. In our laboratory, the use of the Pbf-protected dithioate **113** has been established for ring-closure reactions of this type.[169] Activation of the diol (*S*)-**92** followed by aziridination and *Staudinger* reduction should lead to diamine **112**. The activation should be possible by the conversion of (*S*)-**92** into the corresponding mesylates or tosylates. Via a dihydroxylation of allylglycine **114**, the formation of (*S*)-**92** and its (4*R*)-epimer should be possible. Choosing the matching

AD-mix, a preferred conversion should be observable. Different derivatives of allylglycine **114** have been reported,[170-172] which were all synthesized by *Wittig* methylenation of protected aspartate semialdehyde. Aldehyde **115**, which would be required, should be accessible from L-aspartate methyl ester **116**. Hence, a direct reduction with DIBAL-H might be possible. Another route via the corresponding Weinreb amide could be considered as well, requiring a previous alkaline hydrolysis of the methyl ester and an in-situ-activation of the generated carboxylic acid. For the protection of L-aspartate, Cbz should be suitable for the α-amino group, and TMSE or *tert*-butyl for the acid functionality. This would be consistent with an orthogonal protecting group strategy, required for the envisioned synthesis of peptide structures. In principle, the retrosynthetic strategy (**b**) represents a comparable approach with the synthesis of diol (***S***)-**91** by an asymmetric dihydroxylation as the key step, followed by conversion to the corresponding diamine **118**, and ring-closure towards **117**. The target compound **111** should be accessible via acidic hydrolysis, oxidation of the primary alcohol, and a protection suitable for the synthesis of peptides. The allylic derivative **119**, which would be required, should be accessible from the already mentioned (*R*)-*Garner*'s aldehyde (***R***)-**98**. The preparation of allylic alcohol **120** as a diastereomeric mixture is known, and it uses vinyl bromide in a *Grignard* addition to (***R***)-**98**.[173] The conversion into the corresponding acetate, followed by a reductive, *Tsuji-Trost*-type palladium-catalyzed deoxygenation should then lead to olefin **119**.

4 Results and Discussion

4.1 Synthesis of Different Derivatives of *Garner's* Aldehyde

4.1.1 Preparation of (*R*)- and (*S*)-*Garner's* Aldehyde

Garner's aldehyde **98** is a versatile chiral precursor for the synthesis of amino acid derivatives, and during this work, it was used for the synthesis of 3-hydroxy-L-arginine derivatives and enduracididine precursors in its (*R*)- and (*S*)-configuration. For the preparation, a protocol by *Dondoni* and *Perrone* was adopted.[159] Starting from D-serine (*R*)-121 or L-serine (*S*)-121, respectively, the methyl serinate hydrochloride was prepared by an acid-catalyzed methyl esterification under reflux (Figure 4.1).

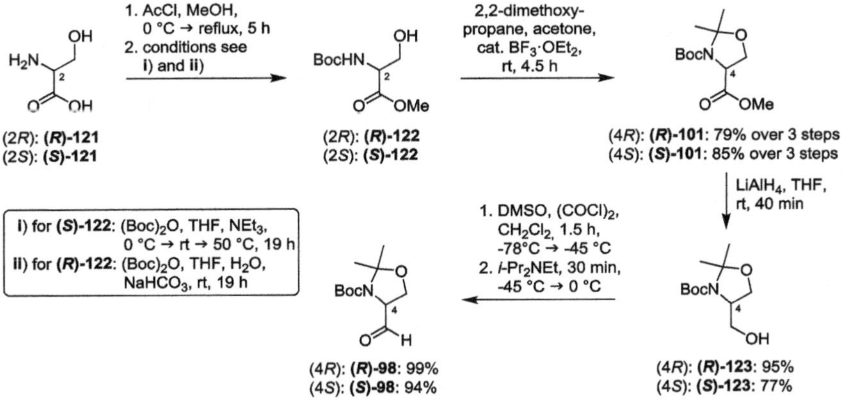

Figure 4.1: Synthesis of (*R*)- and (*S*)-*Garner's* aldehydes (*R*)-98 and (*S*)-98.

Methanolic hydrochloric acid was previously generated by the addition of acetyl chloride to methanol at 0 °C. For the subsequent Boc-protection of the amino group, two different methods were applied. For the preparation of (*S*)-122, the unpurified methyl serinate hydrochloride was suspended in THF. After adding triethylamine, a solution of di-*tert*-butyl dicarbonate in THF was added to yield the *N*-Boc-L-serine methyl ester (*S*)-122 (i). The (*R*)-isomer (*R*)-122 was prepared by following a protocol by *Micale* and coworkers.[174] Instead of triethylamine, saturated aqueous NaHCO₃ solution was used as the base (ii). In both of the cases, products (*S*)-122 and (*R*)-122 were obtained as slightly impure materials in quantitative yields. The second method had the advantage of not needing any inert conditions. Furthermore, the solubility of the hydrochloride was better in the aqueous THF mixture, which most likely resulted in a 'superior' conversion. A possible disadvantage was the partial decomposition of di-*tert*-butyl dicarbonate in the aqueous solution. Isopropylidene protection of compounds **(S)-122** and **(R)-122** by the addition of 2,2-dimethoxypropane and a catalytic amount of boron trifluoride etherate in acetone gave methyl ester **(R)-101** in a yield of 79%, and **(S)-101** in a yield of 85% over three steps. The reduction of **(R)-101** and **(S)-101** by lithium aluminium hydride in THF, followed by *Swern* oxidation with dimethyl sulfoxide and oxalyl chloride in CH_2Cl_2 at -78 °C, delivered the desired *Garner*'s aldehyde derivatives **(R)-98** and **(S)-98** over two steps in good yields of 94% and 72%, respectively.

4.1.2 Preparation of C1'-Deuterated *Garner*'s Aldehyde

The preparation of C1'-deuterated (R)-*Garner*'s aldehyde **99** was achieved analogously. Using lithium aluminium deuteride in the reduction step, methyl ester **(R)-101** was converted into the isopropylidene-protected *N*-Boc-D-[1'-^{2}H$_2$]serinol **100** in a yield of 97%. On a multigram reaction scale, only a yield of 84% was achieved (Figure 4.2). In the ^{1}H NMR spectrum, no signal for the H-1' proton could be detected (Figure 4.2). An analysis of the mass spectrum revealed a deuteration rate of 99%.

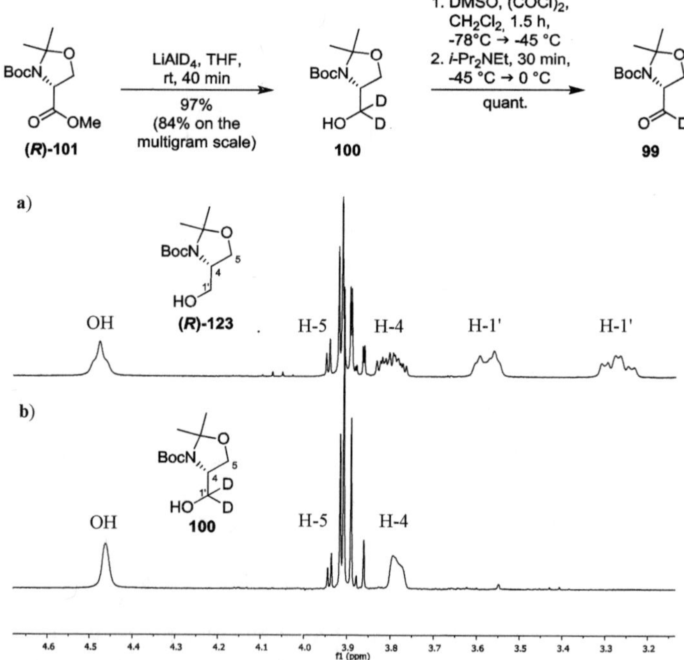

Figure 4.2: Synthesis of [1'-^{2}H]-*Garner*'s aldehyde **99** via a lithium aluminium deuteride mediated reduction step; selected section of the ^{1}H NMR spectra (300 MHz, DMSO-d$_6$, 100 °C) of **a)** the non-deuterated compound **(R)-123** and **b)** the deuterated compound **100**.

Swern oxidation of the deuterated serinol **100** was performed in the same way as the conversion of the non-deuterated compound **(*R*)-123**. The deuterated *Garner*'s aldehyde **99**, which has not been reported in the literature before, was obtained in quantitative yield as a slightly impure material. Due to the possible instability of the material, it was directly converted in the next step without further purification. A small amount of **99** was purified by flash column chromatography to obtain reliable analytical data.

4.2 Precursors for Deuterium Labeled 3-Hydroxy-L-arginine Derivatives

4.2.1 Synthesis of Deuterium-Labeled and Unlabeled TBDMS-Protected Homoallylic Alcohols

The *Grignard* addition of *Garner*'s aldehyde derivatives **(*R*)-98** and **99** was conducted by following a protocol of *Zabriskie* and coworkers.[136] The addition of allylmagnesium chloride to the aldehyde in THF at -80 °C furnished the corresponding homoallylic alcohol as a mixture of two diastereomers in a ratio of 2:1. The homoallylic alcohol **39** was obtained in an overall yield of 84%, starting from **(*R*)-98**. Using deuterated *Garner*'s aldehyde derivative **99**, the *Grignard* addition delivered **124** with a yield of 90% (Figure 4.3).

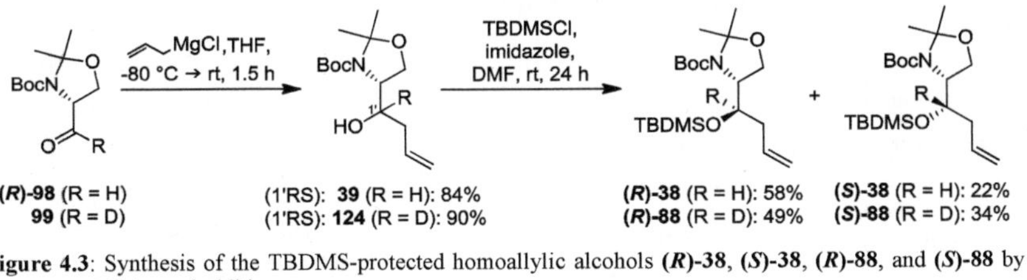

(*R*)-98 (R = H)
99 (R = D)

(1'RS): **39** (R = H): 84%
(1'RS): **124** (R = D): 90%

(*R*)-38 (R = H): 58%
(*R*)-88 (R = D): 49%

(*S*)-38 (R = H): 22%
(*S*)-88 (R = D): 34%

Figure 4.3: Synthesis of the TBDMS-protected homoallylic alcohols **(*R*)-38**, **(*S*)-38**, **(*R*)-88**, and **(*S*)-88** by *Grignard* addition.

Using TBDMSCl and imidazole in DMF, the protection of the homoallylic alcohol **39** as *tert*-butyldimethylsilyl (TBDMS) ether was performed as reported by *Zabriskie* and coworkers.[136] After the protection step, a chromatographic separation of the diastereomers **(*R*)-38** and **(*S*)-38** was possible. On a small scale, overall yields of up to 90% could be obtained. On a multigram scale, an overall yield of around 80% could be observed. This difference can be explained by the problematic chromatographic separation of the diastereomers on a large scale. Using the deuterated derivative **124**, similar yields were obtained (Figure 4.3). *Zabriskie* and coworkers assigned the absolute configuration at the newly formed stereocenter based on NMR data of the Mosher ester derivative of the desilylated alcohol **(*R*)-39**.[136] However, to ensure that this assignment was correct,

M. Büschleb prepared cyclic analogues of both of the desilylated diastereomers (**R**)-**39** and (**S**)-**39**.[169] Using triflic acid anhydride (Tf$_2$O), a spontaneous ring-closure with an inversion of the stereochemical configuration at the C1'-position was observed. The resulting cyclic carbamates (**R**)-**125** and (**S**)-**125** were used for nuclear Overhauser enhancement (nOe) NMR experiments. The 1D nOe ^{1}H NMR studies confirmed the absolute configuration, which had been postulated previously by *Zabriskie* and coworkers (Figure 4.4).[135]

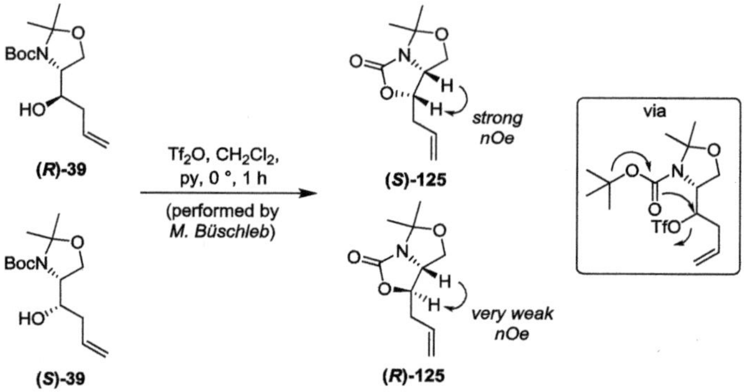

Figure 4.4: Preparation of the cyclic carbamates (**S**)-**125** and (**R**)-**125** for nOe experiments by *M. Büschleb*.[135,169]

4.2.2 Preparation of Precursors for C5-Deuteration

The synthesis towards the precursors (**R**)-**42** and (**S**)-**42** was carried out as reported by *Ducho* and coworkers (see also chapter 2.4.1).[135,160] In almost all reaction steps, slightly improved yields could be obtained (compare Figure 2.13 and Figure 4.5).

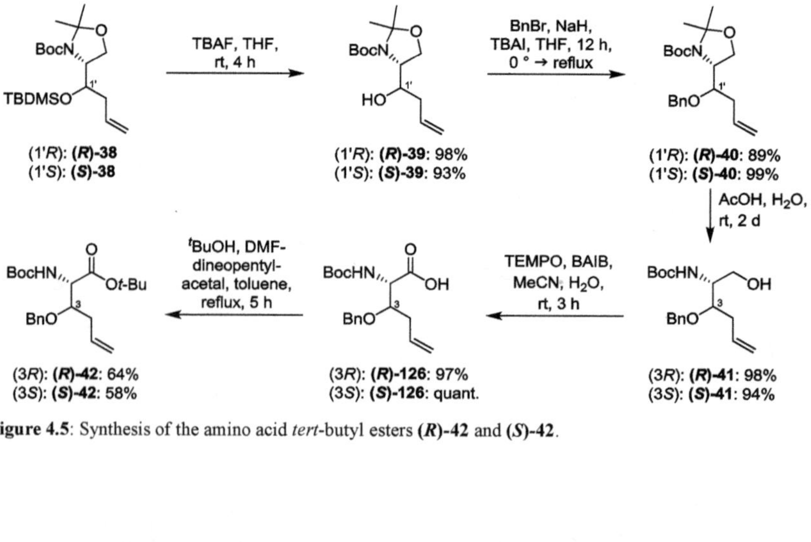

Figure 4.5: Synthesis of the amino acid *tert*-butyl esters (**R**)-**42** and (**S**)-**42**.

4.3 Synthesis of C5-Deuterated 3-Hydroxy-L-arginine Derivatives

4.3.1 Studies for the Introduction of Two Deuterium Labels at the C5-Position

The introduction of a deuterium label at the C5-position was planned by the reduction of **(R)-42** and **(S)-42** with sodium borodeuteride after the ozonolysis reaction. A reductive work-up with dimethyl sulfide should afford the crude aldeyhdes **(R)-96** and **(S)-96**, respectively. However, an equilibrium between **96** and its hemiaminal **127** exists,[161,175-176] making the following reduction difficult. Using an excess of the reducing agent (around 20 eq of NaBH$_4$) had previously been proven to result in a complete conversion.[160] However, following this strategy would only result in one deuterium label at the C5-position. As a second deuterium label could be advantageous for the analysis of feeding experiments, an alternative route for the introduction of a second deuterium label at the C5-position should be investigated. This newly developed strategy was based on an oxidation-reduction sequence (Figure 4.6). A direct oxidation of the ozonolysis product without a reductive work-up was not considered due to safety issues. The combination of potentially reactive ozonolysis products, by-products, and oxidizing agents might easily result in explosive reactions.

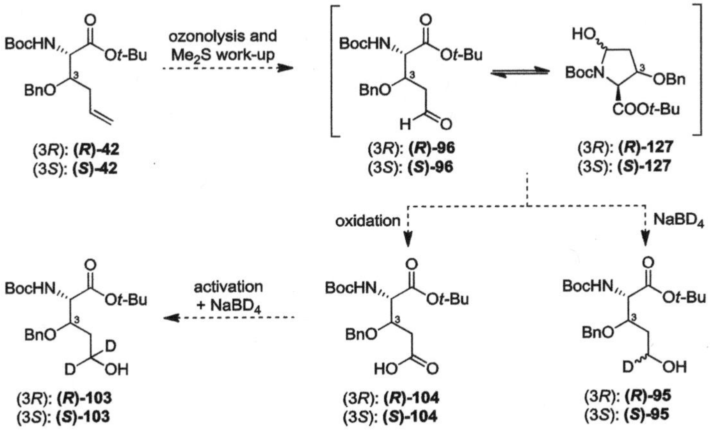

Figure 4.6: Strategy for the introduction of a second deuterium label at the C5-position.

4.3.1.1 Oxidation Attempts Directly from the Hemiaminal

As the equilibrium favors the more stable five-membered ring,[161,175-176] first oxidation experiments were conducted by using *N*-Boc-L-glutamate semialdehyde *tert*-butyl ester **128** as a model compound, which was provided by *O. Ries*.[177] The investigation of different common oxidation methods (for example TEMPO/BAIB, *Pinnick*) showed that only the oxidation to the amide **129** was possible. Stronger oxidation conditions, such as potassium permanganate, only resulted in a decomposition of the material. With respect to the oxidation to the amide **129**, the best results were obtained under *Pinnick* conditions (Figure 4.7).

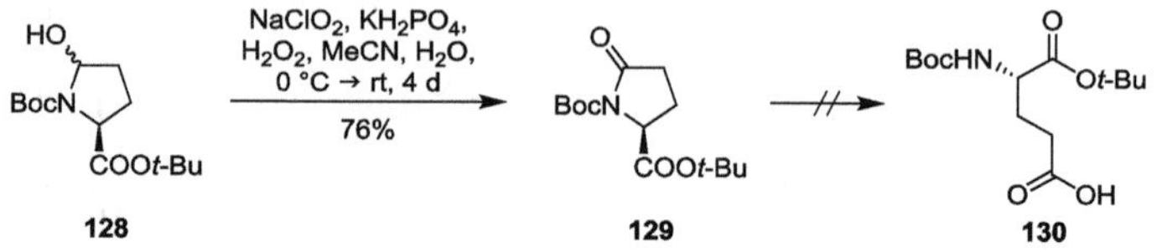

Figure 4.7: *Pinnick* oxidation of the glutamate semialdehyde derivative **128**.

Pinnick oxidation uses sodium chlorite as the oxidizing agent in combination with a scavenger for the hypochlorite formed during the reaction. With hydrogen peroxide as the trapping reagent, a yield of 76% was achieved, but a large excess of sodium chlorite was required (around 40 eq). However, as a saponification towards the corresponding carboxylic acid **130** was not achievable even under harsh conditions, a strategy to circumvent a direct oxidation of the hemiaminal would be desirable.

4.3.1.2 Diboc-Protection Strategy

As the free proton of the Boc-protected amino group was essential for a cyclization, the installation of a second Boc group should prevent the formation of the hemiaminal. A Diboc-protection of compound **(R)-42**, following a protocol by *Corrie* and coworkers,[178] resulted in Diboc-protected derivative **131** with a yield of 85% (Figure 4.8).

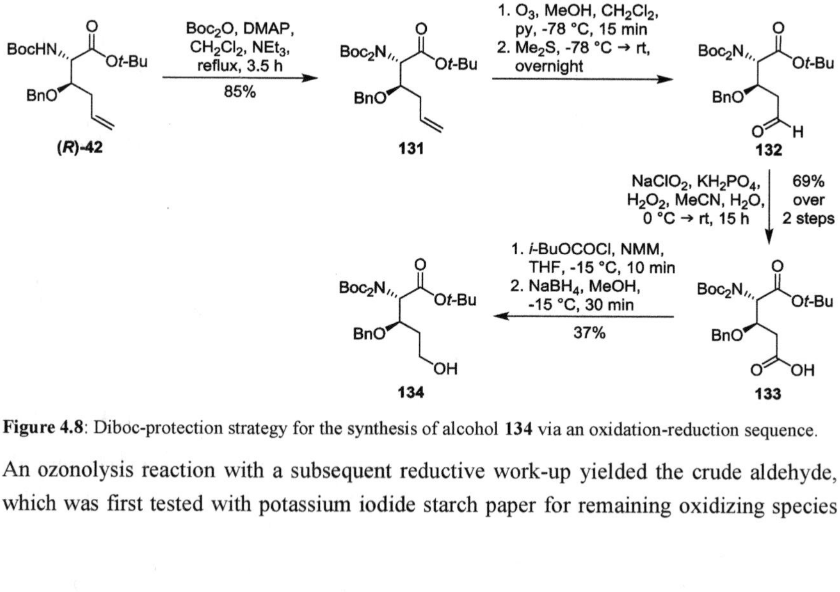

Figure 4.8: Diboc-protection strategy for the synthesis of alcohol **134** via an oxidation-reduction sequence.

An ozonolysis reaction with a subsequent reductive work-up yielded the crude aldehyde, which was first tested with potassium iodide starch paper for remaining oxidizing species before it was oxidized to the corresponding carboxylic acid **133**. For the oxidation, the *Pinnick* conditions which had been developed previously (see chapter 4.3.1.1) were used, achieving a yield of 69% over two steps. For the reduction of the carboxylic acid **133**, activation was required. During an initial attempt, following a protocol by *Miller* and coworkers,[179] an isolation of the product was not possible. The activation was performed by the slow addition of isobutyl chloroformate at -30 °C to a solution of **133** in THF and triethylamine. Subsequently, the activated compound was reduced by sodium borohydride. An activation-reduction protocol, reported by *Malachowski* and coworkers,[180] used *N*-methylmorpholine (NMM) as the base, and the reaction was performed at -15 °C. Test reactions, which consisted in reducing *N*-Boc-glycine under these conditions, resulted in yields of up to 70%. The analogous reduction of the carboxylic acid **133** delivered the alcohol **134** with a yield of only 37%. As the activation-reduction step seemed to be challenging, it was envisioned to minimize an extensive material loss by designing a model compound for optimization attempts.

4.3.1.3 Synthesis of a Model Compound for Optimization Attempts of the Activation-Reduction Sequence

Methyl ester **135** was prepared from commercially available *N*-Boc-glutamic acid 1-*tert*-butyl ester **130** via a protocol by *Hruby* and coworkers (Figure 4.9).[181]

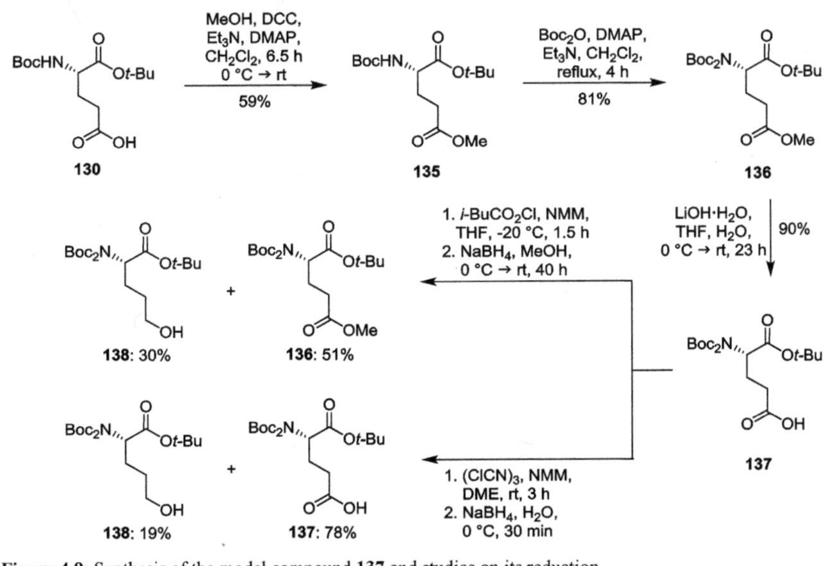

Figure 4.9: Synthesis of the model compound **137** and studies on its reduction.

A subsequent Diboc-protection, which was prepared analogously to the preparation of compound **131** (see chapter 4.3.1.2), and an alkaline hydrolysis with lithium hydroxide delivered the desired model compound **137** in an overall yield of 43% over 3 steps.

Applying the same activation-reduction conditions as for the preparation of alcohol **134**, compound **138** was obtained in a yield of 28%. A closer look at the different fractions of the chromatographic purification revealed that methyl ester **136** was formed as a byproduct in a significant amount (51%). As the reduction in methanol seemed to be problematic, water should be a better choice. A protocol by *Falorni* and coworkers shows the reduction of different protected amino acid derivatives by the activation with cyanuric chloride in dimethoxyethane (DME), followed by a reduction with sodium borohydride in water.[182] Using these conditions, alcohol **138** was obtained in a yield of 19%, and 78% of the starting material **137** could be recovered (Figure 4.9). *Falorni* and coworkers reported the reduction of *N*-Boc-glutamic acid 1-benzylester with a yield of 98%.[182] Despite the second Boc group, this compound is quite similar to the model-compound. The steric hindrance, constituted by the second Boc group, might play a major role during the reaction with the bulky activation reagents and the bulky reducing reagents. A reduction of the carboxylic acid by hydroboration was not considered since borane-d$_3$ (1 M in THF, 0.05 mol, 525 €, *Alfa Aesar*) is rather expensive in comparison with sodium borodeuteride (0.05 mol, 102 €, *Sigma Aldrich*).

If the Diboc-strategy should be applied onto the synthetic route towards 3-hydroxy-L-arginine, a removal of the second Boc group, after having introduced the deuterium labels, would be essential. The steric hindrance, provided by the second Boc group, could result in unpredictable problems during the synthetic route, which has already been established. In this context, a selective cleavage of one Boc group was envisioned before the reaction conditions for the activation-reduction sequence were optimized. As only very small amounts of compound **138** could be isolated, compound **137** was used for initial experiments regarding a selective Boc-deprotection (Figure 4.10).

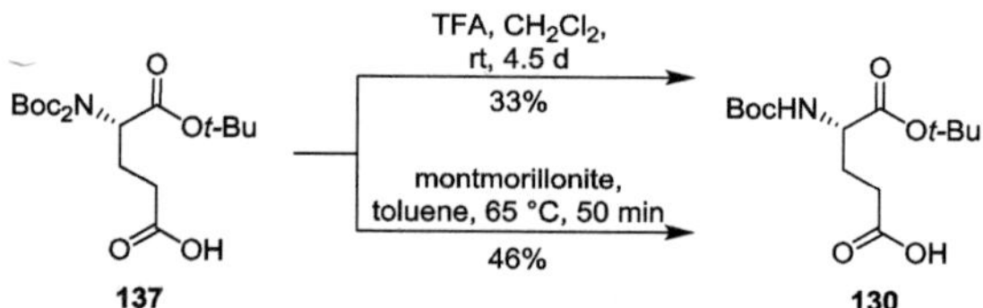

Figure 4.10: Selective Boc-deprotection of compound **137**.

Furthermore, with compound **130** being the product of this reaction, there was a preexisting reference available, which made it much easier to control the reaction by TLC. Using trifluoroacetic acid in CH$_2$Cl$_2$, even after four days, a complete conversion could not be observed. Compound **130** was obtained in a yield of 33%, and only 29% of the starting material **137** could be recovered. These results suggested that the decomposition of the product was as fast as the mono-Boc-deprotection step under these conditions. An attempt using montmorillonite for a selective deprotection resulted in a moderate conversion with a yield of 46%. In this case, no starting material could be recovered. The analysis of the mass spectrum of the crude product showed that, most likely, a Boc-rearrangement took place. This rearrangement delivered the corresponding di-*tert*-butyl ester.

The conducted experiments with model compound **137** showed that the introduction of two deuterium labels at the C5-position should be possible in principle by an oxidation-reduction sequence if a second Boc protecting group is installed. Considering the synthetic difficulties, the additional number of steps, and only moderate to poor yields, an introduction of a second deuterium label did not seem to be reasonable. Based on these studies, it was decided that only one deuterium label should be introduced at the C5-position.

4.3.2 Introduction of a Deuterium Label at the C5-Position

4.3.2.1 Ozonolysis Reaction with Subsequent Reduction

The ozonolysis reaction of **(R)-42** and **(S)-42** furnished the crude aldehydes **(R)-96** and **(S)-96**, respectively, under established conditions. The subsequent reduction was conducted in per-deuterated methanol under inert conditions. At 0 °C, small portions of sodium borodeuteride were added before the reaction mixture was allowed to warm to room temperature. As the equilibrium of the aldehyde **96** with its hemiaminal **127** requires harsh reduction conditions to ensure a complete conversion, a thorough TLC analysis was crucial. Compound **(S)-96** was reduced by adding 18 eq of sodium borodeuteride over 64 h until TLC analysis indicated a nearly complete conversion. The 5-deuterated alcohol **(S)-95** was obtained in a yield of 84% (Figure 4.11).

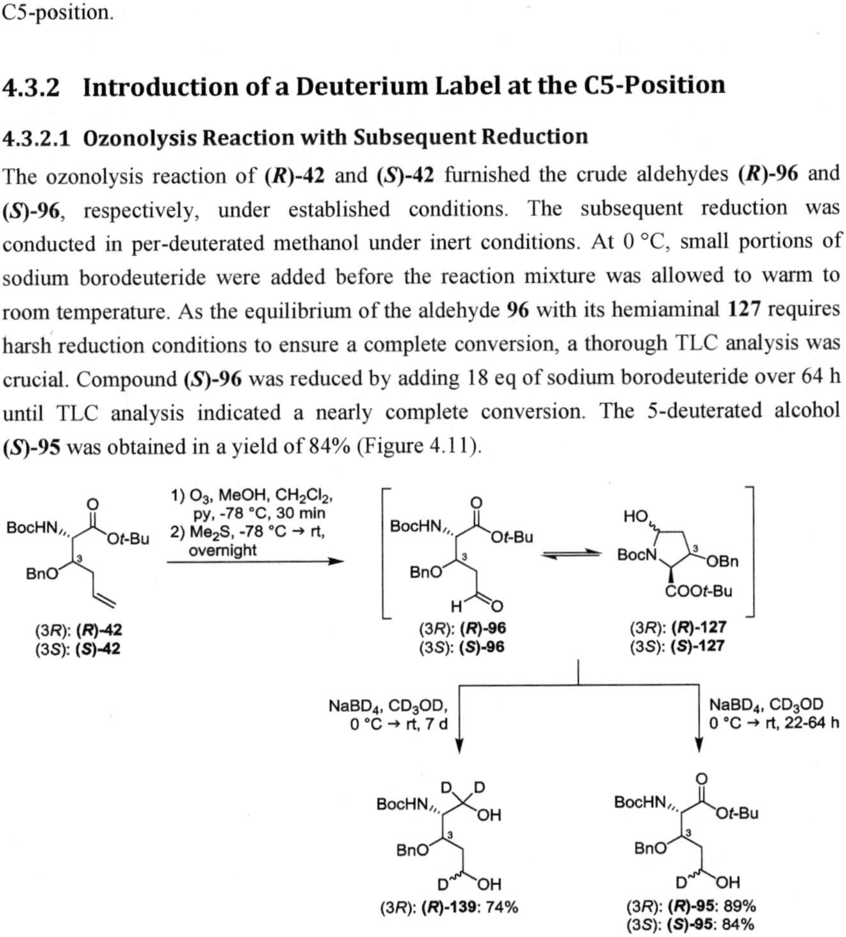

Figure 4.11: Ozonolysis reaction of **(R)-42** and **(S)-42** with subsequent reduction delivered the desired products **(R)-95** and **(S)-95**, or the corresponding byproduct **(R)-139**.

The (*R*)-isomer **(*R*)-96** was reduced by using similar conditions. As TLC analysis indicated a rather slow conversion, 16 eq of sodium borodeuteride were added over 7 days. These long reaction times seemed to favor an over-reduction of the product **(*R*)-95**. It was possible to isolate the diol **(*R*)-139** in a yield of 74%. Direct reductions of amino acid *tert*-butyl esters are not commonly performed due to a poor conversion and an epimerization risk at the C-α-position. However, using stronger reducing agents like lithium aluminium hydride, such reactions are known.[183] By adding an excess of sodium borohydride, a partial reduction of a *tert*-butyl ester was reported by *Dodd* and coworkers.[184] In case of compound **(*R*)-139**, the over-reduction only seemed to be favored by long reaction times. Eventually, the evolvement of local heat in large-scale reactions might be a factor as well. A balanced ratio between the solvent volume and the amount of sodium borodeuteride, which was added, seemed to be essential. Additional methanol-d₄ should be added after adding the first portion of sodium borodeuteride. Using 12 eq of sodium borodeuteride in the reduction of **(*R*)-96** in methanol-d₄, the 5-deuterated isomer **(*R*)-95** could be obtained in a yield of 89% after 22 h.

4.3.2.2 Guanidinylation and Subsequent Deprotection Attempts

The guanidine unit was introduced by a *Mitsunobu* reaction. The required tris-Cbz-guanidine **142** was prepared using a protocol by *Goodman* and coworkers.[185] Starting from guanidine hydrochloride **140**, compound **142** was obtained over two steps in an overall yield of 54% (Figure 4.12).

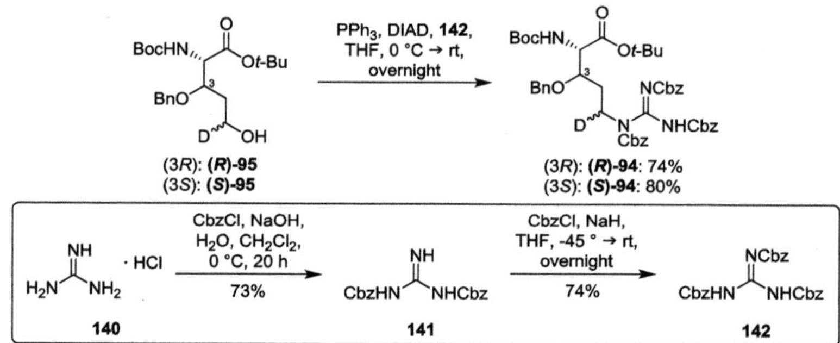

Figure 4.12: Preparation of tris-Cbz-guanidine **142** according to Goodman and coworkers,[185] and the guanidinylation furnishing **(*R*)-94** and **(*S*)-94**.

The guanidinylation of **(*R*)-95** and **(*S*)-95** under *Mitsunobu* conditions delivered the protected 5-deuterated 3-hydroxy-L-arginine derivatives **(*R*)-94** in a yield of 74%, and **(*S*)-94** in a yield of 80%, respectively.

As the deprotection strategy had already been established for the synthesis of both diastereomers of non-deuterated 3-hydroxy-L-arginines **(*R*)-17** and **(*S*)-17**, these reaction steps were not expected to result in major problems. However, using *Pearlman*'s catalyst in the hydrogenation reaction and hydrochloric acid in the subsequent acidic hydrolysis, a double signal set was observed in the ^{13}C NMR spectrum of both compounds **(R)-87** and **(S)-87** (Figure 4.13 **(b)** and Figure 4.14).

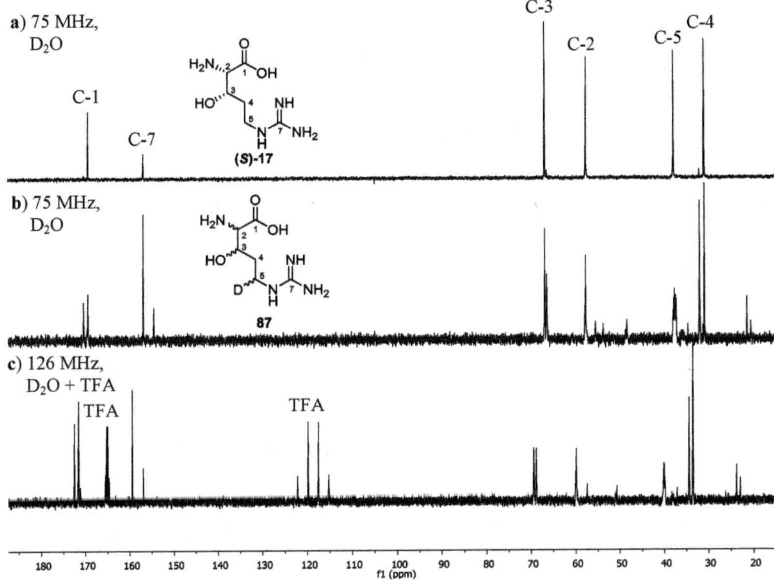

Figure 4.13: Comparison of the ^{13}C NMR spectra of **a)** compound **(S)-17** in D₂O, **b)** compound **87** in D₂O, and **c)** compound **87** in D₂O with the addition of TFA.

To ensure a homogenous protonation pattern, NMR experiments with the addition of trifluoroacetic acid were performed (**c** in Figure 4.13). As the double signal set was also found under acidic conditions, and the measurement of the ^{13}C NMR spectrum at 50 °C gave the same result, it was clear that isomerization had taken place. A reliable interpretation of the ^{1}H and ^{13}C NMR spectra of **143** was quite difficult due to numerous impurities. These observed impurities indicated that the isomerization had taken place during the hydrogenation reaction. Experiments with the earlier-synthesized, non-deuterated compounds **(S)-45** and **(S)-46**[160] proved that this assumption was correct (see **a** in Figure 4.13 and **A** in Figure 4.14). It could clearly be demonstrated that neither the deuterium label nor the acidic hydrolysis was the reason for isomerization. A closer look at the ^{13}C NMR spectra of the 'pure' 3-hydroxy-L-arginine derivatives **(S)-17** and **(R)-17** showed very small, hardly visible, doubled signals (see also **a** in Figure 4.13).

Reactions involving heterogeneous catalysis can be quite challenging. Minimal differences in the composition of a catalyst can result in a major change of reactivity. As purchasing *Pearlman*'s catalyst from different companies did not solve the problem, it is most likely that the batch of *Pearlman*'s catalyst which was used previously had the 'right' composition for the reductive deprotection reaction. In order to find the 'right' conditions, different hydrogenation conditions were investigated. Due to a limited stock of protected 3-hydroxy-L-arginine derivatives, different compounds were used for the screening experiments in order to limit the material loss of valuable compounds.

a) **(S)-45**: R = H
b) **(S)-94**: R = D

a) **46**: R = H: 84%
b) **143**: R = D: quant.

a) **17**: R = H: 96%
b) **87**: R = D: 98%

A

(S)-46

(S)-17

Acidic hydrolysis of the pure (*S*)-isomer **(S)-46** from an earlier-performed hydrogenation reaction resulted in the formation of (3*S*)-3-hydroxy-L-arginine **(S)-17**, containing only minimal impurities of **(R)-17**.

Figure 4.14: Deprotection sequence of the compounds a) (*S*)-**45** and b) (*S*)-**94**; A: the acidic hydrolysis of the diastereomerically pure precursor (*S*)-**46**.

4.3.2.3 Investigation of Hydrogenation Conditions for Cbz- and Benzyl-Deprotection

Performing the hydrogenation step under established conditions using 20 mol% of *Pearlman*'s catalyst over 24 h, a complete conversion could not be observed (Table 4.1, entry 1).

(R)-94: R = D, **(S)-45**: R = H

143: R = D, **46**: R = H

No.	compound	additional conditions	p [bar]	T [°C]	t [h]	conclusion
1	**(R)-94**	-	1	rt	24	No CC, OBn species
2	**(S)-45**	additional 20 mol% cat.	1	rt	48	CC, major impurities
3	**(R)-94**	shaker hydrogenator	3	rt	12	No CC, diff. species
4	**(S)-45**	'bubbling' of H$_2$	1	60	2	No CC
5	**(R)-94**	'bubbling' of H$_2$, RP purification	1	60	7.5	CC, double signal set

Table 4.1: Different hydrogenation conditions using 20 mol% of *Pearlman*'s catalyst in MeOH (RP = reverse phase, CC = complete conversion based on the remaining aromatic signals in the ^{1}H NMR spectra).

By ESI-MS analysis, the benzyl-protected species could be detected (m/z = 438.3 $[M + H]^+$). Doubling the reaction time to 48 h, and increasing the amount of the catalyst that was added (40 mol% in total) resulted in a complete conversion, but major impurities were observed in the ^{1}H NMR spectrum. These observed impurities indicated the formation of side products (entry 2). Considering this, the deprotection of the more stable benzyl group was suggested to be the limiting factor during the hydrogenation. An extended contact time between the substrate and the catalyst could possibly have led to isomerization. That is why it was initially envisioned to reduce the contact time. An attempt to apply more pressure (3 bar H_2) by using a shaker hydrogenator resulted in a mixture of differently deprotected species (entry 3). By bubbling hydrogen through the reaction at 60 °C, the reaction time could be reduced significantly. After 7.5 h, a complete conversion was observed (entry 5). However, a double signal set in the ^{13}C NMR spectrum was detected as well. In contrast, a shortened reaction time of 2 h did not result in a complete conversion (entry 4). 'Forcing' reaction conditions, such as an increased hydrogen pressure, higher temperature, or a more effective hydrogen introduction, did reduce the reaction time significantly. However, these conditions led to the formation of side products. As the activity of *Pearlman*'s catalyst in MeOH under a hydrogen atmosphere seemed to be either too high or too low for a benzyl-deprotection, different solvents, additives, and hydrogen sources were tested (Table 4.2). The addition of acetic acid to the reaction mixture resulted in a loss of reactivity. A complete conversion could only be achieved by adding 60 mol% of *Pearlman*'s catalyst over six days (Table 4.2, entry 1). Performing the hydrogenation in EtOAc instead of MeOH (entry 2), or using transfer hydrogenation conditions (entry 3) required increased reaction times, and these conditions resulted in the formation of side products as well. Even after three to four days, a complete conversion could not be obtained. As the activity of *Pearlman*'s catalyst could not be modified under various conditions, different catalysts were investigated.[186] By the use of palladium on carbon (10 wt%) (entry 4 and entry 7), or palladium black (entry 5), the activity was too low for a complete conversion, even with an increased amount of the catalyst. By performing the hydrogenation with other metal catalyst systems, like *Raney*-nickel in EtOH (entry 6), or zinc with ammonium formate as the hydrogen source in MeOH[187] (entry 8), even worse results were obtained. With *Raney*-nickel, only the starting material could be recovered. By the use of zinc under transfer hydrogenation conditions, a partial Cbz-deprotection could be observed.

Another problem during the investigation of hydrogenation conditions was the purification of the obtained mixtures. Fully deprotected compounds were too polar for normal phase (NP) chromatography while partially deprotected species could only be purified by normal phase chromatography, or they even could not be isolated. The purification by reverse phase (RP) chromatography resulted in the reversed problem. Despite partially deprotected species, such as **144** or **145**, it was not possible to isolate pure compounds.

conditions see table

94: R = D, **45**: R = H   **143**: R = D, **46**: R = H

observed side products

144   **145**

No.	compound	hydrogen source	catalyst	solvent	t	puri-fication	isolated compound	NMR
1	(*S*)-**45**	H_2	Pd(OH)$_2$/C (60 mol%)	MeOH/ AcOH 10:1	6 d	celite	unidentified mixture	major impurities
2	(*R*)-**94**	H_2	Pd(OH)$_2$/C (20 mol%)	EtOAc	3 d	MF	**143**[a], **144** no CC	double signal set
3	(*S*)-**94**	1,4-cyclo-hexadiene	Pd(OH)$_2$/C (40 mol%)	MeOH	4 d	MF + RP-column	no CC	aromatic signals
4	**45**	H_2	Pd/C (60 mol%)	MeOH	4 d	celite	no CC	aromatic signals
5	(*R*)-**94**	H_2	Pd black (40 mol%)	MeOH	40 h	MF	no CC	aromatic signals
6	(*S*)-**94**	H_2	Raney-Ni	EtOH	3 d	MF	(*S*)-**94**	starting material
7	(*R*)-**94**	NH$_4$COOH	Pd/C (60 mol%)	MeOH	19 h	MF + NP-column	no CC	aromatic signals
8	(*R*)-**94**	NH$_4$COOH	Zn (60 mol%)	MeOH	22 h	MF + NP-column	**145** no CC	**145**

Table 4.2: The investigation of hydrogenation conditions (MF = membrane filter, RP = reverse phase, NP = normal phase, CC = complete conversion based on the remaining aromatic signals in the ^{1}H NMR spectra, a: obtained as mixture).

With no isolated side products at hand, the interpretation of the reaction outcome was quite difficult. The final attempt was a 'step-by-step' deprotection strategy (Figure 4.15). Hence, it was envisioned, on the one hand, to reduce the contact time between the catalyst and the product, and on the other hand, to get a different reactivity of the substrate in the hydrogenation reaction. Shortened reaction times under established hydrogenation conditions resulted in a mixture of the benzyl-protected species **145**, and the deprotected species **143** (Figure 4.15). The subsequent acidic hydrolysis gave the corresponding Boc- and *tert*-butyl-deprotected derivatives **146** and **17**. A final hydrogenation step overnight removed the remaining benzyl group. An analysis of the ^{13}C NMR spectra showed again that isomerization had taken place (Figure 4.16). Surprisingly, the mixture, which was obtained by performing the first step overnight (**i**), showed a smaller second signal set than the one that was carried out over 2 h (**ii**).

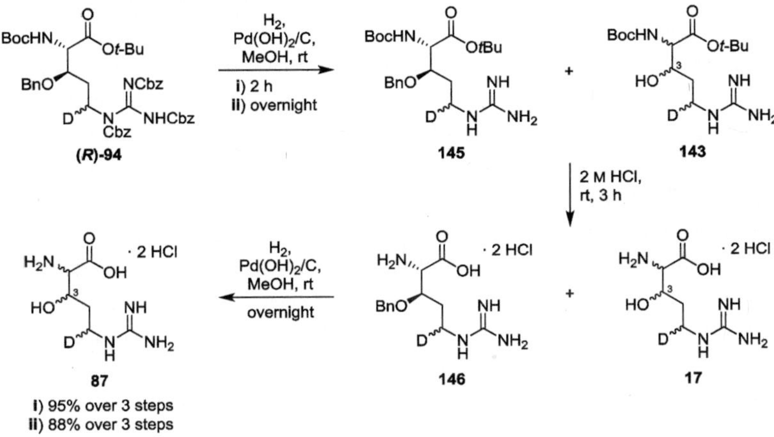

Figure 4.15: The step-wise hydrogenation of (*R*)-94.

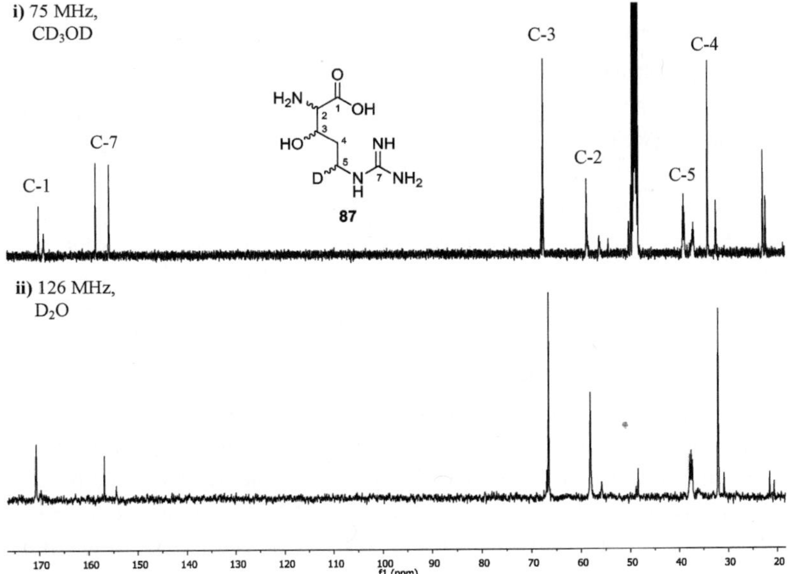

Figure 4.16: ^{13}C NMR spectra of compound **87** with **i**) the first step being performed over 2 h, and **ii**) the first step being performed overnight.

Due to previous considerations, a shortened contact time between the catalyst and the substrate should minimize isomerization. This contrary observation showed the unpredictability of the hydrogenation reaction with little prospect to find the 'right' conditions within a reasonable time, especially since it was not possible to encircle the exact problem.

The interest in the role of trace contaminants in metal-catalyzed reactions has increased over the last years. For example, in a putatively palladium-free, copper-catalyzed *Sonogashira* coupling, ppb levels of palladium proved to be involved most likely.[188] Another example are copper contaminants that play a role in iron-catalyzed cross coupling reactions.[189] Probably, such an unpredictable behavior based on trace contaminants in commercially purchased catalysts is commonly more known as presented by literature since negative results are hard to publish. However, metals are always difficult to free from low levels of contaminants of related metals or other impurities. There is always the chance that impurities are present which give better or different results than experienced in previously performed reactions. In these cases, the best option is, most likely, to find an alternative pathway. As the Cbz-deprotected compound **145** could be isolated before, and as no double signal set could be observed with this compound so far, the protection strategy for the 3-hydroxy group was changed to the *tert*-butyldiphenylsilyl (TBDPS) group.

4.3.2.4 Protection of the 3-Hydroxy Group as *tert*-Butyldiphenylsilyl Ether

For initial attempts to perform a *tert*-butyldiphenylsilyl (TBDPS) protection, diastereomeric mixtures of homoallylic alcohol **39**, which had a different diastereomeric ratio than 2:1 (see chapter 4.2.1), were used. As a consequence, entries 1-4 (Table 4.3) are not necessarily representative for the isolated yields of both diastereomers. TBDPS-protection was carried out analogously to TBDMS-protection (see chapter 4.2.1) by using TBDPSCl and imidazole in DMF (Table 4.3). The yield could be improved by increasing the amount of the TBDPSCl added (entry 3-4). Applying heat did not seem to improve the conversion significantly (entry 2). The factor which seemed to have a larger impact was the reaction time (entry 1 and 3). Although the yield of the reaction could be improved, the purification proved to be difficult. An increased amount of TBDPSCl used in the reaction resulted in purification problems. It was shown that 2-3 eq of TBDPSCl should be enough for satisfying yields (entry 6 and 7). The separation of the diastereomeric mixture proved to be difficult as well (entry 5-7). The major diastereomer **(R)-147** could be isolated in good yields of up to 57% on a small scale (entry 5), 46% on a medium scale (entry 6), and 30% on a large scale (entry 7). However, especially on larger scales, it was difficult to even isolate the pure (*S*)-diastereomer **(S)-147** (see entry 7).

TBDPSCl, imidazole, DMF, additional conditions see table

$$39 \longrightarrow 147$$

(3*R*):   (*R*)-147
(3*R+S*): 147
(3*S*):   (*S*)-147

No.	39 [g]	TBDPSCl [eq]	imidazole [eq]	T	t	yields of the separated diastereomers 147	yield in total
1[a]	0.1	1.5	2.5	rt	5 d	(*R*) 8%, (*S*) 32%, (*R+S*) 6%	46%
2[a]	0.1	1.6	2.0	80 → 100 °C	5 h	(*R*) 2%, (*S*) 2%, (*R+S*) 9%	13%
3[a]	0.1	1.6	2.0	rt	3 d	(*R*) 5%, (*S*) 30%	35%
4[a]	0.1	3.0	4.0	rt	3 d	(*R*) 11%, (*S*) 2%, (*R+S*) 67%	80%
5	0.1	5.0	6.0	rt	3 d	(*R*) 57%, (*R+S*) 37%	94%
6[b]	2.2	3.0	4.0	rt	3 d	1st: (*R*) 40%, (*R+S*) 41% 2nd: (*R*) 6%, (*S*) 4%, (*R+S*) 29%[b]	81%
7	12	2.0 + 0.5	2.0	rt	4 + 1 d	(*R*) 30%, (*R+S*) 58%	88%

Table 4.3: TBDPS-protection of **39** under different conditions (a: different diastereomeric ratio of the starting material, b: two subsequently performed column chromatographies).

Therefore, a strategy analogous to the benzyl-protection could be feasible. Starting with TBDMS-protection and a separation of the diastereomers, followed by a TBDMS-deprotection and TBDPS-protection of each pure diastereomer should solve the problem. As the other steps of this strategy have all been performed successfully, an according synthesis of diastereomerically pure **147** in good yields can be considered accomplished. For the following route, it was decided to investigate the synthetic steps with the TBDPS-protected alcohol (*R*)-**147** until the end of the route, and with (*S*)-**147** for the next steps only. An assignment of the configuration of the diastereomers was done based on the assigned configurations of the desilylated alcohols (*R*)-**39** and (*S*)-**39** (see chapter 4.2.1). TBDPS-deprotection of (*R*)-**147** and (*S*)-**147**, using tetra-*n*-butylammonium fluoride (TBAF) trihydrate in THF, gave the corresponding alcohols (*R*)-**39** and (*S*)-**39**. A comparison of the ^{1}H NMR and ^{13}C NMR spectra proved, analogously to the TBDMS-protected alcohols (*R*)-**38** and (*S*)-**38**, that the faster eluting compound was the one with the (*R*)-configuration.

4.3.2.5 Introduction of a C5-Deuterium Label within the TBDPS-Protection Strategy

After the TBDPS-protection step, the preparation of the protected 3-hydroxy-L-arginine derivative (*R*)-**152** was accomplished analogously to the benzyl-protection route (see chapter 4.2.2, 4.3.2.1 and 4.3.2.2). The first two steps were performed with both of the diastereomers (*R*)-**147** and (*S*)-**147**, providing the alcohols (*R*)-**148** and (*S*)-**148** in yields of

87% and 86%, respectively, and the carboxylic acids (***R***)-**149** and (***S***)-**149** in yields of 93% and 88%, respectively (Figure 4.17).

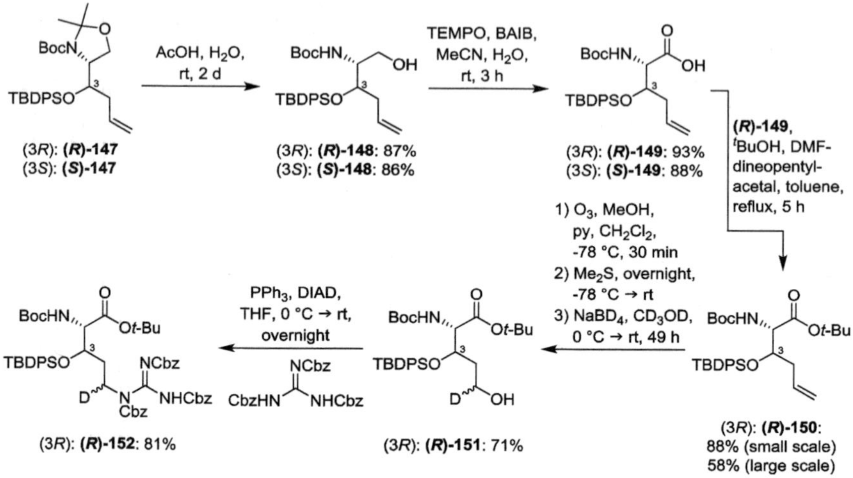

Figure 4.17: Synthesis of the 5-deuterated 3-hydroxy-L-arginine derivative (***R***)-**152**.

A purification of the oxidation product from the TEMPO/BAIB reaction by a basic and an acidic extraction did not succeed in the case of the compounds (***R***)-**149** and (***S***)-**149**. The aqueous solution was directly acidified, and, after extraction, the crude product was purified by column chromatography. Esterification under *Eschenmoser* conditions proceeded surprisingly well on a small scale. The desired product (***R***)-**150** was obtained in a yield of 88%. On a large scale, the yield of 58% was consistent with the yields of previously reported *tert*-butyl esterifications under *Eschenmoser* conditions. After ozonolysis reaction and reductive work-up, the C5-deuterium label could be introduced by using sodium borodeuteride to give (***R***)-**151** with a yield of 71%. The subsequent guanidinylation reaction under established *Mitsunobu* conditions furnished TBDPS-protected 3-hydroxy-L-arginine derivative (***R***)-**152** in a yield of 81%.

4.3.2.6 TBDPS Deprotection

An initial attempt for TBDPS deprotection was carried out analogously to the deprotection of the compounds (***R***)-**147** and (***S***)-**147**, which was done to verify the stereochemical configuration (see chapter 4.3.24). However, by using TBAF trihydrate in THF with a reaction time of 23 h, the removal of one Cbz group and elimination took place (Figure 4.18 and entry 1 in Table 4.4). The side product **153** was obtained in a yield of 35%, and it was characterized by ^{1}H NMR, ^{13}C NMR, ^{1}H,^{1}H COSY, ^{1}H,^{13}C HSQC, and ESI-HRMS. An observed coupling between the H-5 to an NH-signal shows that the Cbz group was removed from NH-6 (Figure 4.18).

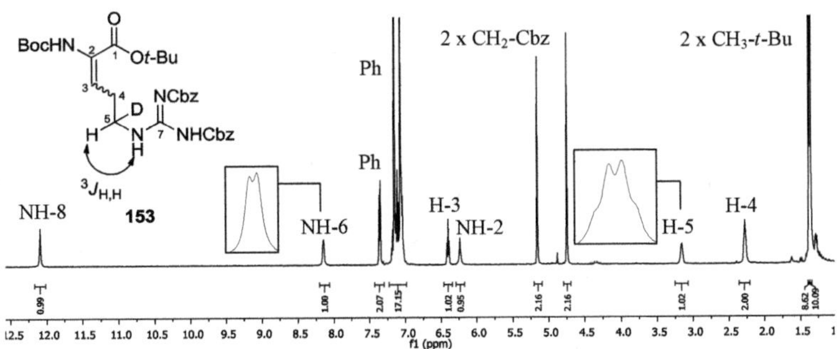

Figure 4.18: ^{1}H NMR (500 MHz, C_6D_6) spectrum of the elimination product **153**.

The ^{13}C NMR shift of C-2 and C-3 were in the range of typical double bonds with δ (C-2) = 126.58 ppm   and δ (C-3) = 129.38 ppm.   Furthermore,   ESI-HRMS   analysis showed a signal at m/z = 620.2827, which was consistent with a calculated mass of $[M_{153}+Na]^+$ with m/z = 620.2801.

Using ammonium fluoride in MeOH as the deprotection agent, only the removal of Cbz groups could be observed. The compounds **155** and **156** were obtained as a mixture (entry 2 in Table 4.4). Further deprotection attempts with TBAF trihydrate under a shortened reaction time (entry 3) and with a TBAF solution (1.0 M in THF, entry 4 and 5) resulted in elimination and a partial removal of the Cbz group as well. In this context, the elimination seemed to be favored by increased reaction times (entry 4). Based on these considerations, it was tried to perform the deprotection under buffered conditions. By the addition of acetic acid,[190] the elimination product **153**, the partially Cbz-deprotected compound **156**, and the starting material **(R)-152** were obtained. (entry 6). Besides these compounds, another side product was isolated. Analysis of this side product by ESI-MS revealed a mass to charge ratio of m/z = 750.3488, which was consistent with a calculated mass of the desired product **(R)-154** with m/z = 750.3455 ($[M_{154}+Na]^+$). As an additional NH proton, which coupled with H-5, could be observed by ^{1}H NMR experiments and a partial Cbz-deprotection could be ruled out since the signals for all of the three Cbz groups were observed, a rearrangement of a Cbz group was proposed leading to **157**. Urethane-protected guanidines are known to cause problems due to rearrangement reactions,[185] especially under basic conditions.[169] A rearrangement of the Cbz group from N^5 to the hydroxyl group would also explain the observation of the elimination product **153**. Such elimination reactions of carbonate derivatives under basic conditions can be exploited for the synthesis of didehydroamino acids. *Chandrasekaran* and coworkers reported a synthesis of didehydroamino acids from carbonate derivatives of serine and threonine using TBAF.[191] A phosphate-buffered reaction only gave a mixture of the partially Cbz-deprotected product **156** and the elimination product **153** (entry 7). A deprotection reaction using hydrogen fluoride pyridine complex[192] resulted in elimination, Cbz-removal, and, most likely, decomposition (entry 8). By using the hydrogen fluoride triethylamine complex, it is possible to ensure less harsh conditions.[192] An according deprotection at room temperature only resulted in a poor conversion (entry 9). Approximately 85% of the unpurified starting material **(R)-152** could be recovered next to 16% of the side product **157**. Performing the reaction under reflux[193] delivered the unpurified, partially Cbz-deprotected species **156** and decomposition products, which could not be identified (entry 10).

No.	reagent	solvent	additive	T	t	isolated side products or starting material
1	TBAF·3H$_2$O (2.5 eq)	THF	-	rt	23 h	**153** (35%)
2	NH$_4$F (10 eq)	MeOH	-	rt	6 d	**155**[a], **156**[a]
3	TBAF·3H$_2$O (2.5 eq)	THF	-	rt	4.5 h	**(R)-152** (43%), **153** (23%)
4	TBAF in THF (2.5 eq)	THF	-	rt	3 d	**153**[a], **155**[a], **156**[a]
5	TBAF in THF (1.1 eq)	THF	-	rt	3 h	**153** (23%)
6	TBAF in THF (1.1 eq)	THF	AcOH	rt	3 h	**(R)-152** (15%), **153**[a], **156** (29%), **157** (17%)
7	TBAF, buffered solution (3.3 M)	THF	KH$_2$PO$_4$ solution	rt	10 d	**153** (29%), **156** (20%)
8	HF·Py (28 eq)	THF	-	0 °C → rt	4 h	**155** (37%), **153** (23%)
9	HF·NEt$_3$ (30 eq)	THF	NEt$_3$	rt	11 d	**(R)-152**[a], **157** (16%)
10	HF·NEt$_3$ (20 eq)	MeCN	NEt$_3$	reflux	7 d	**156**[a], decomposition

Table 4.4: TBDPS-deprotection attempts of **(R)-152** under different conditions (a: obtained as mixture or as impure material).

Cbz-deprotections due to the use of TBAF are commonly known.[194] A global TBDPS- and Cbz-deprotection would be advantageous. However, analogous to the hydrogenation problem, the results obtained showed again how sensitive the deprotected 3-hydroxy-L-arginine derivatives can be. Either the conditions were too smooth for conversion, or too harsh, leading to elimination or decomposition. This observation did not only make the deprotection a challenging task, but it also made the synthesis of 3-hydroxy-L-arginine intriguing. As a TBDPS-deprotection prior to hydrogenation did not seem to be possible due to a rearrangement of the N^5-Cbz group, it was envisioned to remove the Cbz groups first.

4.3.2.7 Hydrogenation of the TBDPS-Protected 3-Hydroxy-L-arginine Derivatives Followed by Global Acidic Hydrolysis

By using 10 mol% of *Pearlman*'s catalyst in MeOH under a reaction time of 1.5 h (Table 4.5, entry 1), an initial attempt to hydrogenate (***R***)-**152** led to the same problems as in previously-performed hydrogenation reactions (see chapter 4.3.2.2). The analysis of the ^{13}C NMR spectrum of the obtained crude mixture of **158** revealed signs of isomerization, or the formation of side products. The deprotection of the crude mixture of **158**, using TBAF trihydrate (**ii**), resulted in purification problems due to the polarity of the putative product **143**. A hydrogenation approach using triethylsilane as the hydrogen source and palladium(II) chloride as the catalyst resulted in purification problems as well (entry 2). Considering these results, deprotection strategies with easily removable reagents were investigated only (entry 3-7, Figure 4.5).

Although the observation that isomerization or the formation of side products had taken place showed that the removal of the benzyl group had not been the essential problem before, further hydrogenation conditions should be investigated since the Cbz groups should be less difficult to remove. Acidic hydrolysis, conducted in 6 M hydrochloric acid, was shown to result in a global deprotection of the Boc group, the *tert*-butyl group, and the TBDPS group (entry 3). However, when the unpurified mixture of **158** was used, a pure diastereomer could not be obtained. Using palladium on charcoal as the catalyst, in a solvent system of EtOH and EtOAc (1:1)[195] and with a subsequent purification by normal phase chromatography (CH$_2$Cl$_2$/MeOH 9:1→8:1), the pure diastereomer (***R***)-**158** was furnished for the first time, and after acidic hydrolysis, the pure diastereomer (***R***)-**87** was obtained (entry 4). Nevertheless, the analysis of the ^{13}C NMR spectrum of the crude hydrogenation product of **158** showed that isomerization or the formation of side products had taken place. As only the isolation of the pure diastereomer (***R***)-**158** was possible by normal phase column chromatography, it was tried to isolate the side products by reverse phase column chromatography (entry 6 and 7), but no compounds could be isolated. In both cases (entry 6 and 7), an analysis of the ^{13}C NMR spectrum of the crude hydrogenation product of **158** indicated the formation of side products. As it could be shown that the acidic hydrolysis of the pure compound (**R**)-**158** resulted in the desired pure diastereomer (**R**)-**87**, a strategy without the necessity of a hydrogenation step should be developed.

No.	hydrogen source	catalyst	solvent	t	purification of i	further deprotection	yield	side products
1	H_2	Pd(OH)$_2$/C (10 mol%)	MeOH	90 min	MF	ii	i) **158**: 87%[a]	yes
2	HEt$_3$Si	PdCl$_2$ (30 mol%)	CH$_2$Cl$_2$ + NEt$_3$ (3.0 eq)	4 h	MF + RP column	-	i) no purification possible	-
3	H_2	Pd(OH)$_2$/C (10 mol%)	MeOH	70 min	MF	iii: 24 h, EtOAc	i) **158**: quant[a] iii) **87**: quant[a]	yes
4	H_2	Pd/C (80 mol%)	EtOH/ EtOAc (1:1)	19 h	MF + NP column	iii: 4 h	i) (**R**)-**158**: 31% iii) (**R**)-**87**: 43%	yes[b]
5	H_2	Pd/C (20 mol%)	EtOH/ EtOAc (1:1)	80 min	MF + NP column	-	i) (**R**)-**158**: 9%	yes[b]
6	1,4-cyclohexadiene	Pd/C (30 mol%)	MeOH	24 h	MF + RP column	-	i) no purification possible	yes
7	H_2	Pd/C (20 mol%)	EtOH/ EtOAc (1:1)	85 min	MF + RP column	-	i) no purification possible	yes

Table 4.5: Hydrogenolysis of (**R**)-**152** under different conditions (i) followed by two different TBDPS deprotection strategies (ii,iii) (a: obtained as mixture, b: although the formation of side products was observed, an isolation of the pure diastereomer (**R**)-**158** was possible, MF = membrane filter, RP = reverse phase, NP = normal phase).

4.3.2.8 Alloc-Protection Strategy for the Guanidine Moiety

Besides tris-Cbz-guanidine **142**, *Goodman* and coworkers also reported the synthesis of tris-Boc-guanidine[185] and bis-Alloc-guanidine **160** with tris-Alloc-guanidine **159** as a side product.[196] As the synthesis and the application of tris-Boc-guanidine had already proven to be problematic,[197] it was envisioned to prepare tris-Alloc-guanidine **159** for the application in guanidinylation reactions. It was tried to change the established conditions of *Goodman* and coworkers for the preparation of bis-Alloc-guanidine **160** in a way that tris-Alloc-guanidine **159** was formed preferentially. *Goodman* and coworkers performed the reaction of guanidine hydrochloride **140** by using allyl chloroformate at 0 °C under basic conditions in a two-phase system with benzyltriethylammonium chloride as a phase transfer catalyst.[196] However, by extending the reaction time to 23 h and warming the reaction to room temperature, only 4% of tris-Alloc-guanidine **159** could be obtained (**i** in Figure 4.19).

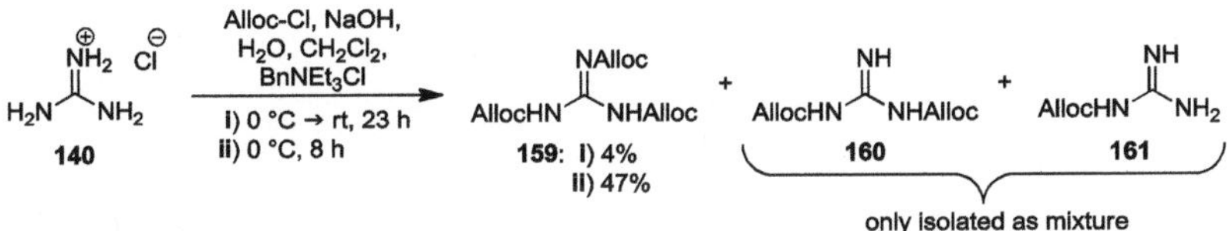

Figure 4.19: Synthesis of tris-Alloc guanidine **159** with mono- and bis-Alloc-guanidine **160** and **161** as side products.

The rest was a mixture of mono- and bis-Alloc-guanidines **160** and **161**. TLC analysis during the reaction showed that, most likely, an extended reaction time and higher temperatures led to the cleavage of the Alloc group. Using allyl chloroformate in excess (6 eq), and performing the reaction at 0 °C for 8 h afforded tris-Alloc-guanidine **159** in a yield of 47% (**ii** in Figure 4.19). Analogous conditions to the preparation of tris-Cbz-guanidine,[185] by the addition of sodium hydride and allyl chloroformate to a solution of bis-Alloc-guanidine in THF, only resulted in a poor conversion. Considering this fact, it did not seem to be advantageous to separate the mixture of mono- and bis-Alloc-guanidines. With enough tris-Alloc-guanidine **159** at hand, it was not intended to improve the reaction conditions of **ii** at this stage (Figure 4.19).

The guanidinylation reaction of (**R**)-**151**, using tris-Alloc-guanidine **159** under *Mitsunobu* conditions, delivered the guanidinylated compound (**R**)-**162** in yields of up to 90% (Figure 4.20). The deprotection of the Alloc group was achieved by the catalysis with tetrakis(triphenylphosphine)-palladium(0) in THF.[177,198] Dimedone was added as a scavenger for the generated allyl species. Thereby, it prevented undesired alkylations, and simultaneously, it regenerated the palladium(0) catalyst. By successive acidic and basic extractions, like described by *Johnson* and coworkers,[198] no product could be isolated because the product seemed to be too unpolar for such a purification strategy. However, an acidic washing step was essential for a proper phase separation. As the remainders of the catalyst could neither be removed by a combined acidic and basic work-up nor by column chromatography, the Alloc-deprotected compound (**R**)-**158** could only be obtained as an impure material. However, a subsequent acidic hydrolysis of compound (**R**)-**158** using 6 M hydrochloric acid led to the pure diastereomer (**R**)-**87** with a yield of 90% over two steps (Figure 4.20). Due to solubility issues of (**R**)-**158** in hydrochloric acid, an ultrasonic treatment, which was repeatedly performed over 24 h, was essential. Without the ultrasonic treatment, only moderate yields of ca. 40% could be obtained (see also Table 4.5, entry 5).

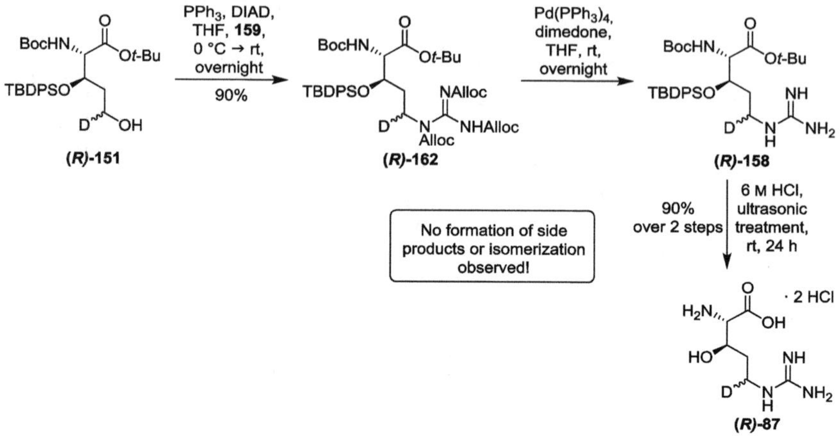

Figure 4.20: Synthesis of the deuterated (3*R*)-3-hydroxy-L-arginine derivative (*R*)-87 via the TBDPS/Alloc-protection strategy.

Due to the novel TBDPS/Alloc protecting strategy, the application of palladium-catalyzed hydrogenolysis, which led to isomerization or side product formation, could be prevented. The newly developed global acidic deprotection step then furnished diastereomerically pure 3-hydroxy-[5-^{2}H]-L-arginine (*R*)-87 in excellent yield.

4.3.2.9 Elucidation of the Isomerization and of the Formation of Side Products During the Palladium-Catalyzed Hydrogenation

The elimination occuring during the attempts to remove the TBDPS protecting group using TBAF was related with a rearrangement of a Cbz group. This rearrangement led to the corresponding carbonate **157**, which underwent elimination under the basic conditions (see chapter 4.3.2.6). However, it could be assumed that the OTBDPS group might also act as a leaving group under certain conditions. *Shibasaki* and coworkers reported a β-elimination of a TBDPS-protected hydroxyl group by TBAF during an attempt to perform a TBDPS-deprotection of an aldehyde derivative. Thus, a conjugated enal was formed.[199] *Wang* and coworkers described transition-metal-catalyzed formations of carbon-carbon double bonds via organozinc reagents and carbonyl compounds by the addition of trimethylsilyl chloride (TMSCl).[200] A mechanism was proposed which involved the formation of a secondary silyl ether, followed by the loss of trimethylsilanol to give the (*E*)-alkene. *Zhang* and coworkers doubted the key role of transition-metal catalysis during this olefination. They proposed TMSCl to be essential as a Lewis acid promoter. However, they assumed that a similar elimination mechanism would be involved.[201] Although, under most conditions, silyloxy groups are more labile to hydrolysis rather than to elimination, a closer look at the literature showed that the problem of the β-elimination of silyloxy groups during total syntheses is known, especially when conjugated systems can be formed. *Evans* and coworkers described a hydrolysis of polyfunctional amides where a base-promoted elimination of the C3-silyloxy group took place.[202] With those examples in mind, it could be proposed that the OTBDPS group acted as a leaving group in the described palladium-catalyzed hydrogenation reactions, leading to the formation of side products, or isomerization. As an isolation of the side products of the hydrogenation reaction was not possible, and as the ^{1}H and ^{13}C NMR spectra of the crude mixtures provided only limited information, putative side products were identified by a detailed analysis of ESI-MS spectra of the crude mixtures (Figure 4.21).

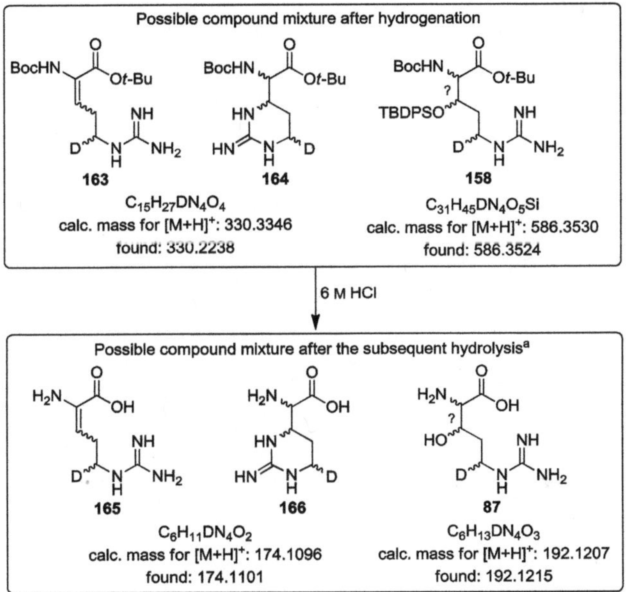

Figure 4.21: The possibly formed side products during the hydrogenation reaction of compound (**R**)-**152**, and their identification by ESI-HRMS (a: the compounds were obtained as their hydrochlorides).

The analysis of the mixture, which was obtained by the Pd/C-catalyzed hydrogenation of TBDPS-protected compound (**R**)-**152** in a solvent mixture of EtOH and EtOAc, showed two significant peaks in the ESI-HRMS spectrum, one at m/z = 586.3524 (100), and one at m/z = 330.2238 (61). The most intense peak at m/z = 586.3524 would be consistent with the Cbz-deprotected, desired product **158**. The other significant peak at m/z = 330.2238 would be consistent with two structural isomers, the elimination product **163**, and capreomycidine derivative **164** formed via a putative ring-closure. As a reliable interpretation of the NMR spectra was not possible, it could not be ruled out that an isomerization of **158** had already taken place, resulting in a diasteromeric mixture. After the acidic hydrolysis, some impurities could be removed by reverse phase chromatography, and different fractions could be obtained. This achievement made the spectroscopic analysis of the resultant mixture slightly easier. ESI-MS revealed signals that would be consistent with compounds **165**, **166**, and **87** (Figure 4.21). A comparison of the ^{1}H NMR spectra of the pure diastereomers (***R***)-**87** (**a** in Figure 4.22 and Figure 4.23) and (***S***)-**87** (**b**) with the obtained mixtures of the hypothetical compounds **165**, **166**, and **87** (**c** and **d**) showed that, besides an isomeric mixture of **87**, a significant amount of more than one side product was definitely obtained. As an analysis of the ^{13}C NMR spectra did not reveal any signals which would be consistent with typical shifts of double bonds, it was proposed that the product **166**, obtained via ring-closure, was formed during the hydrogenation reaction, resulting in up to four possible stereoisomers. However, it could not be entirely ruled out that the elimination product **163** was formed during the hydrogenation as well.

There are two possibilities why the corresponding hydrolyzed species **165** could not be observed after the acidic deprotection step: 1) the purification by reverse phase chromatography was not suitable for this compound, or 2) an acid-catalyzed addition of water had taken place, which led to a diastereomeric mixture of the 3-hydroxyspecies **87**.

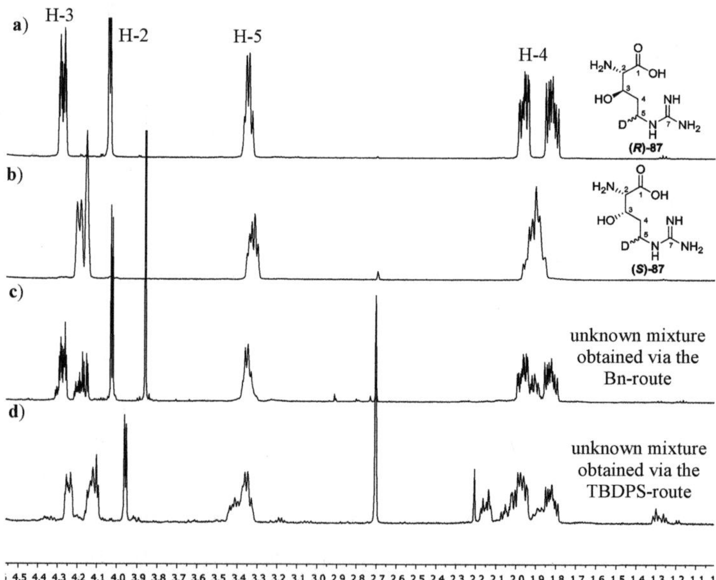

Figure 4.22: Comparison of the ^{1}H NMR spectra (500 MHz, D$_2$O) of the diastereomerically pure compounds (***R***)-**87** and (***S***)-**87** (**a** and **b**) with the mixtures obtained by the hydrogenation reactions with subsequent acidic hydrolysis of the Bn-protected species (**c**), and the TBDPS-protected species (**d**).

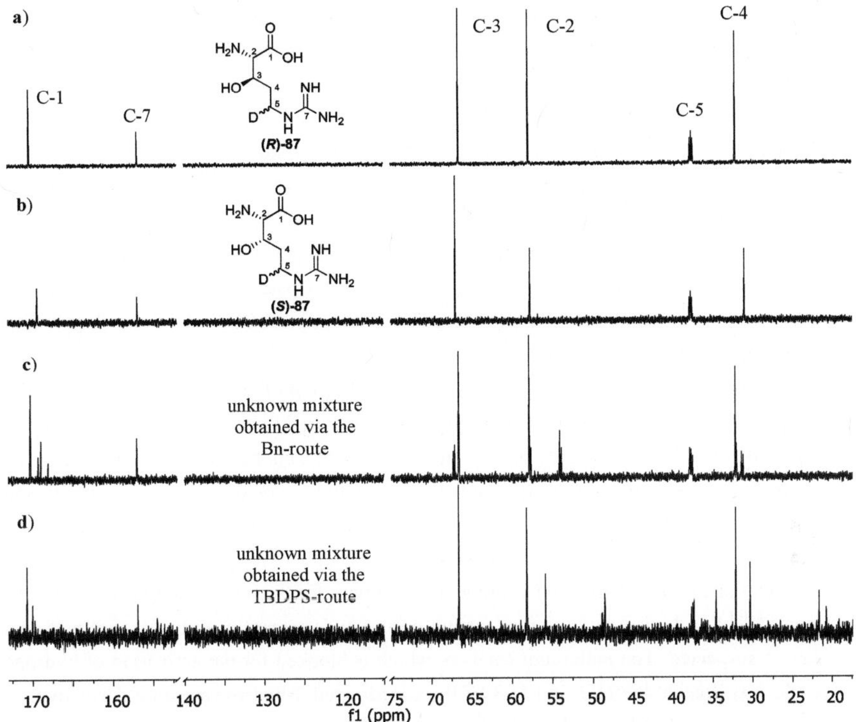

Figure 4.23: Comparison of the ^{13}C NMR spectra (126 MHz, D_2O) of the diastereomerically pure compounds **(R)-87** and **(S)-87** (**a** and **b**) with the mixtures obtained by the hydrogenation reactions with subsequent acidic hydrolysis of the Bn-protected species (**c**), and the TBDPS-protected species (**d**).

As compound **163** should be polar enough for reverse phase chromatography, it was most likely that an acid-catalyzed addition of water had taken place. This would explain the fact that different impurities were observed after the hydrogenation step, and that a clearer picture of the reaction outcome was given after the acidic hydrolysis. To get an idea how such an unusual reaction behavior could occur in palladium-catalyzed hydrogenations, it is useful to have a closer look at the rather unusual functionalities of 3-hydroxyarginine derivatives.

Guanidine, as a functional group found in biologically active compounds, can, for example, mediate transition state binding. In this context, a non-covalent hydrogen bonding of the guanidinium group with anions, such as carboxylates or phosphates, plays an important role.[203] Guanidine itself is categorized as a strong organic base with a pK_a value of its conjugated acid around 13. This high basicity and a remarkable stability are based on the controversially discussed 'Y aromaticity' of the protonated species, which leads to a delocalized π-system.[204] Due to these chemical properties, guanidine is not only interesting as a highly functional moiety of biologically active compounds, but it is also interesting regarding catalysis. Hence, guanidine derivatives were not only elucidated as catalytically active species themselves,[205] but also as ligands for catalysts.[206] Guanidine derivatives, which are used in coordination chemistry, mostly bind via the imino substructure. In addition, bidentate binding via the imino and amino group could be observed, providing a hemilabile system.[207] The coordination of palladium through guanidine was used, for example, in a *Heck* reaction with $PdCl_2$ in a guanidine acid-base ionic liquid.[208] Such a system does not only offer ligands for the stabilization of the activated palladium species, but it also offers a strong base, which favors β-elimination, and polar solvent conditions. Another example is the use as a directing group in palladium-catalyzed C-H functionalization.[209]

Considering the metal binding abilities of guanidine, coordination towards the palladium catalyst during hydrogenation can be assumed. This would keep the 3-hydroxy-L-arginine derivative in a direct proximity to the catalyst. As such a coordination would compete with the hydrogen loading ability of the catalyst, the reactivity would be decreased significantly, resulting in extended reaction times and in an incomplete conversion. In 2008, *Coquerel* and *Rodriguez* reported that activated double bonds can be reduced by a Pd/C-triethylamine system with triethylamine as the source for the newly incorporated hydrogen atoms.[210] For the isomerization occurring during hydrogenation, a similar mode of action can be suspected. The palladium catalyst, which is blocked for the activation of hydrogen, could just 'snatch' the H-2 and H-3 of the coordinated 3-hydroxyarginine derivative. The unselective reattachment of the hydrogen atoms at C2 and C3 would lead to a loss of stereochemical information, and therefore isomerization would occur. In this context, concerted mechanisms as well as non-concerted mechanisms might principally play a role. For the formation of a putative elimination product **163** or the corresponding ring-closure derivative **164**, radical mechanisms might be conceivable leading to a highly unselective reaction outcome. Having all these considerations in mind, a speculative mechanism, which would lead to the observed formation of side products and isomerization during the palladium-catalyzed hydrogenation, was proposed. This hypothetical mechanism is depicted in Figure 4.24. Based on ESI-MS measurements, the same side products could be assumed for the hydrogenation of the benzyl-protected compounds (**R**)-**94** and (**S**)-**94**, and therefore, the same hypothetical mechanism was suggested (Figure 4.24). However, a closer look at the ^1H and ^{13}C NMR spectra of the mixture, which contained the deprotected species **166** and **87**, showed that, presumably, isomerization was favored in the case of the OBn-protected species (Figure 4.22 and Figure 4.23). This mixture (**c**) seemed to be mainly a diastereomeric mixture of (**R**)-**87** and (**S**)-**87** (see **a** and **b**).

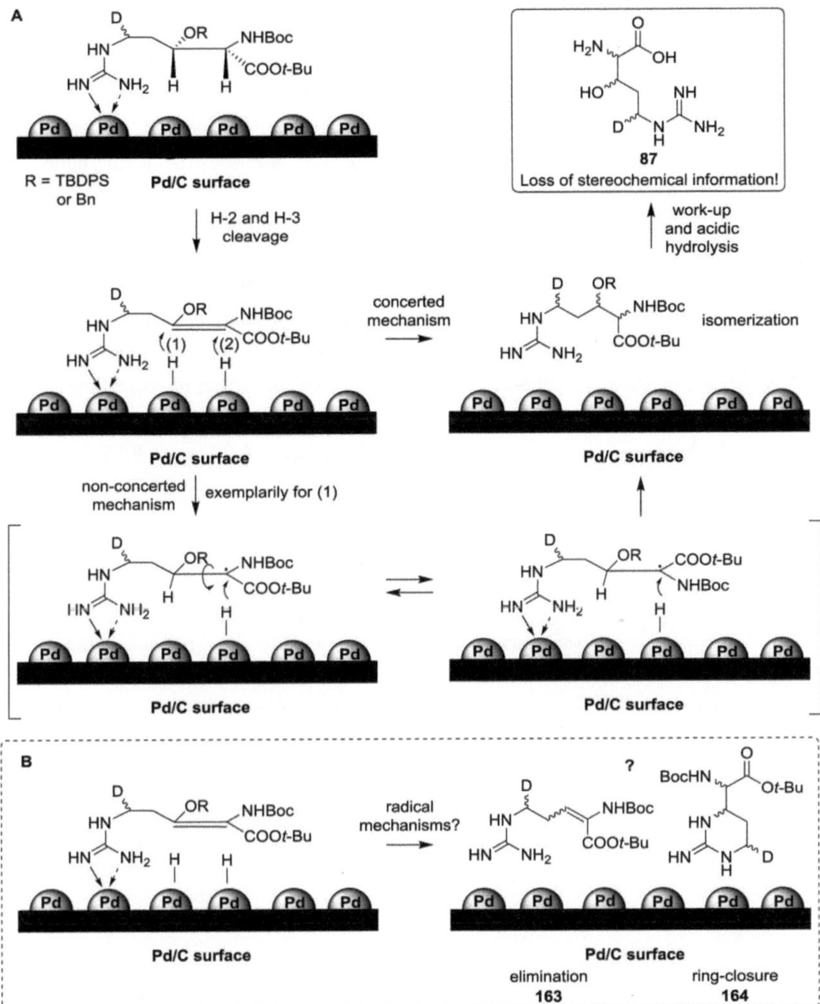

Figure 4.24: The hypothetical reaction mechanism, which could have occurred during the palladium-catalyzed hydrogenation of (**R**)-**152** or (**R**)-**94**/(**S**)-**94**, explaining isomerization (**A**) and the formation of side products (**B**).

In contrast, the mixture obtained by the hydrogenation of the silylated derivative (**R**)-**152** (**d**) clearly presented more side products. This observation would be consistent with the theory that OTBDPS might be the more effective leaving group, especially since the hydroxy group of the benzylated species was partially removed by hydrogenation. However, it could not be ruled out that an elimination of water took place after debenzylation, as elimination or a ring-closure was indicated based on ESI-MS analysis. The proposed reaction mechanism would deeply depend on a coordination of palladium by the 3-hydroxy-L-arginine derivative. An inhibition of this coordination could eventually prevent isomerization and the formation of side products. Without this coordination, the crucial proximity to the catalyst would be missing, and instead, the surface of the catalyst would be loaded by hydrogen. The ability of guanidine to coordinate metals is highly pH-dependent. The protonated guanidinium species would not offer a free bond for the coordination of palladium. In this context, some experiments were carried out using the TBDPS-protected compound (**R**)-**152** (see chapter 4.3.2.10).

4.3.2.10 Hydrogenation Reactions under Acidic Conditions

As the pure diastereomer (**R**)-**158** could only be obtained under hydrogenation conditions using Pd/C (10 wt%) in EtOH and EtOAc and after a chromatographic purification (see entry 4 and 5 in Table 4.5), these conditions were used for further investigations. To see if interactions of the protected compound (**R**)-**152** with the catalyst were already possible without the presence of hydrogen, a solution of (**R**)-**152** and 20 mol% Pd/C in EtOH and EtOAc (1:1) was stirred for 90 min under an atmosphere of argon (Table 4.6, entry 1).

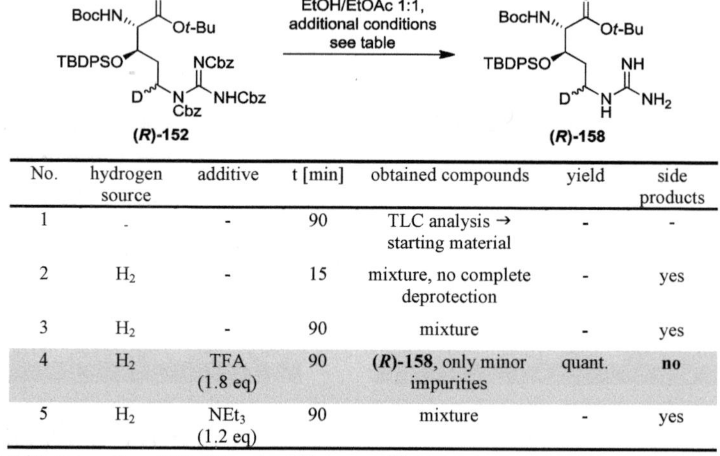

No.	hydrogen source	additive	t [min]	obtained compounds	yield	side products
1	-	-	90	TLC analysis → starting material	-	-
2	H_2	-	15	mixture, no complete deprotection	-	yes
3	H_2	-	90	mixture	-	yes
4	H_2	TFA (1.8 eq)	90	(**R**)-**158**, only minor impurities	quant.	**no**
5	H_2	NEt₃ (1.2 eq)	90	mixture	-	yes

Table 4.6: Hydrogenation of (**R**)-**152** under different conditions.

As expected, TLC analysis did not show any conversion. By changing the argon atmosphere to a hydrogen atmosphere, a conversion could be observed by TLC analysis already after 15 min (entry 2). After the work-up of the reaction, the analysis of the ^{1}H and ^{13}C NMR spectra revealed a mixture of the diastereomeric compounds **158**, side products, and partially Cbz-deprotected species. An extended reaction time resulted in the isolation of a mixture, as it had been observed before (entry 3). By the addition of trifluoroacetic acid (TFA) to the hydrogenation mixture, a quantitative isolation of the pure diastereomer (**R**)-**158** was possible (entry 4). The analysis of the ^{1}H and ^{13}C NMR spectra only revealed minor impurities, but there were not any indications for isomerization or for the formation of side products observed. To investigate if basic conditions could accelerate the formation of side products, a hydrogenation with the addition of triehylamine was performed (entry 5). No significant difference in comparison to the reaction without triethylamine was observed (entry 3). In this context, it could not be ruled out that a coordination of triethylamine instead of (**R**)-**152** 'distorted' the reaction outcome. However, it was proven that the addition of acid to the hydrogenation reaction clearly prevented isomerization and the formation of side products. These results strongly supported the proposed mechanism that is depicted in Figure 4.24. The acidic hydrolysis of the pure diastereomer (**R**)-**158**, which was obtained by hydrogenation, delivered (3*R*)-3-hydroxy-L-[5-^{2}H]arginine (**R**)-**87** in a yield of 77% over two steps (see Figure 4.25).

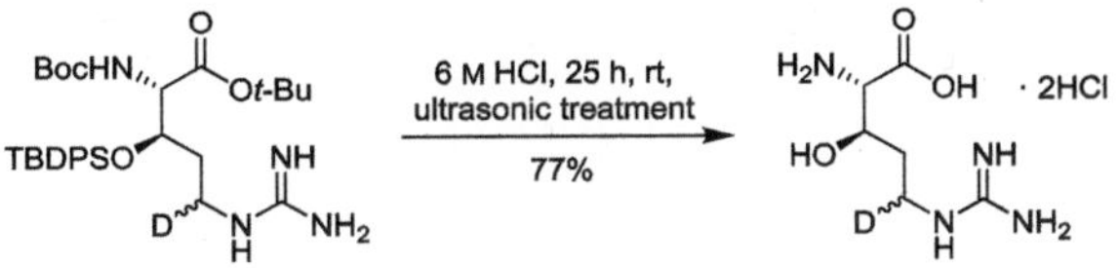

Figure 4.25: Acidic hydrolysis of the pure diastereomer (**R**)-**158** to yield pure (**R**)-**87**.

With the pure diastereomer (**R**)-**158** at hand, the essential role of hydrogenation for isomerization and the formation of side products was investigated. A mixture of (**R**)-**158** and Pd/C in EtOH and EtOAc was stirred overnight under an atmosphere of argon. An addition of triethylamine was expected to ensure the non-protonation of (**R**)-**158**. However, only the starting material could be recovered quantitatively. In this context, it can be proposed that some kind of 'activated' NH species might be essential for isomerization and the formation of side products. Furthermore, the theory of a radical-induced β-elimination might be supported as well.

As the formation of side products was prevented by the addition of acid to the hydrogenation mixture of the TBDPS-protected compound (**R**)-**152**, the hydrogenation of the benzyl-protected species (**R**)-**94** and (**S**)-**94** was also revised under these aspects (Table 4.7). Due to previously performed reactions, *Pearlman*'s catalyst was considered essential for an effective removal of the benzyl protecting group. However, after the addition of TFA, *Pearlman*'s catalyst proved to be too unreactive for a complete Cbz- and Bn-deprotection of compound (**S**)-**94** (Table 4.7, entry 1). This change in reactivity under acidic conditions had already been observed upon the addition of acetic acid (see chapter 4.3.2.3, Table 4.2, entry 1). Surprisingly, using Pd/C in EtOH and EtOAc with the addition of TFA, the formation of the desired diastereomer was observed without obtaining side products or isomerization (entry 2-5). Accordingly, around 40 mol% of the catalyst were required to completely remove the benzyl group. In most of the performed hydrogenations, minor aromatic impurities of the crude product were observed by ^{1}H NMR analysis. This observation indicated that a complete removal of the benzyl group could not be achieved.

However, by reverse phase chromatography after hydrogenation, the purification of the crude products **(R)-143** and **(S)-143** was achieved. A subsequent acidic hydrolysis delivered the pure target compounds **(R)-87** and **(S)-87** in good yields of around 80% over two steps (entry 4 and 5). Performing the reaction in 80% aqueous acetic acid also gave the pure diastereomer **(S)-94**. However, only a yield of 60% was obtained after the acidic hydrolysis (entry 6).

(3R): **(R)-94**
(3S): **(S)-94**

(3R): **(R)-143**
(3S): **(S)-143**

(3R): **(R)-87**
(3S): **(S)-87**

No.	compound	catalyst	solvent	additive	t [h] (i)	result of hydrogenation (i)	yield	side products
1	**(S)-94**	Pd(OH)$_2$/C (20 mol%)	MeOH	TFA (1.8 eq)	15	no complete conversion	-	no
2	**(R)-94**	Pd/C (40 mol%)	EtOH/ EtOAc (1:1)	TFA (1.8 eq)	24	**(R)-143**, minor aromatic impurities	i: quant. ii: 71%	no
3	**(S)-94**	Pd/C (50 mol%)	EtOH/ EtOAc (1:1)	TFA (1.8 eq)	22	pure **(S)-143**	i: quant. ii: 71%	no
4	**(R)-94**	Pd/C (40 mol%)	EtOH/ EtOAc (1:1)	TFA (1.8 eq)	23	**(R)-143**, minor aromatic impurities → RP column	i: 98% ii: 79%	no
5	**(S)-94**	Pd/C (40 mol%)	EtOH/ EtOAc (1:1)	TFA (1.8 eq)	23	**(S)-143**, minor aromatic impurities → RP column	i: quant. ii: 83%	no
6	**(S)-94**	Pd/C (20 mol%)	aq. AcOH (80%)	-	22	**(S)-143**, minor impurities	i: quant. ii: 60%	no
7[a]	**(R)-94**	Pd/C (20 mol%)	HCl (6 M)	AcOH	53	solubility problems	unidentified product	-

Table 4.7: Revision of the deprotection steps using the benzylated compounds **(R)-94** and **(S)-94** (a: combined deprotection conditions).

An attempt to achieve a global deprotection by combining the hydrogenation with the acidic hydrolysis led to unsatisfying results (entry 7). Using 6 M hydrochloride acid as the solvent for compound **(R)-94** resulted in solubility problems. After the addition of acetic acid and an extensive ultrasonic treatment, it was possible to obtain a suspension, which could be stirred under a hydrogen atmosphere. After 53 h, the catalyst was removed, and the reaction mixture was diluted with water prior to evaporation. The analysis of the ^{1}H and ^{13}C NMR spectra showed that all of the protecting groups were removed, but additional signals were observed, which could not be identified, even by a comparison with previously obtained mixtures. No signal for the guanidine-C could be observed in the

^{13}C NMR spectra, and the H-5 signal was shifted significantly to high-field in comparison with the ^{1}H NMR spectra of compounds **(R)-87** and **(S)-87** (Figure 4.26).

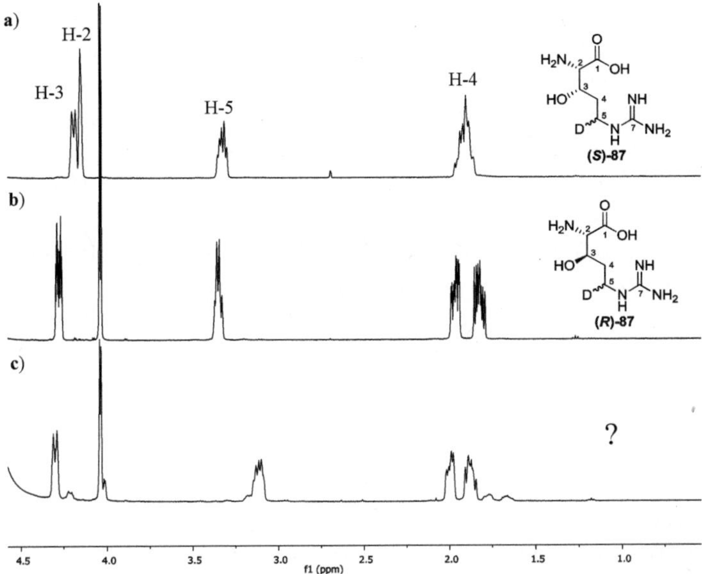

Figure 4.26: Comparison of the ^{1}H NMR spectra (500 MHz, D$_2$O) of a) **(S)-94**, b) **(R)-94**, and c) an unidentified mixture, obtained by hydrogenation in 6 M HCl.

Hypothetically, a cleavage of the guanidine group could be consistent with such observations. However, in contrast, the ESI-MS analysis showed the most intense peak at m/z = 192.12, which would match an expected peak of [M+H]$^+$ of the compound **(R)-87**. An additional peak was found at m/z = 248.19. This would be consistent with a *tert*-butyl esterified derivative. However, as no CH$_3$-signals could be observed by NMR analysis, this observation is especially peculiar. By comparison with the ^{1}H NMR spectra of **(S)-87**, isomerization could be excluded. However, the ^{13}C NMR signals of quaternary guanidine-Cs are sometimes difficult to detect, but performing the hydrogenation in 6 M hydrochloric acid definitely delivered more than one species (see also Figure 4.26). Even if a deprotection in one step would be a nice dodge, it did not seem to be advantageous to elucidate this reaction more deeply with respect to such dubious results.

However, by the addition of TFA, the hydrogenation problem was solved, and after subsequent acidic hydrolysis both pure diastereomers **(R)-87**, and **(S)-87** were obtained in a sufficient amount for the potential use in biosynthetic studies.

4.4 Synthesis of 5'-Deuterated Uridine as an Easily Accessible Nucleoside Building Block

The synthesis of 5'-deuterated uridine **89**, which should be easily accessible, was envisioned in order to obtain a potential probe for feeding experiments. Starting from uridine **1**, a short reduction-oxidation sequence over four steps delivered the desired deuterated derivative **89** (Figure 4.27). The 2'- and 3'-hydroxy functionalities of uridine **1** were protected by an isopropylidene group, following a protocol by *Diederichsen* and coworkers.[211] The protected uridine derivative **71** was oxidized with TEMPO and BAIB under conditions, which had already been established in the synthetic pathway towards 3-hydroxy-L-arginine (see also chapters 4.2 and 4.3). Carboxylic acid **105** was obtained in a yield of 72%. In order to investigate the following reduction and the deprotection sequence, the non-deuterated compounds **71** and **1** were prepared first. An activation by isobutyl chloroformate with the addition of *N*-methylmorpholine (NMM), and a subsequent reduction by sodium borohydride in methanol delivered **71** in a yield of 80%. Due to a loss of material after the aqueous work-up, it was essential to purify the crude mixture by column chromatography without prior extraction steps. Initial attempts to deprotect **71** in a mixture of acetic acid and water required long reaction times, or the application of heat to obtain a yield of around 70%. By the use of trifluoroacetic acid and water (5:1), a yield of 94% could be achieved.

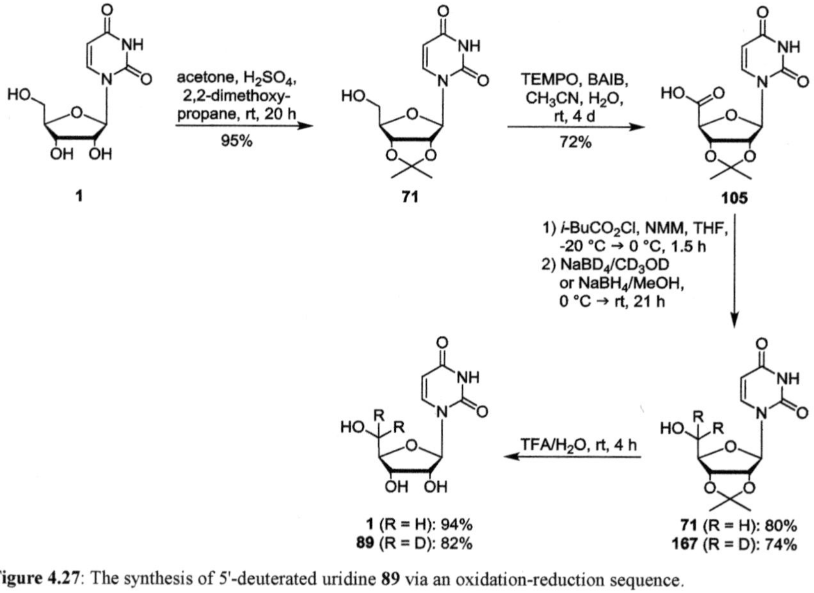

Figure 4.27: The synthesis of 5'-deuterated uridine **89** via an oxidation-reduction sequence.

No.	amount of 105	solvent	NaBD$_4$ [eq]	yield 167	deuteration rate [%] D$_2$: DH : H$_2$
1	100 mg	CD$_3$OD	3	45%	96 : 4 : 0
2	100 mg	CD$_3$OD	5	55%	97 : 3 : 0
3	100 mg	CD$_3$OD	8	51%	97 : 3 : 0
4	100 mg	CH$_3$OH	3	29%	58 : 35 : 7
5	3 g	CD$_3$OD	5	74%	94 : 6 : 0

Table 4.8: Reduction of **105** using sodium borodeuteride. The deuteration rates were calculated based on mass spectra. In this context, the ^{13}C-peak was not considered for the calculations since it would only have a minor impact on entry 4. For the illustration of the relative deuteration rates, a calculation including the ^{13}C-peak was considered negligible.

Analogous conditions were used for the preparation of the deuterated uridine derivative **167**. In the activation-reduction step with sodium borodeuteride, only a yield of 45% could be obtained (Table 4.8, entry 1). By using 5 eq of sodium borodeuteride, the yield could be increased up to 55%, and thereby, a high deuteration rate was obtained (entry 2). The application of more than 5 eq did not lead to further improvements of the reaction outcome (entry 3). As minor impurities were observed in the ^{1}H NMR spectrum of the deprotected species **89**, crystallization attempts were performed additionally to the column-chromatographic purification. However, a purification by crystallization only resulted in minor improvements of the purity, and a significant loss of material. Therefore, the purification by crystallization did not seem to be desirable. As per-deuterated methanol is quite expensive, a deuteration using non-deuterated methanol was investigated (entry 4).

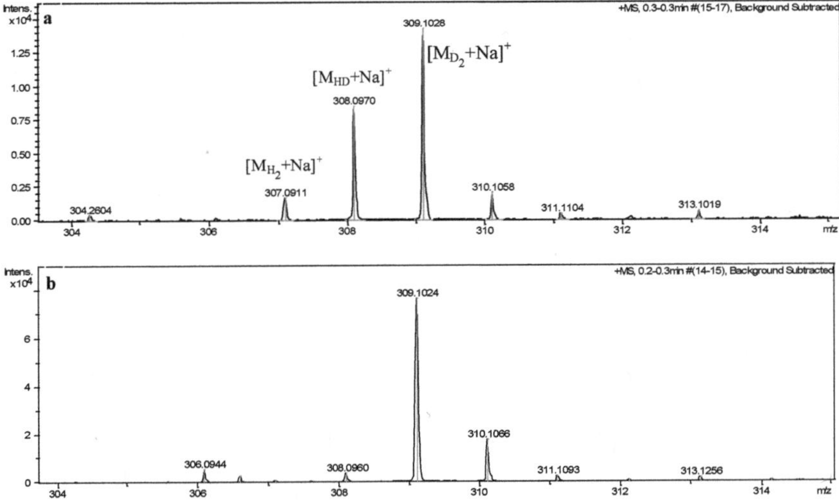

Figure 4.28: Comparison of the mass spectra of **167** after the use of **a**) non-deuterated methanol, and **b**) per-deuterated methanol as the solvent.

However, by the use of sodium borodeuteride in non-deuterated methanol, a deuterium-hydrogen exchange led to unsatisfying rates of deuteration (Figure 4.28). Therefore, using per-deuterated methanol was essential. On a large scale, compound **167** was obtained with an improved yield of 74%. The subsequent acidic hydrolysis delivered [5',5'-^{2}H$_2$]uridine **89** in a yield of 82% with a sufficient purity.

4.5　Synthesis of Different Nucleosyl Amino Acid Derivatives for Biosynthetic Studies

4.5.1　Synthesis of 5'-Deoxygenated Derivatives Comprising an Aminopropyl Linker

The synthesis of the naturally occurring deprotected derivative of (5'S,6'S)-nucleoside building block **86**, and the deprotected derivatives of its isomers **80** and **83**, as well as the synthesis of the 5'-deoxygenated deprotected derivatives of **(S)-77** and **(R)-77** have already been accomplished in our laboratory.[152] In collaboration with the group of *Van Lanen*, phosgene-modified derivatives of these synthetic probes were used as references for the identification of biosynthetic intermediates by the comparison by HPLC and NMR spectroscopy. The results delivered an additional prove for the assigned stereochemistry of the biosynthetic product of a LipK-catalyzed transaldolation, and therefore, these compounds supported the identification of LipK as a L-threonine:uridine-5'-aldehyde transaldolase (see also chapter 2.3.1).[100]

The diastereomers **(S)-90** and **(R)-90**, which comprise an aminopropyl linker, were synthesized in a similar manner. Both of the precursors **(S)-78** and **(R)-78**, which were required for the synthesis of **(S)-90** and **(R)-90**, were provided by *A. Spork*, and their synthesis has already been summarized in chapter 2.5. The acidic global deprotection of related, *tert*-butyl-esterified nucleoside derivatives had already been established by *Ichikawa* and *Matsuda*, using aqueous 80% trifluoroacetic acid.[149] Applying these conditions on the acidic hydrolyses of compounds **(S)-78** and **(R)-78**, the target compounds **(S)-90** and **(R)-90** could be obtained (Figure 4.29)

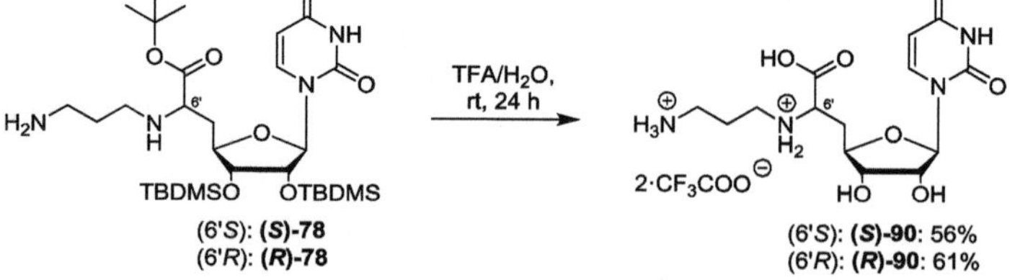

Figure 4.29: Acidic hydrolyses of the provided precursors **(S)-78** and **(R)-78** delivering **(S)-90** and **(R)-90**.

A subsequent purification by reverse phase HPLC (water with the addition of 0.1% trifluoroacetic acid (TFA) as the eluent) and lyophilization delivered the pure compounds (*S*)-90 and (*R*)-90 as their TFA salts in yields of 56% and of 61%, respectively. The stereochemistry of these 5'-deoxygenated nucleoside precursors was introduced via asymmetric hydrogenation by *A. Spork*. The configuration was initially assigned based on empirical results and detailed NMR studies, but the configuration could be proven later on by X-ray crystallography of a synthetic urea derivative.[152]

4.5.2 Synthesis of the Glycosylated Nucleosyl Amino Acid Derivative

4.5.2.1 Synthesis of the Protected Nucleoside Building Block

The preparation of the protected derivative β-75 was reported by *Ichikawa* and *Matsuda* (see also chapter 2.5 and Figure 4.30).[150]

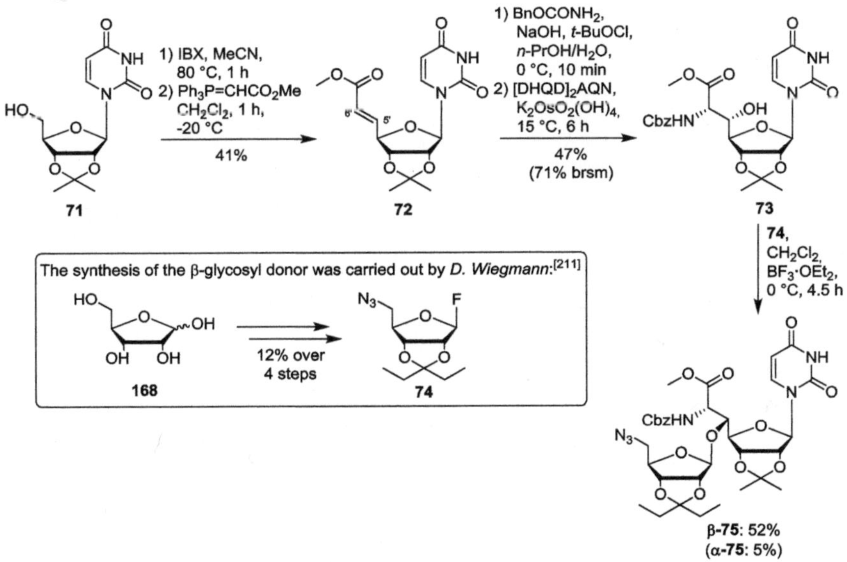

Figure 4.30: Synthesis of the glycosylated nucleoside derivative β-75 following a protocol from *Ichikawa* and *Matsuda*[150,212] (brsm: based on recovered starting material).

·The introduction of the isopropylidene protecting group of the 2'- and 3'-hydroxy function was performed as described in chapter 4.4. Oxidation with 2-iodoxybenzoic acid (IBX) delivered the crude 5'-aldehyde, which was directly used in a *Wittig* reaction with (triphenylphosphoranylidene)acetate. The pure (*E*)-olefin 72 was obtained in a yield of 41% on a small scale. On a large scale, only yields of around 30% could be achieved. Besides the clean product, only impure material could be isolated. The impurities could be identified as triphenylphosphine oxide, which was formed during the reaction. These kind of tertiary phosphine oxides, as a common byproduct of the *Wittig* reaction, are known to be non-trivial to remove.[213] They are 'too organic' to be removed by aqueous work-up, and their polarities can be challenging during silica column chromatography, as they 'bleed' over the column. Especially on a large scale, purification by column chromatography proved to be problematic. As applying the conditions, which were reported by *Ichikawa* and *Matsuda* for the chromatographic purification, did not result in the isolation of the pure compound, strategies of crystallization and precipitation were also investigated, using different mixtures of polar and unpolar solvents. The best results were achieved by diluting the concentrated crude product in a small amount of diethyl ether, adding isohexane to the solution, and completing the precipitation in an ice-bath prior to filtration. In order to remove most of the triphenylphosphine oxide, the precipitation step had to be repeated several times. Thus, a significant loss of material was the result. However, a single precipitation step followed by purification by column chromatography resulted in sufficient purity of the product.

The 5'-hydroxy- and the 6'-amino-function were selectively introduced by *Sharpless* asymmetric aminohydroxylation. The reaction was performed at 0 °C with osmium tetroxide ($K_2OsO_4(OH)_4$), a chiral alkaloid ligand ([DHQD]$_2$AQN), and a sodium *N*-chlorocarbamate salt, which was prepared in situ from benzyl carbamate, sodium hydroxide, and *tert*-butyl hypochlorite. The *tert*-butyl hypochlorite, which was required, was prepared from *tert*-butanol, sodium hypochlorite solution, and acetic acid using a protocol from *Minz* and *Walling*.[214] After its preparation, the *tert*-butyl hypochlorite was stored under an argon atmosphere over calcium chloride at 4 °C in the dark. An initial attempt delivered the amino alcohol **73** in a yield of only 26%, and 30% of the starting material could be recovered (Table 4.9, entry 1). The stereochemistry of **73** was assigned based on the comparison of the ^{1}H and ^{13}C NMR with the ones reported by *Matsuda* and coworkers, and they were found to be identical.[215] It is known that the temperature is a sensitive factor during asymmetric aminohydroxylations. *Sharpless* reported that a temperature above 25 °C led to a considerable increase in diol formation as well as to a decline in selectivity. Further studies actually showed that the reaction temperature was crucial. By performing the reaction in a water bath at room temperature, the yield could be increased to 36% (entry 2). As the starting material could be recovered, it was tried to increase the conversion by using 40 mol% of the catalyst and of the ligand. However, the yield decreased, and an isolation of the starting material was not possible (entry 3). Experiments by *Lohray* and coworkers implicated osmium(VIII) species as a source of diol contaminant.[216] With a stoichiometric amount of osmium tetroxide and of the chloramine salt, they nearly obtained a quantitative yield of the diol, but they could not isolate any aminoalcohol. This observation indicated that an increased amount of the catalyst, which was used, would result in a worse chemoselectivity and therefore, a reduced yield. These results might imply that the 15 mol% of the catalyst and of the ligand, which were reported by *Ichikawa* and *Matsuda*, were already the 'optimal' amount for this system. By performing the aminohydroxylation in a water bath at 15 °C, the desired amino alcohol **73** was obtained in a yield of 47%, and 34% of the starting material could be recovered (entry 4), thus leading to a yield of 71% based on recovered starting material (brsm).

No.	[DHQD]$_2$AQN/ K$_2$OsO$_4$(OH)$_4$	T	t [h]	yield 73	recovered starting material 72
1	0.15 eq	rt	2	26%	30%
2	0.15 eq	water bath at rt	3	36%	27%
3	0.4 eq	water bath at rt	3	28%	-
4	0.15 eq	water bath at 15 °C	6	47%	34%

Table 4.9: Different conditions for the asymmetric aminohydroxylation of **72**.

As a sufficient amount of the aminoalcohol **73** had already been prepared, it was not planned to improve the reaction any further. Asymmetric aminohydroxylations were reported to be quite sensitive to several factors.[217] Reasons for not reaching the yield of 96%, which was reported by *Ichikawa* and *Matsuda*, might be traces of contaminants in the reagents, in the catalyst, or in the solvents. For example, the contamination by minor traces of triphenylphosphine oxide could have affected the subsequent asymmetric aminohydroxylation.

Under established conditions at -30 °C, the following β-selective ribosylation of **73** with the ribosyl donor **74** yielded the glycosylated derivatives **β-75** and **α-75**. A chromatographic separation of the isomers delivered **β-75** in a yield of 43%, and **α-75** in a yield of 10%. By performing the reaction at 0 °C and by varying the interval and the amount of boron trifluoride added (3 equivalents in total), up to 52% of the β-isomer **β-75** and 5% of the α-isomer **α-75** could be isolated. An improved conversion with an increased total yield seemed to create relatively more α-isomer **α-75**, and was thereby complicating the chromatographic separation of the isomers and decreasing the isolated yield of **β-75**.

4.5.2.2 Deprotection Strategies

Attempts to synthesize naturally occurring muraymycins showed that the (5'*S*,6'*S*)-β-hydroxy amino acid motif was quite unstable under basic or acidic conditions.[152] One reason might have been the basic unprotected 6'-amino group. Thus, it was envisioned to remove the N6'-Cbz protecting group and to reduce the azide functionality of the ribose functionality in one step at the end of the synthetic route. Transfer hydrogenation using palladium on charcoal as the catalyst and 1,4-cyclohexadiene as the hydrogen source is an established method in our laboratory for these kinds of reductions.[152] Using hydrogen gas, the 5,6-double bond of the nucleobase might be reduced as well. Another advantage of mild conditions for the hydrogenation is the use of easily removable reagents. Filtration and the removal of the solvent should be sufficient to obtain the pure compounds. Therefore, purification by HPLC, prior to the hydrogenolysis reaction, was envisioned. As an unprotected carboxylic acid function could result in purification problems due to its polarity, the isopropylidene- and isopentylidene protecting groups of the hydroxyl functionalities should be removed first (see Figure 4.31).

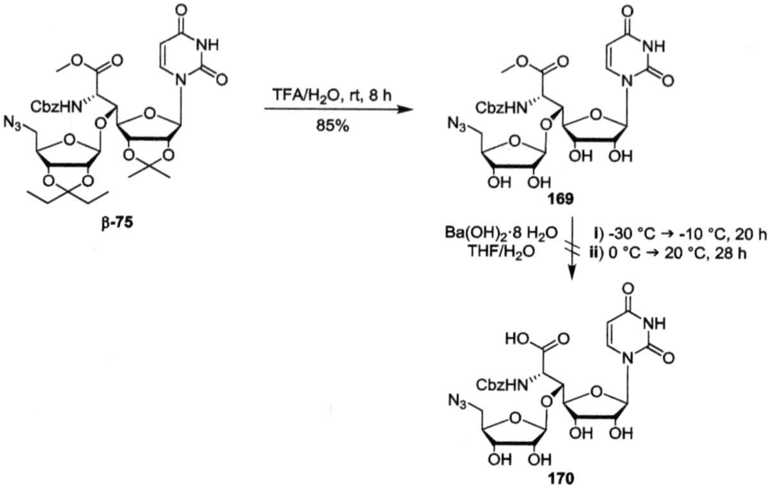

Figure 4.31: Initial deprotection attempts of compound **β-75** by acidic hydrolysis followed by alkaline saponification.

The acidic hydrolysis of **β-75** with aqueous 80% trifluoroacetic acid delivered compound **169** in a yield of 85%. By a subsequent alkaline hydrolysis, followed by column-chromatographic purification (CH$_2$Cl$_2$/MeOH 9:1 with 0.5% AcOH), only 49% of the starting material could be recovered (**i** in Figure 4.31). The conditions of the alkaline hydrolysis were adopted from a protocol of *Ichikawa* and *Matsuda*, which they used for the saponification of similar methyl esters.[150] They reported a removal of the methyl ester by the use of barium hydroxide octahydrate in aqueous THF (THF/H$_2$O 4:1) at -30 °C. However, as the reaction mixture was frozen under these conditions, more THF had to be added, and the reaction mixture was slowly warmed until the reaction mixture thawed (up to -10 °C). As only the starting material could be recovered, it was tried to perform the reaction at a higher temperature in order to obtain a better conversion and to avoid solubility problems. The reaction was performed at 0 °C, and it was then warmed overnight to 20 °C because TLC analysis only showed a poor conversion at 0 °C. After purification by column chromatography, 10% of the starting material could be recovered (**ii** in Figure 4.31). As neither the product nor any side products could be isolated, it was assumed that decomposition had taken place. The analysis of the ^{1}H NMR spectrum of the crude product revealed a mixture of the starting material and decomposition products. The

ESI⁻-MS spectrum of the crude product showed a signal at m/z = 607.17, which would be consistent with the desired product ([M-H]⁻ = 607.16). The analysis of the ^{1}H NMR spectrum of the concentrated aqueous layer from the aqueous work-up step showed that the carboxylic acid might have been extracted into the aqueous phase despite the acidic work-up. However, a purification of the crude mixture, which was obtained from the aqueous layer, and further improvement of the reaction conditions did not seem to be feasible. *Ichikawa* and *Matsuda* reported that the basic hydrolysis of such methyl esters could be difficult.[149] Only by the treatment with barium hydroxide octahydrate, they could isolate the free carboxylic acid. Therefore, an investigation of different bases for the hydrolysis of the methyl ester **169** was not envisioned. As a β-elimination of the sensitive glycosyl unit might be the problem, it was tried to perform an alkaline hydrolysis of the methyl ester prior to glycosylation. A subsequent introduction of a benzylester might be advantageous, as it should be removable under hydrogenolytic conditions. However, by the hydrolysis with lithium hydroxide, it was not possible to isolate the product (Figure 4.32). TLC analysis showed a complete conversion of the starting material, and the analysis of the ^{1}H NMR spectrum of the crude product showed a diverse mixture of, most likely, decomposition products.

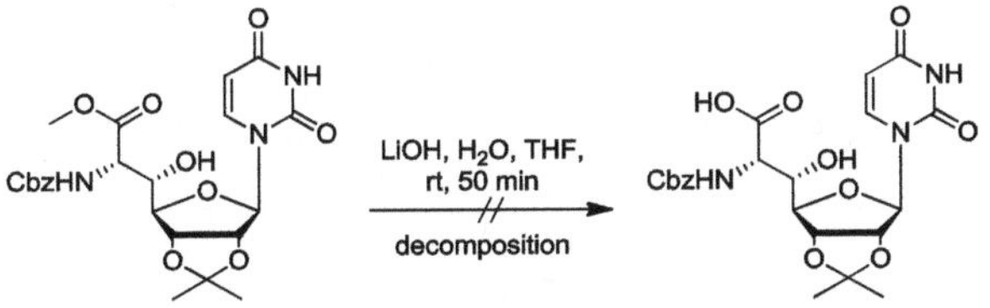

Figure 4.32: Attempt to perform an alkaline hydrolysis of **73** prior to glycosylation.

As compound **73** seemed to be quite sensitive to decomposition under basic conditions, it was envisioned to introduce an ester functionality by the help of which alkaline hydrolysis could be avoided. As mentioned above, a benzyl ester might be advantageous due to the possibility of hydrogenolytic removal. A *Wittig* reaction under the established conditions of the 5'-aldehyde derivative, which was prepared freshly from **71**, was carried out by the use of commercially available benzyl (triphenylphosphoranyliden)acetate. The desired olefin **172** was obtained in an excellent yield of 94% (Figure 4.33).

In contrast to the *Wittig* reaction with the methyl ester derivative, no purification problems with triphenylphosphine oxide arose using the benzyl derivative of the phosphorous ylide. The reason for this could be the differences in the polarities of the compounds **72** and **172**. The benzyl ester **172** is more lipophilic in comparison to the methyl ester **72**. This is reflected by the R$_f$-values of the compounds. Benzyl ester **172** showed a R$_f$-value of 0.27 in isohexane/EtOAc 3:1, while the methyl ester **72** showed a R$_f$-value of 0.27 in isohexane/EtOAc 1:3. The more polar solvent mixture, which was required for the elution of the methyl ester **72**, also eluted triphenylphosphine oxide. The benzyl ester **172** could be purified under quite unpolar conditions, resulting in an uncomplicated removal of triphenylphosphine oxide. For *Sharpless* asymmetric aminohydroxylation of the benzyl ester **172** under established conditions, it was not possible to isolate a pure compound. TLC analysis only revealed one spot, and ESI-MS measurements showed the most intense peak at $m/z = 604.20$, which would be consistent with the desired aminoalcohol **173a** with a calculated mass of m/z ($[M+Na]^+$) = 604.19. For the analysis of the ^{1}H NMR spectrum, a clear interpretation of the signals was not possible. However, in the ^{13}C NMR spectrum, a double signal set could be observed with a ratio of around 1:3. Thus, it is suggested that the asymmetric aminohydroxylation of the benzyl ester **172** is not proceeding selectively under the established conditions and that an isomeric mixture was obtained. As the interpretation of the ^{1}H NMR and ^{13}C NMR spectra did not allow a distinct identification of signals due to overlapping signals, it could not be ruled out that other side products had been isolated as well. However, TLC analysis and ESI-MS measurements indicated that only the 'pure' isomeric mixture had been isolated. Based on this assumption, the yield for the isomeric mixture **173** was calculated, resulting in a yield of 29% (see Figure 4.33).

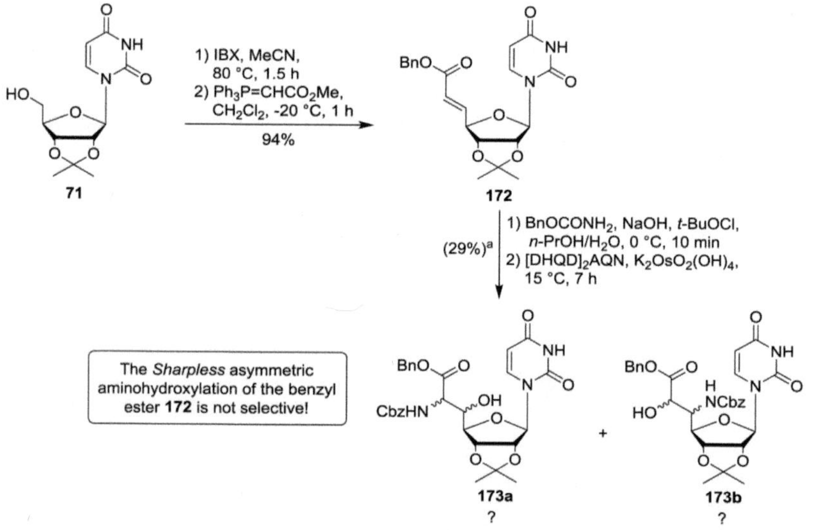

Figure 4.33: *Wittig* reaction for the synthesis of the benzyl ester **172**, and the subsequent asymmetric aminohydroxylation (a: the yield was calculated based on the assumption that only a isomeric mixture of the amino alcohol **173** was isolated).

As mentioned previously, asymmetric aminohydroxylations can be quite sensitive to minor changes in reaction conditions. A comprehensive study of ligand-substrate interactions in phthalazine-based ligand systems was reported by *Janda* and coworkers.[217-218] They proposed a distorted trigonal bipyramide as the active complex with the oxygen atoms in equatorial positions, and the nitrogen atoms of the ligand system and the nitrogen source in axial positions. With such a proposed geometry of the OsO_3N_2 species, the regioselectivity would be strongly affected by the mode in which the olefin binds to the catalyst. An unsymmetrically substituted olefin can be oriented in two ways within the offered binding cleft of the ligand-catalyst system. Thus, ligand-substrate interactions should play an important role for the orientation in which the olefin binds to the catalyst. Consequently, steric demands as well as electronic properties of the substrate might have a major impact on the regioselectivity. *Panek* and coworkers reported that these kinds of interactions play a role in systems with anthraquinone-derived ligands as well.[219] They showed that the *Sharpless* asymmetric aminohydroxylation of aryl ester substrates resulted in a reversal of the regioselectivity, compared to the corresponding alkyl ester derivatives. As a possible reason for this alteration in regioselectivity, they proposed π-complex interactions between the osmium bond to the alkaloid ligand, the aryl moiety of the ester, and the anthraquinone ring of the ligand. Such considerations could explain the difference in selectivity of the asymmetric aminohydroxylation of the benzyl ester **172** in comparison to the asymmetric aminohydroxylation of the methyl ester **72**. As not only the selectivity would have to be optimized towards the matching regioisomer **173a**, but also a reliable assignment of the configuration of the 'right' isomer ought to be done, it did not seem to be promising to invest more time and effort into the elucidation of the asymmetric aminohydroxylation with the benzyl ester as the substrate. Instead, further experiments for the deprotection of the methyl ester **β-75** were carried out.

Alkaline hydrolysis of compound **β-75** with barium hydroxide octahydrate at room temperature and subsequent purification by column chromatography (CH_2Cl_2/MeOH 9:1 with 0.5% AcOH) furnished compound **174** as impure material (Figure 4.34). However, the analysis of ^{1}H NMR, ^{13}C NMR, ^{1}H,^{1}H COSY, ^{1}H,^{13}C HSQC spectra, and ESI-MS analysis clearly showed that the desired product was obtained. The ^{1}H NMR spectrum of an isolated side product was consistent with a previously discussed β-elimination. However, due to impurities of the obtained material, a clear identification was not possible. As a purification of the unprotected carboxylic acid **174** could not be achieved, it was envisioned to perform the acidic hydrolysis with the impure material, and to perform a purification by HPLC then. The acidic hydrolysis of **174** using aqueous 80% trifluoroacetic acid delivered the desired product **170**. An initial purification attempt by HPLC under gradient conditions with a mixture of acetonitrile and water with 0.1% triflouroacidic acid (see chapter 7) resulted in the isolation of **170**, showing only minor impurities (**a** in Figure 4.35). Two attempts to further purify **170** by isocratic conditions (see also chapter 7) did not result in significant improvements of purity (**b** and **c**, Figure 4.35). After the third HPLC run (**c**), NMR measurements were carried out in D_2O in order to exclude the possibility of a partial formation of the methyl ester. However, looking at the ^{1}H NMR spectra (**a-c**), no differences could be observed despite the solvent-dependent signal shifts.

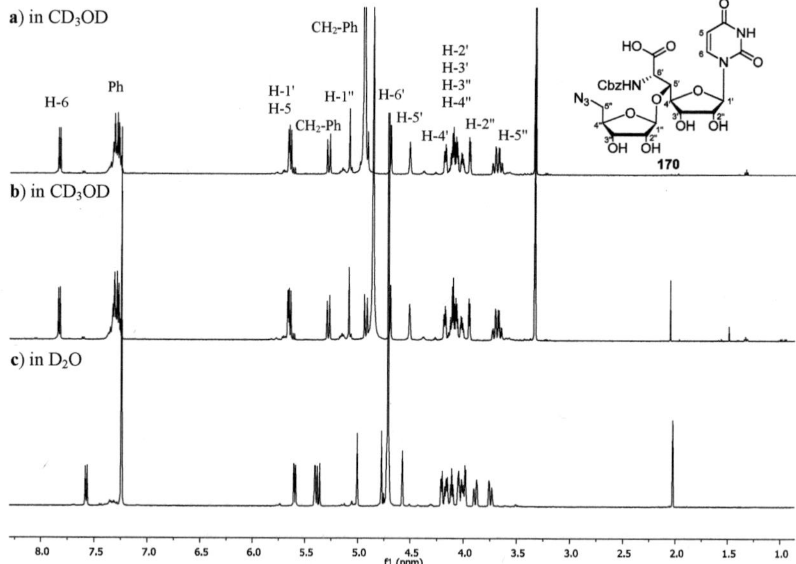

Figure 4.34: Deprotection of the nucleoside building block **β-75** in three steps (a: obtained as impure material).

Figure 4.35: Comparison of the ¹H NMR spectra of **170** after **a)** the first HPLC purification applying gradient conditions (500 MHz, CD₃OD), **b)** after the second HPLC purification applying isocratic conditions (500 MHz, CD₃OD), **c)** after the third HPLC purification applying isocratic conditions (500 MHz, D₂O).

Including the three subsequent HPLC purifications, the nucleoside derivative **170** was obtained in a yield of 36% over two steps. Considering the partial decomposition due to a β-elimination during the alkaline hydrolysis and the repeatedly performed HPLC purification steps after acidic hydrolysis, the moderate yield, which was obtained, was satisfying. As an additional peak was not observed in the HPLC chromatogram of **170**, and ESI-MS measurements did not indicate the formation of another side product, it is suggested that, next to the desired (5'*S*,6'*S*)-derivative **170**, the (5'*R*,6'*R*)-isomer or the regioisomers of the corresponding α-hydroxy β-amino acid derivative were formed in minor amounts during the *Sharpless* asymmetric aminohydroxylation. These isomers could possess the same retention time under the quite polar conditions which were necessary for elution. Therefore, a complete purification by HPLC reverse phase chromatography was not feasible. The transfer hydrogenation under the already mentioned conditions delivered the globally deprotected nucleoside building block **12** in quantitative yield.

For biosynthetic studies, pure compounds and especially isomerically pure compounds are essential. However, with only minimal traces of the putative isomers and the awareness of these impurities, it should be possible to get reliable data from biosynthetic studies using the obtained material of **12** with an estimated purity of ca. 95%.

4.6 Synthesis of Enduracididine Precursors via Catalytic Key Steps

4.6.1 Strategy via *Wittig-Horner* Reaction Followed by Asymmetric Hydrogenation as the Catalytic Key Step

4.6.1.1 Synthesis of the Phosphonate Building Block

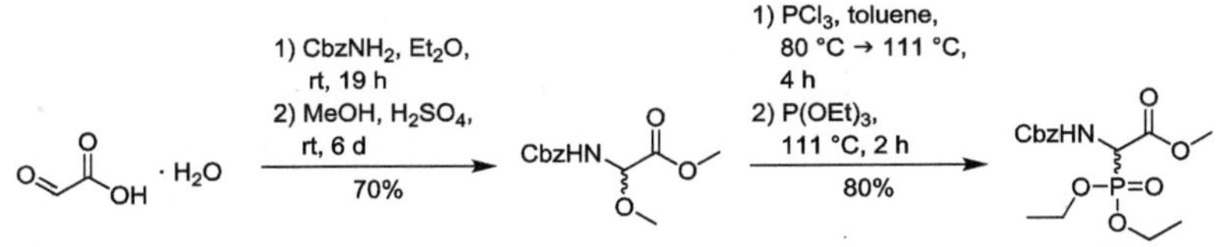

Figure 4.36: Synthesis of phosphonate **176**.

Phosphonate **176** was synthesized following a protocol from *Schmidt* and coworkers and *Zoller* and coworkers (Figure 4.36).[166-167] A solution of glyoxylic acid monohydrate **175** in diethyl ether was treated with benzyl carbamate before methanol and sulfuric acid were added. The protected glycine derivative **110** was obtained in a yield of 70%. The reaction with phosphorous trichloride in refluxing toluene delivered an α-chlorinated derivative. In an *Arbuzov* reaction, this freshly generated derivative was directly converted into the phosphonate building block **176** by the addition of triethyl phosphite with a yield of 80%.

To obtain the desired TMSE ester **109**, a protocol from *Toone* and coworkers was applied (Table 4.10).[168]

1) LiOH, THF, H$_2$O, rt, 3 h

2) 4-DMAP, CH$_2$Cl$_2$, TMSE-OH, EDC, rt, 2-16 h

176 → **109**

No.	TMSE-OH [eq]	coupling reagent	t [h]	yield	remarks
1	1.0	EDC	2	64%	-
2	1.3	EDC	2	70%	-
3	1.0	DIC	2	6%[a]	-
4	1.0	EDC	16	41%	molecular sieves 3 Å
5	1.0	EDC	16	36%[a]	-
6	1.2	EDC	16	67%	large scale

Table 4.10: Two-step transesterification of **176** delivering the phosphonate building block **109** (a: loss of material due to crystallization attempts).

Using lithium hydroxide, the alkaline hydrolysis of methyl ester **176** yielded the free carboxylic acid in a quantitative yield. The obtained carboxylic acid was directly esterified with 2-(trimethylsilyl)-ethanol (TMSE-OH). The activation of the carboxylic acid was carried out by using 4-dimethylaminopyridine (4-DMAP) and 1-ethyl-3-(3-dimethylaminopropyl)carbodiimide hydrochloride (EDC) in dichloromethane. In an initial attempt, a yield of only 64% could be achieved (Table 4.10, entry 1). Hence, different conditions were investigated in order to optimize the esterification step. An excess of 2-(trimethylsilyl)-ethanol resulted in a slightly improved yield (entry 2). Using *N,N'*-diisopropylcarbodiimide (DIC) as the coupling reagent, a poor yield of only 6% was achieved. As the yield of esterification reactions might strongly depend on dry conditions, molecular sieves 3 Å were added to the reaction mixture (entry 4), but compared to entry 5, no significant optimization was observed. Another problem seemed to be the purification of the TMSE ester **109**. After purification by column chromatography, a discrepancy between the obtained mass of the crude and the purified product was observed. A comparison of the ^{1}H NMR spectrum of the crude product with the ^{1}H NMR spectrum of the purified product showed no impurities which could explain such a loss of material. Crystallization attempts under different conditions did not result in any applicable alternative for the purification of **109**. The analysis of the ^{31}P NMR spectrum of a crude product mixture, which was obtained by directly removing the solvent, did not reveal any phosphorous containing species, which could explain the loss of material. Therefore, it could be suggested that a partial decomposition had taken place during the purification by column chromatography, resulting in a loss of material and in a decreased yield. On a large scale applying the improved conditions, TMSE ester **109** could be obtained in a yield of 67% (entry 6).

4.6.1.2 *Wittig-Horner* Reaction Towards the Didehydroamino Acid

By a *Wittig-Horner* reaction, the (*S*)-*Garner's* aldehyde **(*S*)-98** (see chapter 4.1) and the phosphonate **109** were coupled to provide olefin **108** as a mixture of its (*Z*)- and (*E*)-isomers (Table 4.11). The phosphonate **109** was deprotonated using potassium *tert*-butylate (KO^{t}Bu) or potassium hexamethyldisilazane (KHMDS) at -78 °C. The reaction with the aldehyde **(*S*)-98** and a subsequent warming to room temperature overnight yielded **108** as its (*Z*)- and (*E*)-isomers **108a** and **108b**. As the configuration of the isolated diastereomers was initially unclear, the faster eluting isomer with R_f (cyclohexane/Et$_2$O) = 0.40 should be named **108a** and the slower eluting isomer with R_f (cyclohexane/Et$_2$O) = 0.33 should be named **108b** from here on. A putative assignment is discussed later on.

No.	base	base [eq]	phosphonate **109** [eq]	aldehyde (*S*)-98 [eq]	isolated overall yield	diastereoselectivity **108a:108b**[b]
1	KO*t*-Bu	1.05	1.0	2.2	28%[a]	15:85
2	KO*t*-Bu	1.05	1.0	1.1	9%[a]	66:34
3	KO*t*-Bu	1.05	1.0	1.0	8%[a]	-
4	KO*t*-Bu	0.9	1.0	1.0	59%	52:48
5	KHMDS	1.1	1.3	1.0	45%	74:26

Table 4.11: *Wittig-Horner* reaction for the synthesis of the enduracididine precursor **108** (a: no complete separation of **108a** and **108b** by column chromatography could be achieved; b: determined from isolated yields of the pure isomers **108a** and **108b**).

The separation by column chromatography proved to be quite challenging. Isocratic conditions resulted in an improved efficiency of separation, compared to gradient conditions. By choosing cyclohexane and diethyl ether in a ratio of 2:1 as eluent, the best results were obtained (entry 4 and 5). Initial attempts of the *Wittig-Horner* reaction only resulted in poorly separable (*E*)/(*Z*)-mixtures (entry 1-3). In order to prevent a possible isomerization after the coupling reaction, the amount of the base was reduced. By using 0.9 eq of KO^{t}Bu and by applying the improved chromatographic conditions, the diastereomers could be separated for the first time (entry 4). Although the yield of 59% was acceptable, the obtained diastereoselectivity was not sufficient. In order to obtain a diasteroselectivity which would be comparable with the ones of the reported (*Z*)-selective *Wittig-Horner* reactions,[153,220] the use of KHMDS as base was investigated (entry 5). As an excess of the phosphonate **109** should be easier to remove by column chromatography than an excess of the aldehyde (*S*)-**98**, 1.3 equivalents of phosphonate **109** were used. The use of KHMDS should show if KOH contaminants within the KO^{t}Bu could have possibly led to the poor diasteroselectivities. The separation of the diastereomers showed that the diasteroselectivity could be improved, but that the yield was slightly decreased compared to entry 4. A closer look at the other isolated fractions, which were obtained by the purification of reaction 5 by column chromatography, revealed that, most likely, decomposition had taken place. This decomposition would not only explain the moderate yield, but potentially also the rather poor diasteroselectivity.

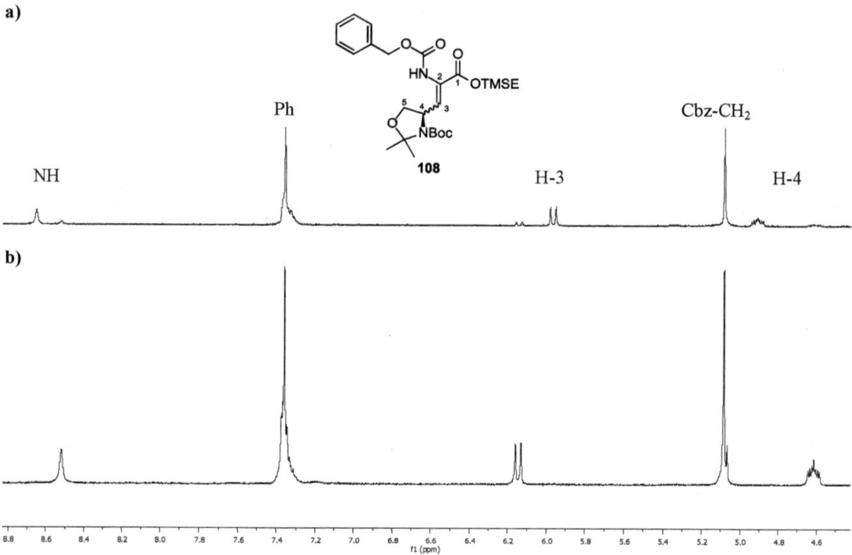

Figure 4.37: Cut offs of the ^{1}H NMR spectra (301 MHz, DMSO-d$_6$, 100 °C) of the didehydroamino acids a) **108a** and b) **108b**.

It was attempted to identify the obtained diastereomers according to *Mazurkiewicz* and coworkers, who reported an empirical rule for the identification of such (*E*)- and (*Z*)-didehydroamino acids by comparing the β-CH and the α-NH shifts in the ^{1}H NMR spectra recorded in CDCl$_3$.[221] According to *Mazurkiewicz* and coworkers, the β-CH and the α-NH of (*Z*)-didehydroamino acids were both further upfield-shifted than the corresponding signal of the respective (*E*)-isomer ($\delta_{\beta\text{-CH}}(Z) < \delta_{\beta\text{-CH}}(E)$ and $\delta_{\alpha\text{-NH}}(Z) < \delta_{\alpha\text{-NH}}(E)$, respectively). However, due to signal broadening and the formation of rotamers in the CDCl$_3$ spectrum, no reliable interpretation could be achieved. The assignment of the signals for the characterization of compounds **108a** and **108b** was carried out for NMR spectra measured in DMSO-d_6 at 100 °C (see Figure 4.37). Looking at the DMSO-d_6 NMR spectra of **108a** and **108b**, the following was found: $\delta_{\text{H-3}}$(**108a**) < $\delta_{\text{H-3}}$(**108b**) and δ_{NH}(**108a**) > δ_{NH}(**108b**). As the *Wittig-Horner* reaction is known to be suitable for the synthesis of (*Z*)-configured didehydroamino acids in very good selectivities,[153,220] it could be proposed that the major diastereomer **108a** (entry 4 and 5) was the (*Z*)-isomer. In this context, the rule of *Mazurkiewicz* would found to be true for the β-CH signals, but it would be violated for the α-NH signals. This observation would be consistent with studies carried out by *A. Spork*, who reported that the α-NH signals of (*Z*)-configured nucleosyl didehydroamino acids were further downfield-shifted than the corresponding signals of the respective (*E*)-isomers in CDCl$_3$ ($\delta_{\alpha\text{-NH}}(Z) > \delta_{\alpha\text{-NH}}(E)$).[152,154] In these cases the assumed configurations were clearly proven by ^{1}H nOe studies.[152,154] Therefore, it could be proposed that the empirical rule with $\delta_{\beta\text{-CH}}(Z) < \delta_{\beta\text{-CH}}(E)$ and $\delta_{\alpha\text{-NH}}(Z) > \delta_{\alpha\text{-NH}}(E)$ reported by *A. Spork* might also be applicable in DMSO-d_6 at 100 °C, and that the major diastereomer **108a** was the (*Z*)-configured diastereomer.

4.6.1.3 Asymmetric Hydrogenation of the Didehydroamino Acid 108a

Knowles and coworkers reported the asymmetric hydrogenation of (*Z*)-didehydroamino acids in high yields and stereoselectivities using chiral rhodium catalysts.[164] The corresponding (*E*)-isomers only gave unsatisfying results. These results were confirmed in our group by the asymmetric hydrogenation of nucleosyl amino acid derivatives.[152] By the use of (*S,S*)-Me-DUPHOS-Rh **177**, the asymmetric hydrogenation of a (*Z*)-configured didehydroamino acid should lead to the L-configured amino acid. The putative (*Z*)-configured didehydroamino acid **108a** was dissolved in methanol under the absolute exclusion of air, and it was stirred with the catalyst **177** under a hydrogen atmosphere (Figure 4.38). Even after 7 days, TLC analysis did not show any conversion. After a column-chromatographic purification, only the starting material was recovered in 81%.

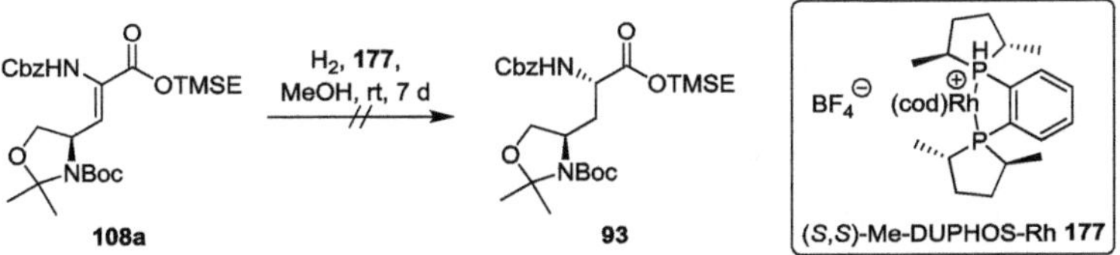

Figure 4.38: Attempt to synthesize the enduracididine precursor **93** by asymmetric hydrogenation.

It could not be ruled out that the assignment of the hypothetical configuration of the compounds **108a** and **108b** was wrong. The hydrogenation of the (*E*)-didehydroamino acid would not lead to a conversion or to a poor conversion. Furthermore, an asymmetric hydrogenation using another catalyst was not considered since most other catalysts are known not to give sufficient conversions with *N*-urethane-protected didehydroamino acid esters. However, as the *Wittig-Horner* reaction gave poor stereoselectivities, and as an optimization of the reaction conditions did not seem to be feasible, it was not promising to conduct further studies of the asymmetric hydrogenation or of the *Wittig-Horner* reaction. Therefore, a new strategy for the synthesis of enduracididine precursors was developed.

4.6.2 Strategy via *Sharpless* Asymmetric Dihydroxylation as the Catalytic Key Step

4.6.2.1 Synthesis of Allylglycine as the Precursor for the Asymmetric Dihydroxylation

Initially, a protection strategy analogous to the *Wittig-Horner* reaction pathway was envisioned, with a Cbz protecting group for the amino functionality and a TMSE protecting group for the carboxylic acid. Starting from L-aspartic acid **178**, a 4-selective methyl esterification was performed following a protocol from *Cox* and coworkers using thionyl chloride.[222] After crystallization, the methyl ester **180** was obtained in a yield of 77% (Figure 4.39).

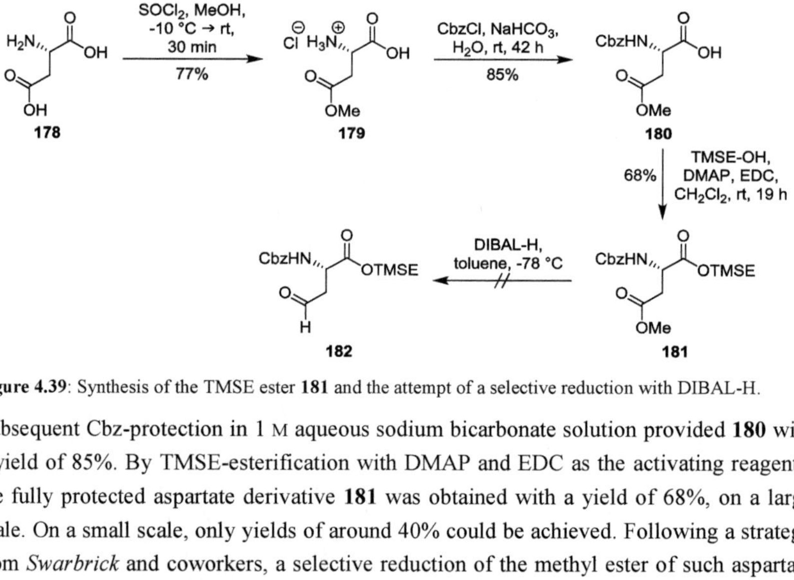

Figure 4.39: Synthesis of the TMSE ester **181** and the attempt of a selective reduction with DIBAL-H.

Subsequent Cbz-protection in 1 M aqueous sodium bicarbonate solution provided **180** with a yield of 85%. By TMSE-esterification with DMAP and EDC as the activating reagents, the fully protected aspartate derivative **181** was obtained with a yield of 68%, on a large scale. On a small scale, only yields of around 40% could be achieved. Following a strategy from *Swarbrick* and coworkers, a selective reduction of the methyl ester of such aspartate derivatives should be possible by the use of DIBAL-H.[223] A temperature of -78 °C and short reaction times were, thereby, essential to selectively obtain the corresponding aldehyde. However, using short reaction times, no conversion could be observed. Extending the reaction time or rising the temperature led to a decomposition of the TMSE ester. As the TMSE ester functionality might not only be too sensitive for this kind of reactions, but as it might also be too sensitive for the following reaction steps, a different ester group was introduced. *Tert*-butyl esterification under *Eschenmoser* conditions furnished compound **183** in a yield of 59% (Figure 4.40). Attempts to obtain the aldehyde **184** via a direct selective reduction with DIBAL-H led to a yield of only up to 12%, depending on the reaction conditions (route **i**). Besides the desired product **184**, the starting material and decomposition products were obtained. Due to the poor yields, a different strategy via the corresponding *Weinreb* amide **185** was investigated (route **ii**). Using lithium hydroxide, the alkaline hydrolysis of **183** gave the corresponding free carboxylic acid in a yield of 90%. Activation and subsequent coupling with *N,O*-dimethylhydroxylamine hydrochloride (DMHH) delivered the *Weinreb* amide **185**. Various coupling reagents were used with triethylamine as base (see Figure 4.40).

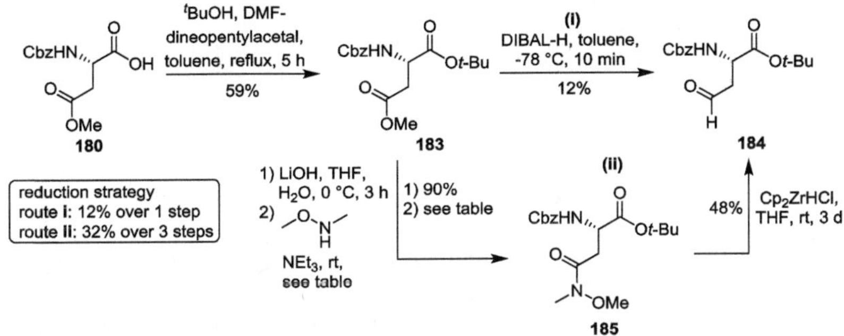

Reaction conditions for the formation of the *Weinreb* amide **185**

No.	coupling reagent [eq]	NEt₃ [eq]	DMHH [eq]	t [d]	yield **185**
1	PyBOP (1.0)	2.0	1.5	4	46%
2	PyBOP (1.0)	1.0	3.0	5	53%
3	PyBOP (1.4)	1.4	2.9	6	73%
4	HOBt/EDC (1.1)	2.2	1.1	4	31%

Figure 4.40: Synthesis of the aspartate semialdehyde **184** via two different reduction strategies: i) directly with DIBAl-H, ii) via the *Weinreb* amide **185** as an intermediate; and different conditions used for the synthesis of the Weinreb amide **185**.

After the optimization of the reaction conditions, especially PyBOP-derived active esters gave good results in the coupling with DMHH (entry 1-3). Using 2.9 eq DMHH, 1.4 eq PyBOP, and 1.4 eq triethylamine, the yield could be increased up to 73%. The use of HOBt in the presence of EDC as the coupling reagent (entry 4) only led to poor yields in comparison to entry 1-3. The reduction of the *Weinreb* amide **185** by the use of *Schwartz's* reagent (Cp₂ZrHCl) afforded the desired semialdehyde **184** in a yield of 48%. Thus, the synthesis of **184** could be achieved with a yield of 32% over three steps (route **ii**). In contrast, the direct reduction with DIBAL-H gave a yield of only 12% (route **i**). Although the reduction strategy with the *Weinreb* amide **185** as an intermediate gave better results than the direct reduction, the yield might be improvable by applying other strategies. It could be promising to reduce the methyl ester **183** to the alcohol, followed by a selective oxidation to the aldehyde. An oxidation might be possible by using *Dess-Martin* periodinane or by applying *Swern* conditions. However, as more promising results were obtained in another synthetic strategy for alkene precursors, which should be used in the asymmetric dihydroxylation (see chapter 4.6.2.2), it was not envisioned at this point to further optimize the reaction conditions for the synthesis of aldehyde **184**.

By *Wittig* methylenation of the semialdehyde **184**, olefin **186** could be obtained (Figure 4.41), following a protocol from *Young* and coworkers.[172] After deprotonation of methyltriphenylphosphonium bromide to the corresponding ylide by potassium hexamethyldisilazane (KHMDS), aldehyde **184** could be converted to the desired allyl glycine derivative **186** with a yield of 64%.

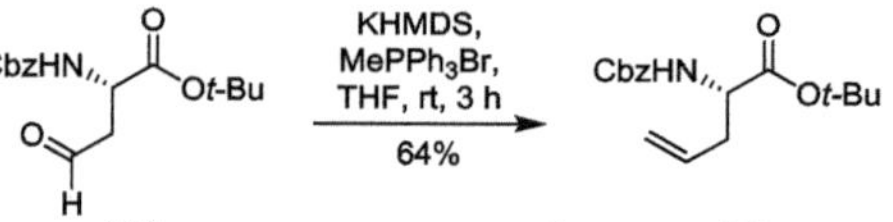

Figure 4.41: *Wittig* methylenation of **184** with protected allyl glycine **186** as the product.

4.6.2.2 Synthesis of a *Garner*'s Aldehyde Derived Olefin as the Precursor for the Asymmetric Dihydroxylation Step

Applying a protocol by *Garner* and *Park* for the *Grignard* addition of vinylmagnesium bromide to (*R*)-*Garner*'s aldehyde (***R***)-**98**,[173] allyl alcohol **120** could be obtained in a yield of 76% (Figure 4.42). The ratio of the diastereomeric mixture, which was obtained, was determined to be 2:5, based on the analysis of the ^{1}H NMR spectrum.

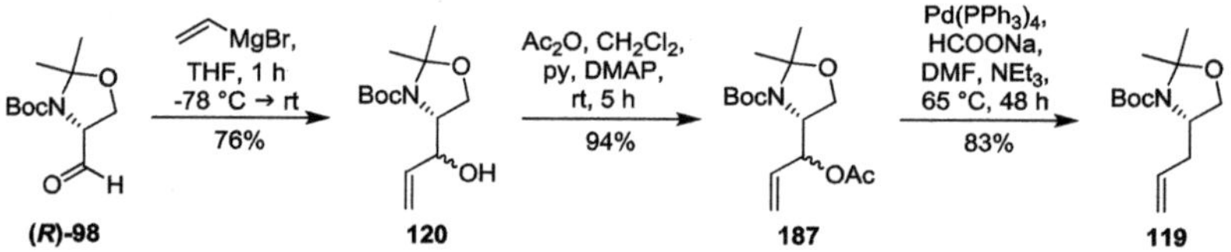

Figure 4.42: Synthesis of the olefin **119** from (*R*)-*Garner*'s aldehyde (***R***)-**98**.

As allyl alcohol derivative **120** was deoxygenated, a separation of the diastereomers was not necessary. The deoxygenation could be achieved via allyl acetate derivative **187** as an intermediate. Therefore, a protocol by *Miyano* and coworkers was applied.[224] By adding pyridine and DMAP, the esterification with acetic anhydride gave the allyl acetate **187** as a diastereomeric mixture with a yield of 94%. By subsequent *Tsuji-Trost*-type palladium-catalyzed reductive deoxygenation with sodium formate as the hydride source, olefin **119** was obtained in a yield of 83%. A temperature of 65 °C, long reaction times (3 days), and the absolute exclusion of air were essential for a good conversion.

4.6.2.3 *Sharpless* Asymmetric Dihydroxylation

The synthesis of the diol **(S)-91** was performed according to the asymmetric dihydroxylation developed by *Sharpless*.[225] Based on mechanistic considerations by *Sharpless* and using his mnemonic device, the formation of the (S)-configured diol **(S)-91** should be preferred by the use of AD-mixα (see Figure 4.43).

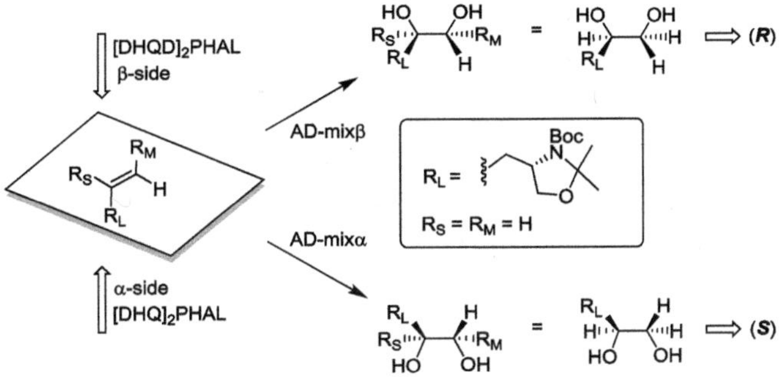

Figure 4.43: Asymmetric dihydroxylation of **119** according to Sharpless.[225]

In order to study a possible substrate-controlled stereoselectivity, the reaction was carried out by using AD-mixα and AD-mixβ separately, and the results were compared. To a solution of AD-mixα or AD-mixβ, respectively, in a mixture of *tert*-butanol and water (1:1), olefin **119** was added at 0 °C. Using AD-mixα, the diastereomeric mixture was obtained with an overall yield of 78%. With AD-mixβ an overall yield of 91% was obtained (Figure 4.44).

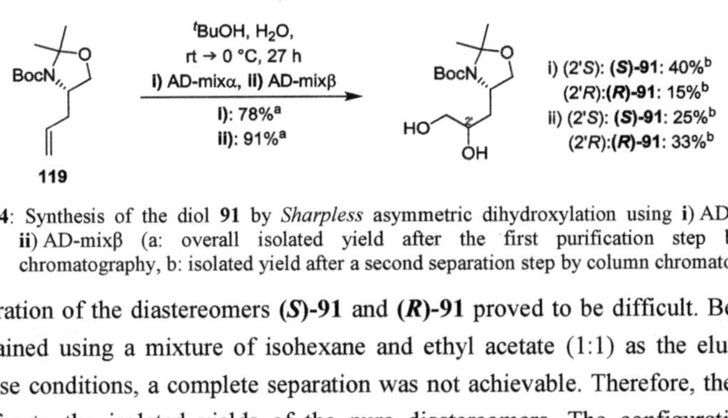

Figure 4.44: Synthesis of the diol **91** by *Sharpless* asymmetric dihydroxylation using i) AD-mixα, and ii) AD-mixβ (a: overall isolated yield after the first purification step by column chromatography, b: isolated yield after a second separation step by column chromatography).

The separation of the diastereomers **(S)-91** and **(R)-91** proved to be difficult. Best results were obtained using a mixture of isohexane and ethyl acetate (1:1) as the eluent. Even under these conditions, a complete separation was not achievable. Therefore, the reported ratios refer to the isolated yields of the pure diastereomers. The configurations were assigned according to *Sharpless'* mnemonic device. Using AD-mixα, the diastereomers were isolated in a ratio of **(S)-91**/**(R)-91** = 5:2. With a yield of 40%, the formation of the putative (S)-isomer **(S)-91** was preferred. Using AD-mixβ led to a ratio of **(S)-91**/**(R)-91** = 2:3. These results were consistent with the reported theory of *Sharpless*.

Regarding the future use as enduracididine precursors, the stereochemistry of the diol **(*S*)-91** or of a corresponding derivative has to be determined by experimental methods as well. Crystallization and X-ray structure analysis, or NOESY experiments of suitable derivatives could be adequate methods in this context.

4.6.2.4 Initial Attempt for the Activation of Diol (*S*)-91 by Conversion into the Corresponding Mesylate

In order to synthesize diamine **118**, a conversion into the corresponding azide would be necessary, which should be reducible to furnish **118**. The synthesis of the azide would require a prior activation of the diol **91**. Therefore, diol **91** should be converted into its corresponding mesylate (Figure 4.45).

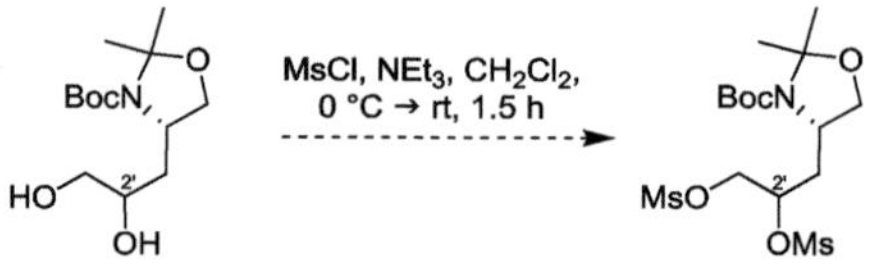

(2'*S*): **(*S*)-91**
(2'*R*): **(*R*)-91**

(2'*S*): **(*S*)-188**
(2'*R*): **(*R*)-188**

Figure 4.45: Attempted synthesis of mesylates **(*S*)-188** and **(*R*)-188**.

In an initial attempt, the desired products **(*S*)-188** and **(*R*)-188** could putatively be obtained using methanesulfonyl chloride and triethylamine in CH_2Cl_2. TLC analysis and the analysis of the ESI-MS spectra of the putative product **(*R*)-188** were consistent with the formation of the corresponding dimesylate. However, by analysis of the ^1H NMR spectra, the formation of side products could definitely be observed. Due to the instability of mesylates, it could be proposed that a partial decomposition had taken place. Therefore, the mesylates should be converted directly into their azides to prevent decomposition.

A possible problem in the developed synthetic strategy could be a ring-closure via the Boc group, due to the activation of the diol **91** (figure 4.46).

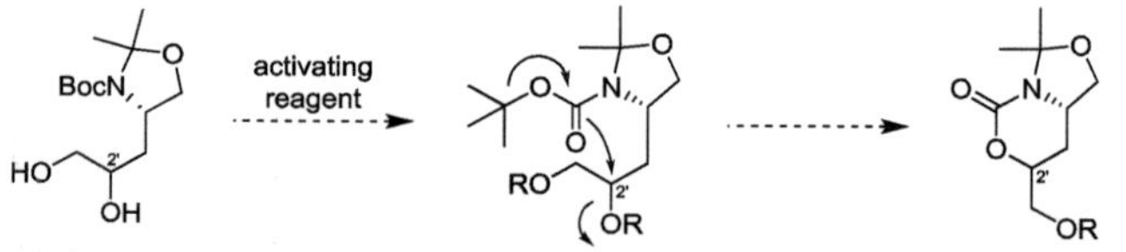

(2'*S*): **(*S*)-91**
(2'*R*): **(*R*)-91**

(2'*S*): **(*S*)-189**
(2'*R*): **(*R*)-189**

(2'*R*): **(*R*)-190**
(2'*S*): **(*S*)-190**

Figure 4.46: Hypothetical ring-closure reaction of the activated diol **189** (OR = leaving group).

In this context, a cyclization analogously to the preparation of the cyclic carbamates **(*R*)-125** and **(*S*)-125**, which were used for nOe experiments (see also Figure 4.4 in chapter 4.2.1),[169] might be conceivable. However, as mesylation seemed to be in principle possible, the developed route via the *Garner*'s aldehyde derived diol **(*S*)-91** seemed to be promising. Further investigations were not carried out as part of this thesis.

4.7 Fermentation Methods and Initial Studies on Muraymycin Biosynthesis

4.7.1 Muraymycin Production by Fermentation of *Streptomyces* sp. NRRL30471 and Its Confirmation by LC-MS

In order to obtain samples for a long-term-storage, several glycerol stocks were prepared from the lyophilized cells of NRRL30471, which were obtained from the Agricultural Research Service Culture Collection from NRRL. This way, a sufficient stock for future fermentation processes was generated. The fermentation was carried out according to the patent issued to the *Wyeth Holdings* Corporation.[158] The production of muraymycins within the secondary metabolism of the organism strongly depends on the conditions of the fermentation. With most conditions, multiple muraymycin derivatives are produced in different quantities. Optimized conditions for the production of a single muraymycin in excess would simplify the procedures of identification, isolation, and characterization. This would be especially important for feeding experiments. Example 5 in the patent describes that, under BPM21 conditions (see chapter 7.3.1), muraymycin C1 and C5 are produced in quantities of approximately 67 mg/L and 4.5 mg/L, respectively.[158] Therefore, cultures were prepared using these conditions. The sample preparation for the LC-MS analysis, which started with an inoculation from the glycerol stock, is described in Figure 4.47. The purification of larger amounts was carried out on a Diaion® WT01S ion exchange resin instead of solid phase extraction (SPE) columns.

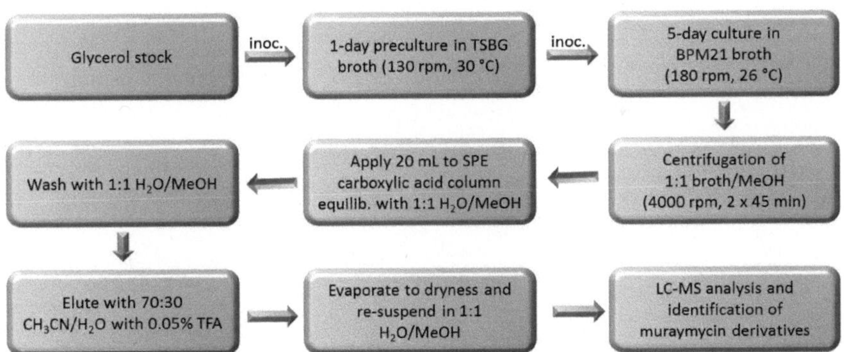

Figure 4.47: Preparation of LC-MS samples (for conditions of tryptic soy broth with glucose (TSBG) and BPM21 broth see chapter 7.3.1; SPE = solid phase extraction, inoc. = inoculation).

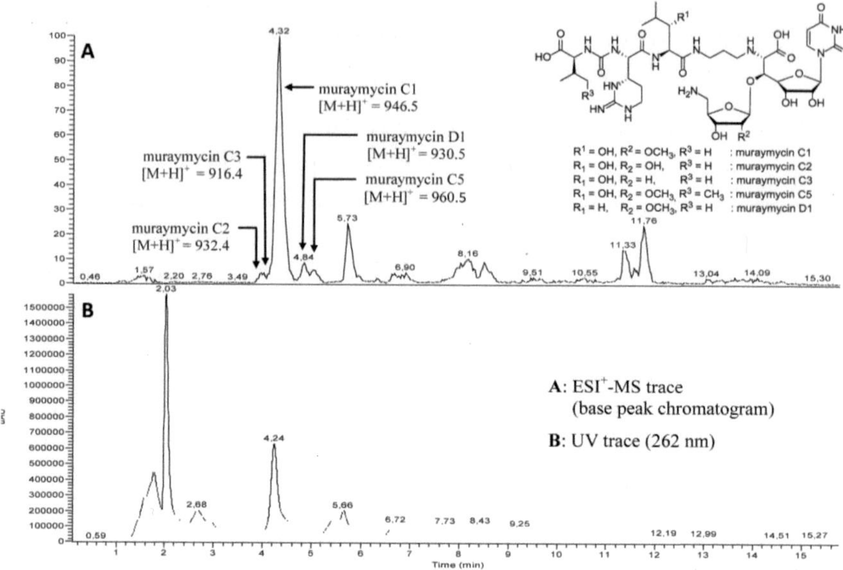

Figure 4.48: Confirmation of the production of muraymycins by LC-MS: A) the ESI⁺-MS trace as the base peak chromatogram and B) the UV trace at 262 nm.

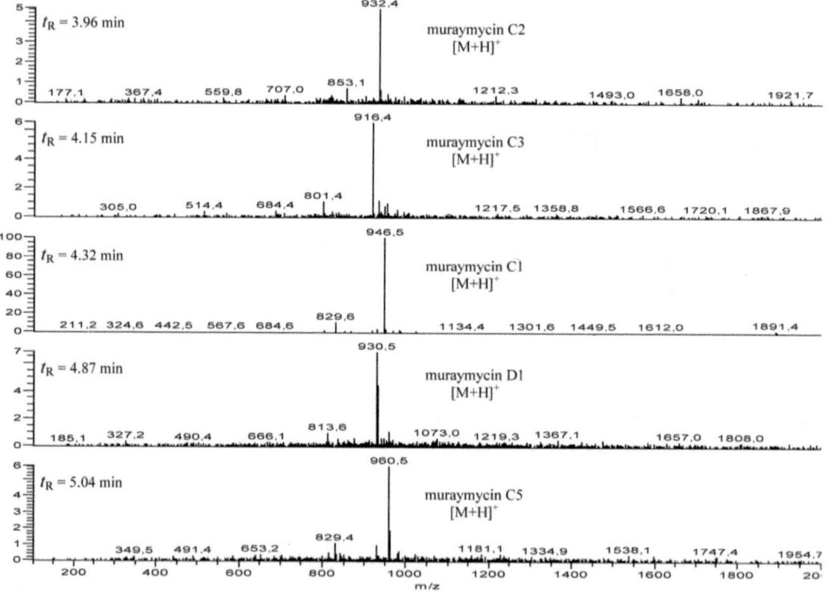

Figure 4.49: Confirmation of the production of muraymycins by LC-MS: ESI⁺-MS spectra of the muraymycin derivatives C2, C3, C1, D1, and C5 (t_R = retention time).

The confirmation of the production of muraymycins was achieved by LC-MS analysis (see Figure 4.48 and Figure 4.49). Using a linear gradient from 10% acetonitrile with 0.1% formic acid and 90% water with 0.1% formic acid to 100% acetonitrile with 0.1% formic acid over 22 min with a flow of 0.8 mL/min (RP C_{18} column) gave a sufficient separation for the detection, and the conditions made the identification of different muraymycin derivatives possible. The presence of the uracil moiety allowed for UV monitoring at 262 nm. In accordance with the patent,[158] muraymycin C1 was detected as the major compound. With retention times between 4.0 and 5.0 min, the muraymycin derivatives C2, C3, C5, and D1 could be detected as well (Figure 4.48 and Figure 4.49). The retention times and the detected mass-to-charge ratios were in good accordance with the calculated and reported values of the muraymycin derivatives C1, C2, C3, C5, and D1 (Table 4.12).

muraymycin derivative	$[M+H]^+$ found	$[M+H]^+$ reported	$[M+H]^+$ calcd.	t_R (min) found[a]	t_R (min) reported
C2	932.4	932.4	932.4	4.0	7.4
C3	916.4	916.3	916.4	4.2	7.5
C1	946.5	946.4	946.4	4.3	7.5
D1	930.5	930.4	930.5	4.8	9.3
C5	960.5	960.4	960.5	5.0	9.8

Table 4.12: Retention times (t_R) and mass-to-charge ratios $[M+H]^+$ which were found in comparison with the calculated and reported values of the muraymycins C1, C2, C3, C5, and D1 (a: different HPLC conditions than reported in the patent were used).[158]

The relative quantities of the different muraymycin derivatives, which were obtained, differed from experiment to experiment. This fact indicated the sensitivity of the secondary metabolism of the *Streptomyces* strains towards minimal changes in the culturing conditions. Furthermore, a detailed investigation of the work-up conditions for the fermentation broth showed that a purification by solid phase extraction (SPE) was essential. Otherwise, the concentration of muraymycin derivatives within the concentrated crude mixture was too low for a proper detection by LC-MS (**A** and **B** in Figure 4.50). In order to obtain a sufficient amount for detailed LC-MS measurements, it was possible to use the 3 mL SPE columns for more than the indicated sample volume. It was shown that up to 20 mL of the methanolic fermentation broth could be applied to an SPE column without a loss of detection-quality. In this manner, it was also possible to make muraymycin derivatives detectable which were produced as minor compounds, such as muraymycin C3. If more than 20 mL of the methanolic fermentation broth were applied, the quality of the LC-MS sample deteriorated due to other impurities (**C** and **D** in Figure 4.50). Overloading of the SPE column most likely was the reason for this observation. The LC-MS chromatograms of such samples were comparable to the LC-MS chromatograms of samples for which a purification by SPE was not performed.

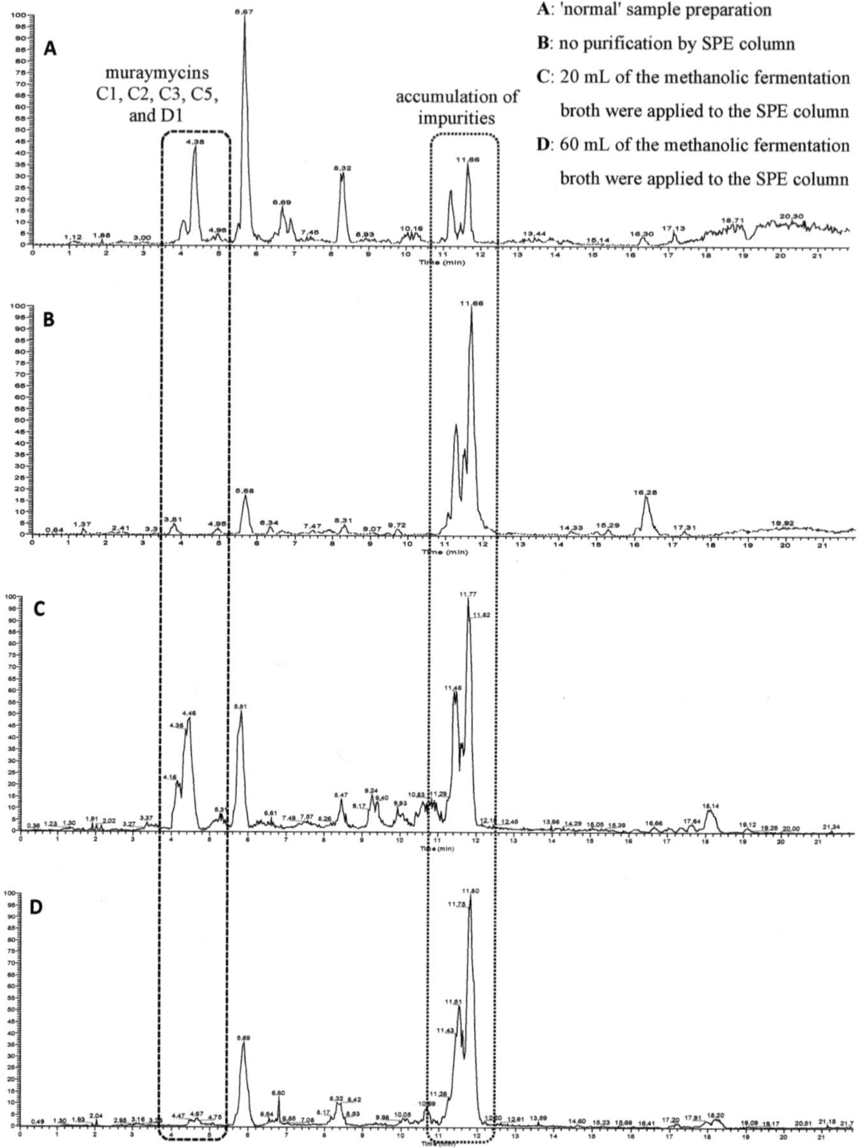

Figure 4.50: Evaluation of the purification conditions for the preparation of LC-MS samples by a comparison of the base-peak chromatograms of differently purified samples.

Choosing the culturing conditions which were reported to produce the most active compound, muraymycin A1 (example 4 in the patent, BPM23A conditions), a detection of muraymycin derivatives was not possible. However, it was demonstrated that the

Streptomyces sp. strains could be cultured in our laboratory. Furthermore, the conditions for the production of muraymycin antibiotics were verified by using BMP21 conditions, and the purification conditions for the preparation of LC-MS samples could be improved.

4.7.2 The Fragmentation Pattern of Muraymcin Derivatives Obtained by LC-MS/MS

Tandem mass spectrometry (MS/MS) might be a helpful tool for the analysis and the interpretation of feeding experiments. Combined with liquid chromatography, it does not only allow the identification of different compounds within the sample, but it also offers information about the structure and the composition of the molecule. Individual fragmentation patterns of the different muraymycin derivatives, which are obtained from feeding experiments, can reveal and verify the specific positions or fragments where the isotope label is located. This could provide additional information for biosynthetic pathways. Therefore, LC-MS/MS analysis should be performed additionally to the LC-MS analysis. Using analogous conditions, MS/MS spectra from the muraymycin derivatives C1, C2, and C5 could be obtained (Figure 4.52–4.54). As the concentration of the derivatives C3 and D1 was too low within the samples, and as an overlapping with the signals of the other muraymycin derivatives occurred, a clear fragmentation pattern could not be observed for these compounds.

In Figure 4.51–4.54, the fragmentation patterns of the muraymycin derivatives are shown which are consistent with the obtained mass-to-charge ratios. The most intense peak, which was observed, is consistent with a cleavage of the valine residue (C1 and C2), or the isoleucine residue (C5), respectively. Therefore, commercially available, isotopically labeled valine derivatives might be attractive compounds for feeding studies in order to prove successful isotope incorporation.

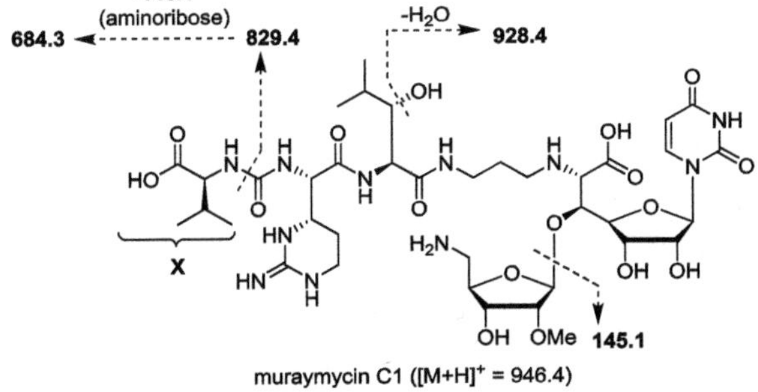

muraymycin C1 ([M+H]$^+$ = 946.4)

Figure 4.51: Principle fragmentation pattern of muraymycins occurring in ESI$^+$-MS/MS experiments, exemplified for muraymycin C1.

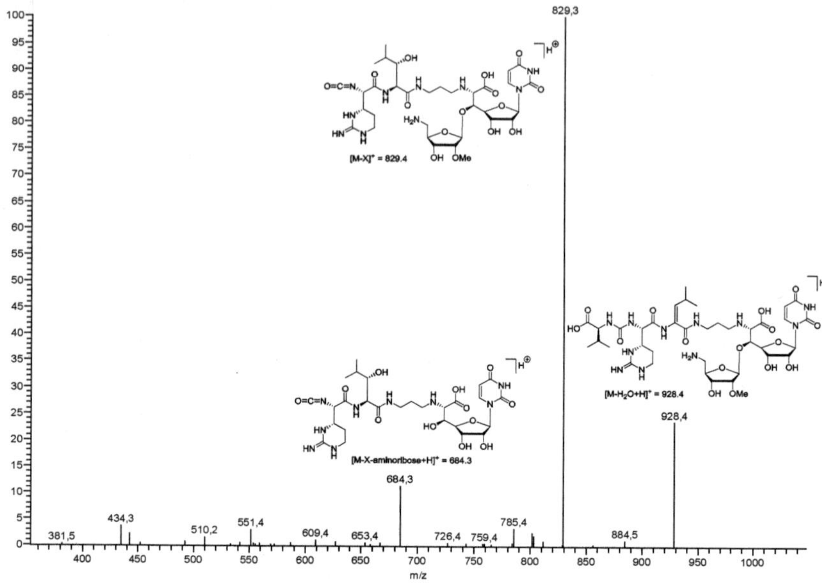

Figure 4.52: Fragmentation pattern of muraymycin C1, obtained by LC-MS/MS with t_R = 4.81 min.

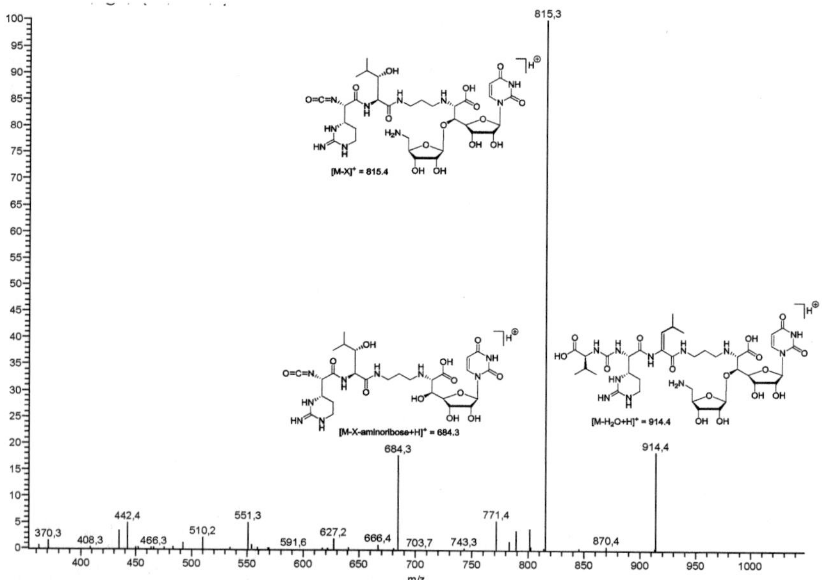

Figure 4.53: Fragmentation pattern of muraymycin C2, obtained by LC-MS/MS with t_R = 4.48 min.

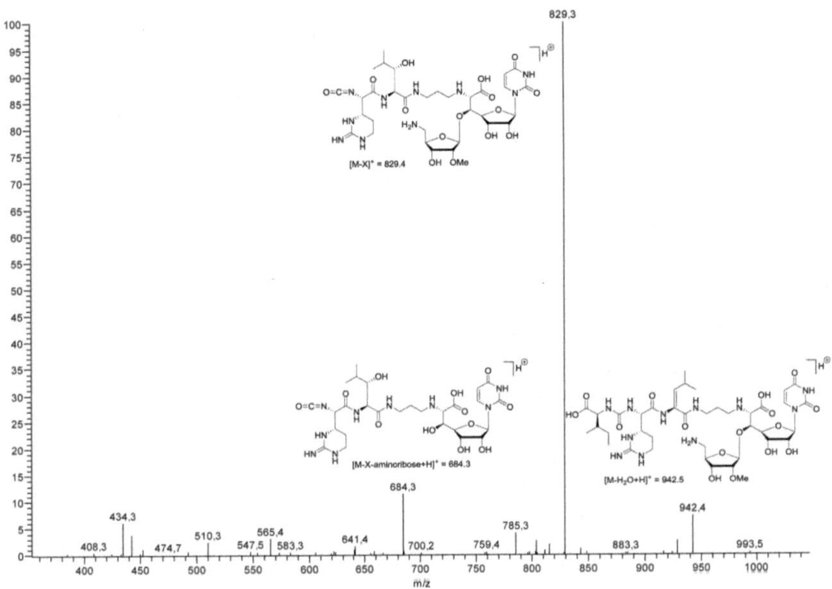

Figure 4.54: Fragmentation pattern of muraymycin C5, obtained by LC-MS/MS with t_R = 5.58 min.

4.7.3 Attempts to Isolate Muraymycin Derivatives by HPLC

As LC-MS experiments clearly showed the possibility of obtaining different muraymycin derivatives at different retention times, a separation by HPLC should be possible. Therefore, it was envisioned to also apply the LC-MS methods for purification by HPLC, and to optimize the conditions for isolation attempts. As muraymycin C1 was found to be the major compound, an isolation of this derivative might provide a sufficient amount for NMR analyses. Furthermore, it could be used as a reference for synthetically produced muraymycin derivatives.

In initial attempts, a gradient from 98% water with 0.1% TFA and 2% acetonitrile to 2% water with 0.1% TFA and 98% acetonitrile on a semi-preparative RP C_{18} column was used. Using these conditions, a separation was not possible. However, a separation from other impurities could be achieved. The purification of 628 mg of the crude mixture (**A** in Figure 4.55) gave 30 mg (**B**, t_R = 17-30 min) and 42 mg (**C**, t_R = 30-40 min) of pre-purified muraymycin mixtures. An enrichment of muraymycin C5 in fraction **C** could be observed. Based on these initial attempts for the isolation of pure compounds, a modification and an optimization of the HPLC conditions might make a separation possible. However, due to a shortness of time, further experiments were not carried out.

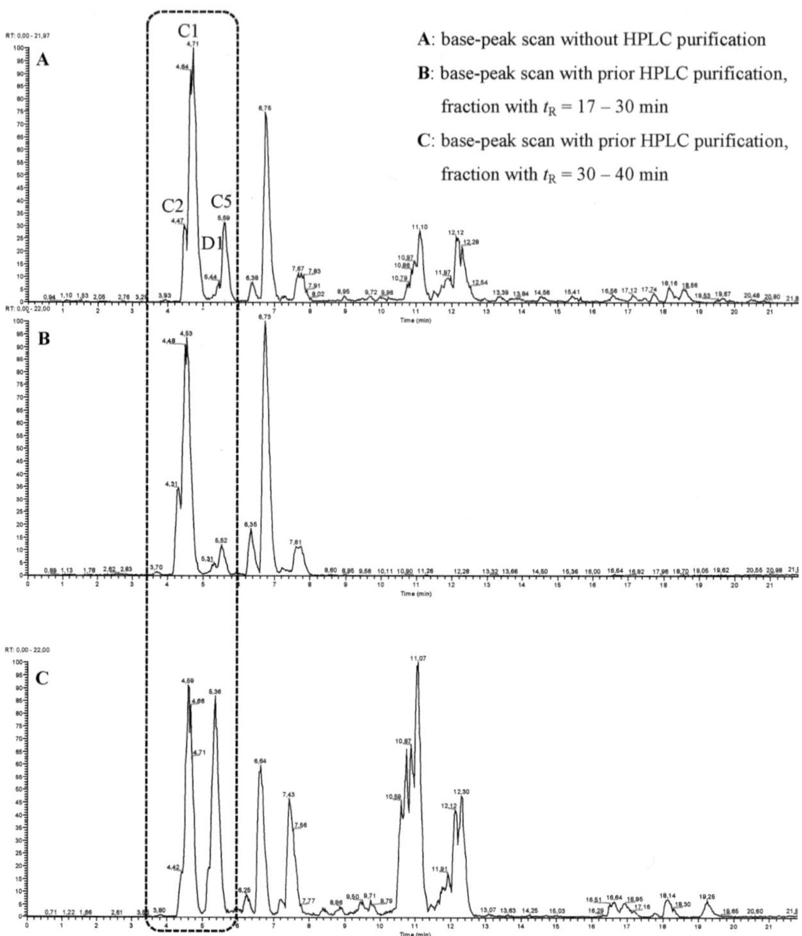

Figure 4.55: Comparison of the base-peak scans with HPLC purification (**B** and **C**) and without HPLC purification (**A**).

4.7.4 Initial Feeding Experiments

For initial feeding experiments, the synthesized [5',5'-^{2}H$_2$]uridine **89** (see chapter 4.4) and commercially purchased [1-^{13}C]glycine were used since they were proposed to be potential intermediates within the biosynthesis of muraymycins (see chapter 2.3.1). As it was shown that the production of muraymycins is quite sensitive to small changes in culture conditions, fermentation broths with unlabeled uridine **1** and glycine **9** were prepared as references. Different pulse feeding conditions were used (see Table 4.13). After purification, LC-MS analysis was performed as described previously. However, labeled compounds could not be detected, and a significant change of intensity of the isotopic signals could not be observed. First, it was considered that insufficient feeding strategies were used (importance of time or time-interval), that the concentration of the labeled compounds within the fermentation broth was too low, or that the compounds were consumed in the primary metabolism of the organism. As it did not seem to be feasible to find the right conditions within a reasonable time without the experience needed, no further feeding experiments were carried out.

compound	c [mg/mL]	addition after 1 d	addition after 2 d	addition after 3 d
uridine **1**	7	0.5 mL	0.5 mL	0.5 mL
[5',5'-^{2}H$_2$]uridine **89**	15	1.0 mL	1.0 mL	1.0 mL
	15	0.5 mL	0.5 mL	0.5 mL
glycine **9**	100	1.0 mL	1.0 mL	1.0 mL
[1-^{13}C]glycine	100	1.0 mL	1.0 mL	1.0 mL

Table 4.13: Different pulse feeding conditions for feeding experiments.

However, recent studies revealed that neither uridine **1** nor glycine **9** are starting materials for the biosynthesis of related nucleoside antibiotics (see chapter 2.3.1).[98,100] These reports would be consistent with the results of the performed feeding studies. However, feeding experiments with a positive result would be necessary to verify the negative results of the feeding studies which were performed with [5',5'-^{2}H$_2$]uridine **89** and [1-^{13}C]glycine. For the future, it seems promising to perform further experiments with other labeled compounds, using the developed pulse feeding conditions. As labeled 3-hydroxy-L-arginine is too valuable for further test-feeding experiments, it seems advisable to purchase labeled compounds commercially. Suitably labeled compounds for further test-feedings might be the amino acids which are incorporated: L-valine (see chapter 4.7.2), L-arginine, L-leucine, or L-threonine. The commercial availability of the isotopically labeled derivatives of these compounds was reviewed, and the costs for 1 g were compared since a sufficient amount of the compounds should be available for further reliable test-feeding studies. It was found that only ^{13}C-labeled L-valine is obtainable with reasonable costs ([1-^{13}C]-L-valine, 1 g, 315.50 €, *Sigma Aldrich*). Another commercially available, inexpensive labeled compound for test-feeding experiments could be ^{13}C-labeled sodium bicarbonate (NaH^{13}CO$_3$, 1 g, 101.00 €, *Sigma Aldrich*). This compound was used to elucidate the biosynthetic formation of the ureido-bond motif of syringolins.[111] Therefore, [1-^{13}C]-L-valine and ^{13}C-labeled sodium bicarbonate seem to be promising compounds for the conduction of further experiments. Feeding studies which use these labeled derivatives might not only help to find suitable feeding conditions, but they might also give intriguing insights into the formation of muraymycins in *Streptomyces* sp. However, the fact that neither [5',5'-^{2}H$_2$]uridine **89** nor [1-^{13}C]glycine were incorporated are already strongly indicating that, analogously to related nucleoside antibiotics, not uridine, but uridine monophosphate (UMP) acts as an biosynthetic precursor,[98] and that not glycine, but threonine is involved in the formation of the glycyl-uridine motif.[100]

5 Conclusion

The central topic of this work was the formation of unusual amino acid structures by biocatalytic and catalytic methods. For the elucidation of the biosynthetic pathway of *Streptomyces*-produced muraymycins and related natural products, a 'biosynthetic tool kit' was compiled and prepared. This does not only comprehend the completed synthesis of deuterium-labeled compounds, such as (3*R*)- and (3*S*)-3-hydroxy-[5-^{2}H]-L-arginine and other potential intermediates with a nucleosidic structure, but it also includes the development of fermentation methodology including the possibility to detect the muraymycin derivatives by LC-MS. Furthermore, different synthetic routes with catalytic key steps for the preparation of the non-proteinogenic amino acid enduracididine were investigated. A strategy with a *Sharpless* asymmetric dihydroxylation as the key step provided promising precursors bearing the potential to be converted into enduracididine building blocks.

For the synthesis of (3*R*)- and (3*S*)-3-hydroxy-[5-^{2}H]-L-arginine **(R)-87** and **(S)-87**, and of corresponding precursors towards C3-labeled 3-hydroxy-L-arginine, the homoallylic alcohols **39** and **124** are essential key structures (Figure 5.1). *Grignard* addition to the literature-known *Garner*'s aldehyde **(R)-98**, which was prepared from D-serine **(R)-121** in a yield of 74% over five steps, created the C1'-stereocentre and made both of the epimers of **39** accessible.

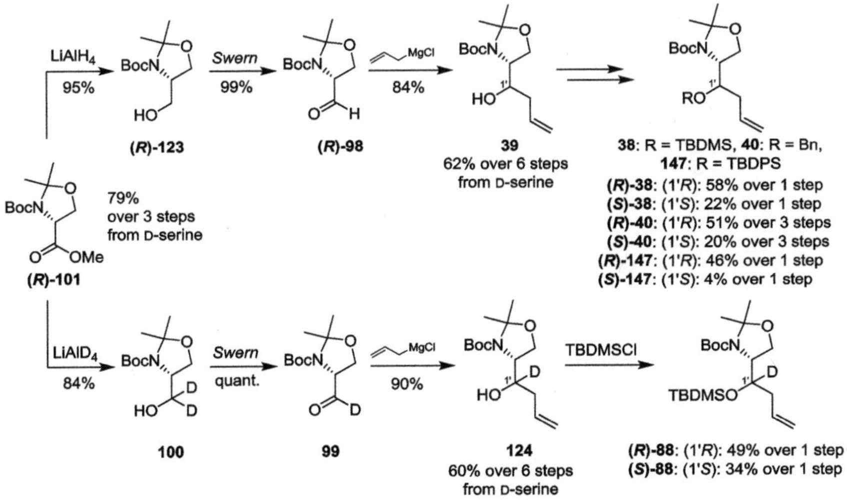

Figure 5.1: Synthesis of the homoallylic alcohols **(R)-38**, **(S)-38**, **(R)-40**, **(S)-40**, **(R)-88**, **(S)-88**, **(R)-147**, and **(S)-147** as the key structures for different synthetic strategies.

The C1'-deuterium label of **124** was introduced by reduction of the methyl ester **(R)-101** to the deuterated alcohol **100**.

By introducing a suitable 3-hydroxy protecting group, the obtained diastereomers of **39** and of **124** could be separated. In this context, the TBDMS-protected derivatives could be separated most effectively, providing 58% of **(R)-38** and 22% of **(S)-38**. The corresponding deuterium-labeled derivatives **(R)-88** and **(S)-88** were obtained in yields of 49% and of 34%, respectively. Starting from compound **39**, the benzyl-protected homoallylic alcohols **(R)-40** and **(S)-40** could be obtained over three steps in yields of 51% and of 20%, respectively. Separation of the diastereomers was achieved via the TBDMS-protected derivatives **(R)-38** and **(S)-38**, followed by deprotection and subsequent benzyl protection. The separation of the TBDPS-protected derivatives **(R)-147** and **(S)-147** delivered isolated yields of 46% and 4%, respectively. On a large scale, the pure (S)-isomer **(S)-147** could not be isolated. Therefore, the strategy used for the benzyl-protected derivatives **(R)-40** and **(S)-40** might be a good alternative for the synthesis of both of the TBDPS-protected diastereomers **(R)-147** and **(S)-147** as well.

Based on the differently protected homoallylic alcohols **40** and **147**, three different approaches towards the target compounds (3R)- and (3S)-3-hydroxy-[5-^{2}H]-L-arginine **(R)-87** and **(S)-87** could be established: (i) via a Benzyl/Cbz protecting group strategy, (ii) via a TBDPS/Cbz protecting group strategy, and (iii) via a TBDPS/Alloc protecting group strategy (Figure 5.2).

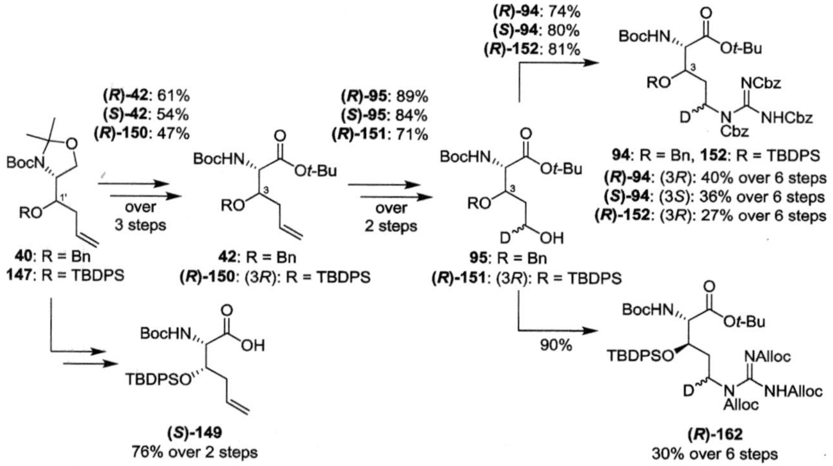

Figure 5.2: Synthesis of the protected 3-hydroxy-L-arginine precursors **(R)-94**, **(S)-94**, **(R)-152**, and **(R)-162**.

The synthesis of the (S)-diastereomer via the TBDPS-protection strategy was only carried out towards the carboxylic acid **(S)-149**. The reason for that was a limited availability of the homoallylic alcohol **(S)-147** due to a challenging separation of the diastereomers

(R)-147 and **(S)-147** by column chromatography. Therefore, only the first steps were investigated for the (S)-isomer. The complete route based on the TBDPS-protected derivatives was investigated using the corresponding (R)-isomers though. Starting from the homoallylic alcohol **(S)-147**, carboxylic acid **(S)-149** was obtained in a yield of 76% over two steps (Figure 5.2). Over three steps, derivatives **(R)-40**, **(S)-40**, and **(R)-147** were converted into their corresponding *tert*-butyl esters **(R)-42**, **(S)-42**, and **(R)-150** with yields of 61%, 54%, and 47%, respectively. The introduction of the deuterium label could be achieved by an ozonolysis reaction with a reductive work-up and a subsequent reduction with sodium borodeuteride. Attempts to develop a strategy for the introduction of two deuterium labels at the C5-position failed. The obtained deuterated alcohols **(R)-95**, **(S)-95**, and **(R)-151** were then guanidinylated. Using tris-Cbz-guanidine as the guanidinylation reagent, the compounds **(R)-94**, **(S)-94**, and **(R)-152** were accessible. Using tris-Alloc-guanidine in the *Mitsunobu* reaction of **(R)-151**, the derivative **(R)-162** was synthesized. The differently protected 3-hydroxyarginine derivatives were synthesized in yields of 30-40% over six steps. Starting from the homoallylic alcohols **(R)-40** and **(S)-40**, the benzyl- and Cbz-protected 3-hydroxyarginine derivatives **(R)-94** and **(S)-94** were obtained over six steps in yields of 40% and 36%, respectively. The TBDPS- and Cbz-protected derivative **(R)-152** and the TBDPS- and Alloc-protected derivative **(R)-162** were obtained over six steps in a yield of 27% and 30%, respectively.

The deprotection of the Cbz-protected derivatives **(R)-94**, **(S)-94**, and **(R)-152** by palladium-catalyzed hydrogenation unexpectedly proved to be problematic due to isomerization and the formation of side products. However, by detailed investigations of the deprotection reactions of the TBDPS-protected derivative **(R)-152**, it was not only possible to identify a substrate-catalyst coordination as the problem, but it was also possible to develop a strategy to avoid isomerization and the formation of side products.

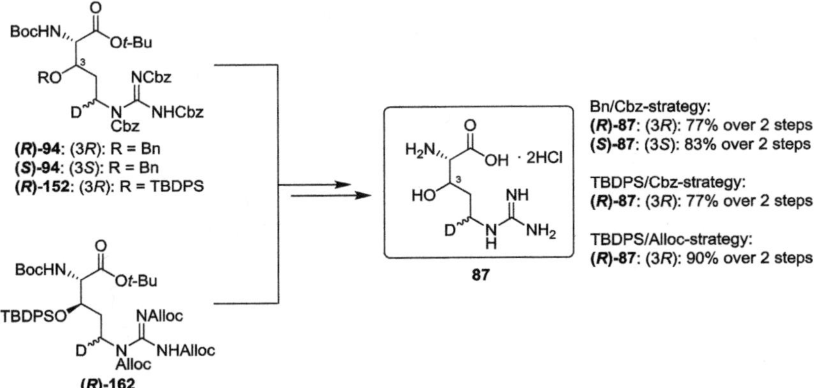

Figure 5.3: Deprotection of **(R)-94**, **(S)-94**, **(R)-152**, and **(R)-162** yielding the pure diastereomers **(R)-87** and **(S)-87**, respectively.

The use of Pd/C with the addition of trifluoroacetic acid delivered the pure compounds. Using the benzyl-protected derivatives, an additional reverse-phase column chromatography was necessary. Via the Bn/Cbz-strategy, both of the diastereomers (3*R*)- and (3*S*)-3-hydroxy-L-arginine **(R)-87** and **(S)-87** could be obtained from **(R)-94** and **(S)-94** over two steps in yields of 77% and of 83%, respectively (Figure 5.3). The TBDPS/Cbz-strategy furnished **(R)-87** in a yield of 77% over two steps. An alternative strategy via the TBDPS- and Alloc-protected derivative **(R)-162** delivered (3*R*)-3-hydroxy-L-arginine **(R)-87** in a yield of 90% over two steps. Comparing the different protection strategies from the homoallylic alcohols over eight steps, the obtained yields are quite similar with 31% and 30% via the Bn/Cbz-route, 21% via the TBDPS/Cbz-route, and 27% via the TBDPS/Alloc-route. However, due to the experienced problems during the palladium-catalyzed hydrogenation, and due to the harsh conditions of the acidic hydrogenation, which have to be applied to prevent an obvious substrate-catalyst coordination, the deprotection of the Alloc groups seems to be more reliable. As the utility of a process lies in its reproducibility, the Alloc/TBDPS protecting strategy should be favored for future syntheses.

However, both of the pure C5-deuterium-labeled diastereomers of 3-hydroxy-L-arginine **(R)-87** and **(S)-87** were synthesized. The synthesis of **(R)-87** was achieved over three alternative protection strategies. Furthermore, the optimized synthetic route via the TBDPS/Alloc protecting strategy now offers a solid synthesis for 3-hydroxy-L-arginine derivatives. Therefore, the corresponding C3-deuterated diastereomers of 3-hydroxy-L-arginine **(R)-97** and **(S)-97** should be easily accessible via this developed route. Additionally, the synthetic problems aroused during the palladium-catalyzed hydrogenation gave new insights into the complex behavior of the deprotected 3-hydroxy-L-arginine derivatives. In this context, a strong pH-dependence was shown. Having these results in mind, it can be assumed that both of the deuterated derivatives of 3-hydroxy-L-arginine **(R)-87** and **(S)-87** are not only valuable substances for the elucidation of the biosynthetic pathway of muraymycins, but that they might also be interesting substrates for other synthetic and biosynthetic studies. The non-deuterated 3-hydroxy-L-arginine derivatives **(R)-17** and **(S)-17** had already been provided to *Schofield* and coworkers, and they proved to be valuable standards in the elucidation of an oxygenase-catalyzed ribosome hydroxylation.[137]

Another deuterated compound which was synthesized was [5',5'-^{2}H$_2$]uridine **89**. The deuterium label at the C5'-position could be introduced by an oxidation-reduction strategy (Figure 5.4). Starting from uridine **1**, the corresponding carboxylic acid **105** was obtained over two steps in a yield of 68%. Activation and a subsequent reduction furnished the deuterated uridine derivative **167** with a yield of 74%. Acidic deprotection gave the desired

[5',5'-^{2}H$_2$]uridine **89** in a yield of 82%. Overall, [5',5'-^{2}H$_2$]uridine **89** was synthesized in a yield of 41% over four steps starting from uridine **1**.

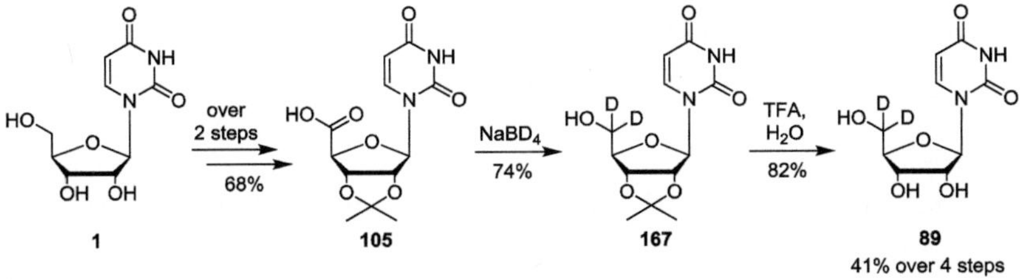

Figure 5.4: Synthesis of [5',5'-^{2}H$_2$]uridine **89**.

For the elucidation of common biosynthetic pathways within the formation of high-carbon nucleoside antibiotics, different nucleosidic building blocks were synthesized. The 5'-deoxygenated derivatives **(R)-90** and **(S)-90**, which comprise an aminopropyl linker, were obtained by the acidic hydrolysis of the compounds **(R)-78** and **(S)-78** in yields of 53% and of 64%, respectively (Figure 5.5). The precursors **(R)-78** and **(S)-78**, which were required, were provided by *A. Spork*.[152]

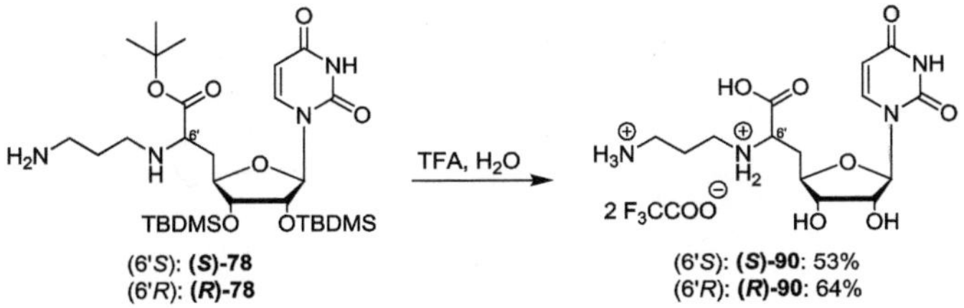

Figure 5.5: Synthesis of the 5'-deoxygenated nucleoside building blocks **(R)-90** and **(S)-90**.

The glycosylated nucleoside building block **12** was synthesized based on a strategy by *Ichikawa* and *Matsuda*.[150] They reported the synthesis of the protected derivative **β-75**. Over five steps, **β-75** was obtained in a yield of 10% (figure 5.6). Especially the purification of the *Wittig*-product **72** (see chapter 4.5.2.1) and problems with the subsequent aminohydroxylation resulted in moderate yields, which were not comparable to the ones reported by *Ichikawa* and *Matsuda*.

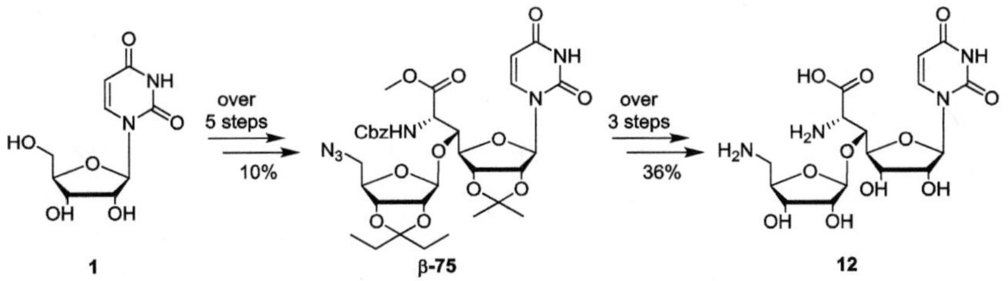

Figure 5.6: Synthesis of the glycosylated nucleoside building block **12**.

Furthermore, the aminohydroxylation, which had been reported to be selective, seemed to result in a minimal contamination by isomeric products, or an undesired formation of the diol. Due to the moderate reproducibility of the *Ichikawa*-route towards β-**75**, the development of an alternative route might be interesting for future studies. However, deprotection and reduction delivered the desired nucleosidic building block **12** in a yield of 36% over three steps. Overall, the glycosylated compound **12** was obtained in a yield of 4% over eight steps (Figure 5.6).

All of the three synthesized nucleosidic building blocks (*R*)-**90**, (*S*)-**90**, and **12** will be used in collaboration with the group of *Van Lanen* (College of Pharmacy, University of Kentucky, Lexington, USA) for the elucidation of the biosynthetic pathway of high-carbon nucleoside antibiotics. It is a particularly intriguing question if glycosylation or an attachment of the side chain occurs first.

The synthesis of precursors towards the non-proteinogenic amino acid enduracididine was achieved by employing a catalytic key step. In an initial attempt, the strategy used was a *Wittig-Horner* reaction with a subsequent asymmetric hydrogenation (Figure 5.7).

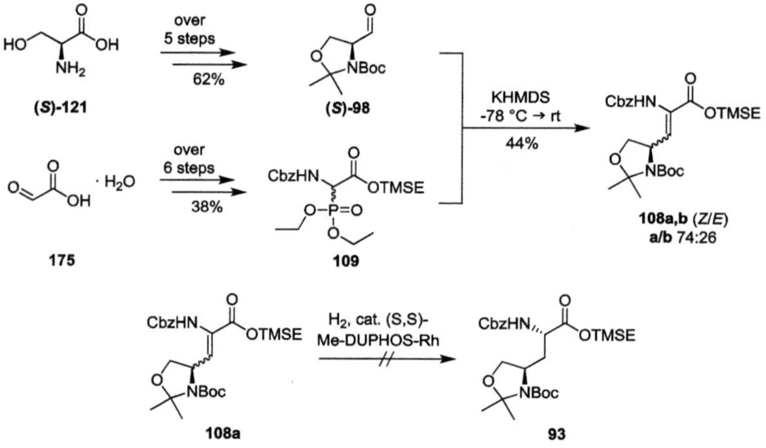

Figure 5.7: Attempted synthesis of the enduracidine precursor **93** via an asymmetric hydrogenation of olefin **108a**.

The required building blocks, the (*S*)-*Garner*'s aldehyde (*S*)-**98** and the phosphonate **109**, were synthesized from D-serine (*S*)-**121** with a yield of 62% over five steps, and from glyoxylic acid monohydrate **175** with a yield of 38% over six steps, respectively. The *Wittig-Horner* reaction, which was performed under different conditions, showed unusually poor diastereoselectivities. Optimization attempts gave the didehydroamino acid **108** in a yield of 44% and in a ratio of 74:26. Since the ^{1}H NMR spectra of the pure diastereomers **108a** and **108b**, which were measured in deuterated chloroform, were not interpretable, the empirical rule of *Mazurkiewicz* could not be applied for the assignment of the configuration. It could only be assumed that the major diastereomer **108a** contained a (*Z*)-double bond, as this kind of *Wittig-Horner* reaction is known to be (*Z*)-selective. In addition, ^{1}H NMR spectra in DMSO-d$_6$ also hinted towards the major product being (*Z*)-configured based on the chemical shift of β-CH signals. An attempted stereoselective asymmetric hydrogenation towards the enduracididine precursor **93** was not successful. As the *Wittig-Horner* reaction had already proven to be problematic, further studies were not envisioned at that time. Instead, an alternative strategy was developed.

As an alternative strategy, *Sharpless* asymmetric dihydroxylation should serve as the catalytic key step. Therefore, two different synthetic routes were investigated. The first one depended on the allylglycine derivative **186** as the olefin for the dihydroxylation step. In three successive steps, L-aspartic acid **178** was converted into the protected L-aspartate derivative **183** in a yield of 39% (Figure 5.8). Selective reduction with DIBAL-H over one step gave the corresponding aspartate semialdehyde **184** with a yield of only 12%. By a reduction strategy via the corresponding *Weinreb* amide **185**, aldehyde **184** was obtained with a yield of 32% over three steps. The *Wittig* methylenation of **184** furnished the allylglycine derivative **186** with a yield of 64%. Starting from L-aspartic acid **178**, allylglycine derivative **186** was obtained in a yield of 8% over seven steps. As the synthesis of the allylglycine derivative **186** only gave poor yields, the focus for the elucidation of the asymmetric dihydroxylation reaction was placed within the second approach. Therefore, it was not envisioned to optimize the synthetic route via the allylglycine **186**.

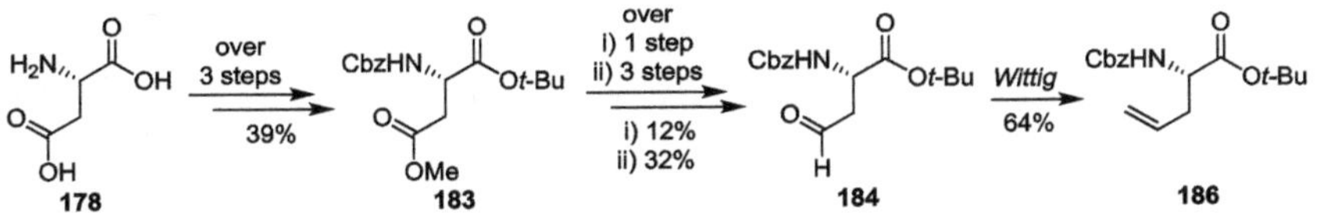

Figure 5.8: Synthesis of the allylglycine derivative **186**.

The second synthetic route was based on the *Garner*'s aldehyde-derived olefin **119** as the substrate for the asymmetric dihydroxylation. As described previously, *Garner*'s aldehyde (***R***)-98 was obtained from D-serine in a yield of 74% over five steps. *Grignard* addition followed by a deoxygenation step delivered the olefin **119** in a yield of 59% over three steps (Figure 5.9). Using AD-mixα for the *Sharpless* asymmetric dihydroxylation of the olefin **119**, the (*S*)-configured diol (***S***)-91 was obtained in a yield of 40%. The diastereomers (***S***)-91 and (***R***)-91 could be isolated in a ratio of (*S*)/(*R*) = 5:2. The assignment of the configuration was performed based on *Sharpless*' mnemonic device. The diastereomeric ratio of (*S*)/(*R*) = 2:3, which was obtained with AD-mixβ, was consistent with these theoretical considerations. The diol (***S***)-91 was synthesized from D-serine with a yield of 17% over nine steps (Figure 5.9).

Figure 5.9: Synthesis of diol (*S*)-91 via *Garner*'s aldehyde (*R*)-98.

Furthermore, an initial attempt to activate the diol groups of (*S*)-91 and (*R*)-91 for azide introduction showed that mesylation should be feasible. A potential ring-closure did not seem to have occurred. These results showed that the diol (*S*)-91 is a promising precursor for the synthesis of enduracididine-derived building blocks.

Besides the synthetic work, fermentation methods for the cultivation of muraymycin-producing *Streptomyces* sp. were established. The culturing process was validated, and the production of muraymycins was proven by the identification of the muraymycin derivatives C1, C2, C3, C5, and D1 by LC-MS analysis. Furthermore, the purification of the methanolic fermentation broth could be optimized, and first attempts to isolate the major compound C1 by purification with semi-preparative HPLC delivered promising results. By the application of MS/MS techniques, the fragmentation patterns of the muraymycin derivatives C1, C2, and C5 were elucidated. Using [5',5'-^{2}H$_2$]uridine **89** and commercially available [1-^{13}C]glycine, initial feeding studies did not result in the incorporation of a deuterium label within the muraymycin derivatives. However, this result is consistent with recent studies on the biosynthesis of related high-carbon nucleoside antibiotics, which were published after the feeding experiments had been carried out. It was reported that not uridine, but uridine monophosphate (UMP) acts as a biosynthetic precursor, and that not glycine, but threonine is involved in the formation of the glycyl-uridine motif.[98,100] These new insights and the completed synthesis of both diastereomers of 3-hydroxy-L-arginine (*R*)-87 and (*S*)-87, combined with the acquired biosynthetic and analytical methods, constitute a comprehensive biosynthetic toolkit with promising prospects for future experiments (Figure 5.10).

Overall, biocatalytic and catalytic approaches for the formation of unusual amino acid structures were developed. With the obtained results and target structures, principle methods for sustainable biocatalytic and catalytic processes were established.

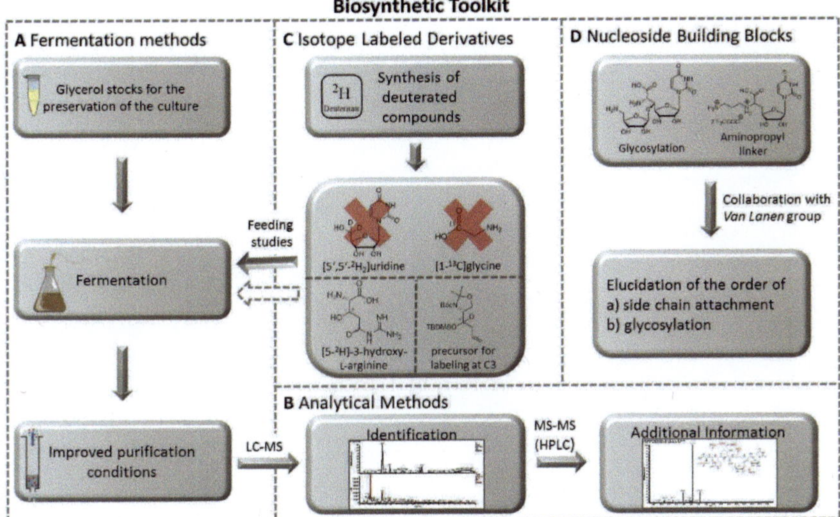

Figure 5.10: The compiled toolkit for the elucidation of the biosynthesis of muraymycin antibiotics.

6   Outlook

One primary objective should be to continue the initiated studies on muraymycin biosynthesis. As discussed in chapter 4.7.4, ^{13}C-labeled L-valine and ^{13}C-labeled sodium bicarbonate are promising compounds for further test-feeding experiments. They are commercially available, and they are relatively inexpensive compared to other labeled amino acids, such as L-arginine, L-leucine, or L-threonine. Using these labeled derivatives, feeding studies might not only help to find suitable feeding conditions, but they might also give intriguing insights into the formation of muraymycins in *Streptomyces* sp. Furthermore, the 'negative results' which were obtained by feeding experiments with [5',5'-^{2}H$_2$]uridine **89** and [1-^{13}C]glycine could be verified.

With verified feeding conditions at hand, further feeding experiments with the pure C5-deuterium labeled diastereomers of 3-hydroxy-L-arginine **(R)-87** and **(S)-87** can be carried out. These experiments should show whether 3-hydroxy-L-arginine acts as an intermediate in the biosynthesis of muraymycins, and, if so, which of the diastereomers is converted. With the knowledge of the relevant diastereomer, the corresponding C3-deuterium labeled diastereomer of 3-hydroxy-L-arginine **(R)-97** or **(S)-97** can be synthesized (Figure 6.1). The deuterium labeled precursors **(R)-88** and **(S)-88**, which are required for the synthesis of **(R)-97** or **(S)-97**, have already been synthesized in sufficient amounts. The optimized synthetic route via the TBDPS/Alloc protection strategy should give access to the desired compounds without the occurrence of any isomerization reaction.

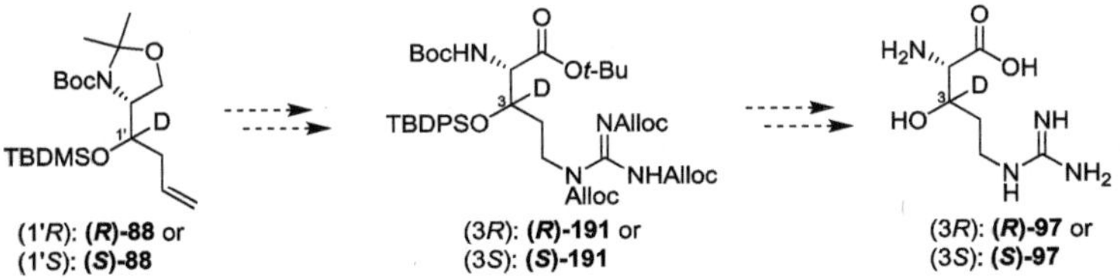

Figure 6.1: Possible synthesis of the C3-deuterated 3-hydroxy-L-arginine derivatives **(R)-97** or **(S)-97**.

With the C3-deuterated diastereomers of 3-hydroxy-L-arginine **(R)-97** and **(S)-97** at hand, a detailed investigation of the stereochemical course of the biosynthetic pathway will be feasible. In this manner, a possible epimerization reaction at the C3-position could be elucidated. As Mur15 and Mur16 were proposed to be involved in the formation of epicapreomycidine, in vitro studies using these enzymes will be crucial.

In the context of the oxygenase-catalyzed arginyl hydroxylation of ribosomal protein Rpl16 reported by *Schofield*,[137] there is a rising interest in 3-hydroxy-L-arginine as building block for solid-phase peptide synthesis (SPPS). Considering this, the development of an efficient synthesis towards a SPPS building block of 3-hydroxy-L-arginine, which would be based on the established synthetic route reported in this work, would offer intriguing prospects.

Due to the moderate reproducibility of the *Ichikawa*-route (see chapter 4.5.2.1) towards the protected nucleoside derivative **β-75**, the development of a different strategy might be interesting. As the aminohydroxylation not only seemed to produce minor amounts of isomeric contaminants, but also resulted in moderate yields, an alternative strategy for the introduction of the 5'-hydroxy and 6'-amino function is desirable. The generation of the corresponding epoxide, followed by a S_N2-like epoxide opening (as described in chapter 2.5) might be suitable in this context. As the alkaline hydrolysis of the methyl ester resulted in β-elimination and decomposition, a modification as *tert*-butyl ester is conceivable. In this context, a glycosylation of the *tert*-butyl ester under mild conditions has to be developed.

The synthesized nucleosidic building blocks **(R)-90**, **(S)-90**, and **12** will be used for biosynthetic studies in a collaboration with the *Van Lanen* group (College of Pharmacy, University of Kentucky, Lexington, USA). These studies might give important insights into common biosynthetic pathways for the formation of high-carbon nucleoside antibiotics, and they will provide the information if glycosylation or attachment of the side chain occur first. In this context, an important long-term goal is to obtain new nucleoside antibiotics by engineered biosynthesis and semisynthesis.

For the synthesis of an enduracididine-based building block for peptide syntheses, diol **(S)-91** is a promising precursor. Activation by mesylation has already been proven to be feasible. A subsequent direct azide introduction, followed by *Staudinger* reduction, should lead to diamine **118**. Guanidinylation should give the corresponding ring-closure product, which can be converted into building block **111** by an acidic hydrolysis of the isopropylidene protecting group, oxidation and suitable protecting group manipulations (Figure 6.2).

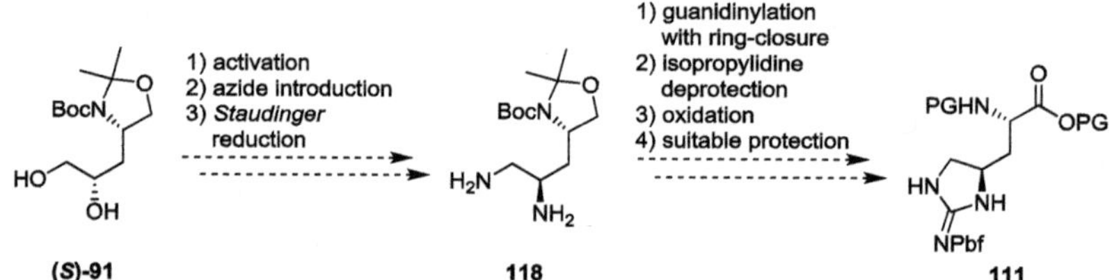

Figure 6.2: Possible synthetic route towards enduracididine-derived building block **111**.

If this synthetic route should not lead to the desired building block **111**, the analogous pathway using allylglycine **186** could be an alternative. *Sharpless* asymmetric dihydroxylation should be feasible under the analogous conditions which were used for the synthesis of **(S)-91**. The obtained diol **(S)-192** should be convertible into the diamine **193** by the application of an analogous strategy. A guanidinylation reaction with ring-closure should then lead to enduracididine precursor **194** (Figure 6.3).

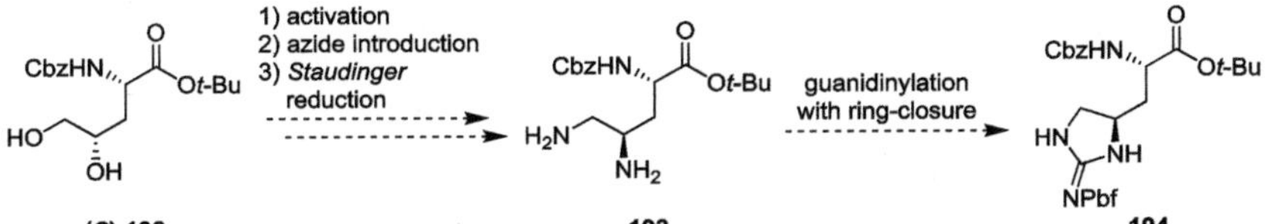

Figure 6.3: Alternative synthetic route towards enduracididine-derived building block **194**.

With respect to the synthetic strategy via *Wittig-Horner* reaction and subsequent asymmetric hydrogenation, a variation of the substrates for the *Wittig-Horner* reaction might be interesting. In this context, aldehyde **196** might be an intriguing candidate (Figure 6.4). The *Wittig-Horner* reaction of **196** with a suitable phosphonate building block **197** should directly lead to didehydroamino acid **198**. Aldehyde **196** might be synthesized starting from L-2,3-diaminopropionic acid **195**. By suitable protecting group manipulations, the amino acid should be convertible into the corresponding *Weinreb* amide. Guanidinylation with ring-closure could then possibly lead to the corresponding 5-membered ring derivative, which could then be selectively reduced to the aldehyde **196**. Another possibility would be to proceed via the corresponding alcohol with a final selective oxidation. Within this strategy, the stability of the aldehyde **196** might be challenging. Furthermore, the selectivity of the *Wittig-Horner* reaction might be unsatisfying, considering the reaction outcome using *Garner*'s aldehyde **(S)-98**.

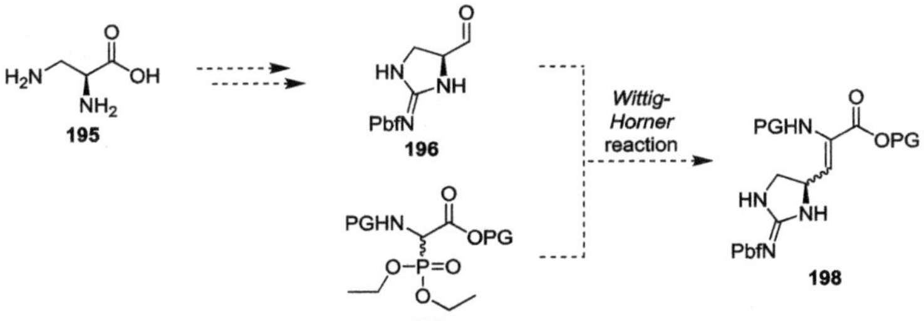

Figure 6.4: Possible synthesis of the didehydroamino acid **198**.

Regarding the proposed mechanism for the palladium-catalyzed hydrogenation reaction of the Cbz- and TBDPS-protected 3-hydroxyarginine derivative **(R)-152**, an additional investigation by LC-MS might be interesting. Reaction conditions which favor a ring-closure reaction could be elucidated. New insights into the mechanism leading to ring-closure might give new inspirations for the design of a biomimetic ring-closure reaction. In this context, the hydroxyl protecting group might play a key role. Furthermore, an investigation of asymmetric hydrogenation reactions might be carried out. These experiments could show if i) ring-closure also occurs mediated by other metals and catalyst systems, and ii) if a selective ring-closure reaction might in principle be feasible.

7 Experimental

7.1 General Methods

7.1.1 General Working Techniques

Ozone was generated using a *Fischer* ozone generator model 502 or model 500. Reactions involving oxygen and/or moisture sensitive reagents were carried out under an atmosphere of argon using anhydrous solvents. The glass equipment used for these reactions was dried by heating prior to use. Anhydrous solvents were obtained as indicated in chapter 7.1.4. For reactions at low temperatures (T ≤ 0 °C), suitable freezing mixtures (ice in water, dry ice or liquid nitrogen in acetone) or a cryostat were used. Lyophilization was carried out using an Alpha2-4 unit of *Christ*.

7.1.2 Starting Materials and Reagents

Chemicals were purchased from *Sigma-Aldrich, Alfa Aesar, ABCR, VWR, Roth, Chempur, Grüssing, Acros Organics, Merck, GL Biochem, Deutero, Appli Chem* and *Fluka* in p.a. quality or ACS reagent grade und used without further purification. *N*-Boc-L-glutamate semialdehyde *tert*-butyl ester **128**, which was used as model compound for oxidation experiments, was provided by O. Ries.[177] The precursors (***S***)-78 and (***R***)-78, which were used in the synthesis of 5'-deoxygenated nucleosyl derivatives, were provided by A. Spork.[152] The β-glycosyl donor **74** was provided by D. Wiegmann.[212]

7.1.3 Solvents

The following solvents were purchased in technical quality and distilled prior to their use (reactions without inert conditions, extractions, chromatography).

Dichloromethane (CH$_2$Cl$_2$): distilled.

Diethyl ether (Et$_2$O): dried over KOH, distilled on a rotary evaporator and stored over KOH.

Ethyl acetate (EtOAc): distilled.

Isohexane: distilled.

n-Hexane: distilled.

Petroleum ether: distilled, boiling range 35/70.

All other solvents were of p.a. quality. Distilled water was used throughout, except for RP (reverse phase) column chromatography or where it is indicated otherwise. In these cases, bidistilled ultra-pure water (Milli-Q) was used.

7.1.4 Anhydrous Solvents

For reactions under inert conditions, the following dry solvents were used.

Acetone: absolute, degassed and stored over molecular sieves (4 Å).

Acetonitrile (MeCN): dried and purified by an *MBraun* solvent purification system (MB-SPS-800), degassed and stored over molecular sieves (4 Å).

Dichloromethane (CH$_2$Cl$_2$): dried over CaH$_2$, distilled under inert conditions and stored over molecular sieves (3 Å).

N,N-Dimethyl formamide (DMF): absolute, degassed and stored over molecular sieves (4 Å).

Dimethyl sulfoxide (DMSO): absolute, degassed and stored over molecular sieves (4 Å).

Ethyl acetate (EtOAc): absolute, degassed and stored over molecular sieves (4 Å).

Methanol (MeOH): absolute, degassed and stored over molecular sieves (3 Å).

Pyridine: dried over CaH$_2$, distilled under inert conditions und stored over molecular sieves (4 Å).

Tetrahydrofuran (THF): dried over sodium/benzophenone, distilled under inert conditions and stored over molecular sieves (4 Å).

Toluene: HPLC grade, degassed and stored over molecular sieves (4 Å).

Triethylamine (NEt$_3$): absolute, degassed and stored over molecular sieves (4 Å).

7.1.5 Chromatography

Column chromatography: Column chromatography was carried out on silica gel 60 (0.040-0.063 mm, 230-400 mesh ASTM, *VWR*) under flash conditions.

Reverse phase (RP) column chromatography: RP column chromatography was carried out on RP silica gel 90 C$_{18}$ (0.040-0.063 nm, *Fluka*) under flash conditions.

Thin layer chromatography (TLC): TLC was performed on aluminium plates precoated with silica gel 60 F$_{254}$ (VWR). The visualization of the spots was carried out using UV light (254 nm) where appropriate and/or staining under heating (KMnO$_4$ staining solution: 1 g KMnO$_4$, 6 g K$_2$CO$_3$ and 1.5 mL 5 % NaOH (aq) (w/v), all dissolved in 100 mL H$_2$O; ninhydrin staining solution: 0.3 g ninhydrin, 3 mL AcOH, all dissolved in 100 mL 1-butanol; vanillin/sulfuric acid staining solution: 4 g vanillin, 25 mL concd H$_2$SO$_4$, 80 mL AcOH, and 680 mL MeOH).

High-performance liquid chromatography (HPLC): All HPLC methods were carried out on a standard system by *Hitachi* with the UV/vis detector L-7400 (detection at 260 nm), the mixer L-7614 and the pump L-7150. Analytical HPLC was performed with a *VWR* LichroCart®Purospher® RP18e column (5 μm, 4 x 125 mm) and with a correspondent pre-column (4 x 4 mm), also purchased from *VWR*. Semi-preparative HPLC was performed with a *VWR* LichroCart®Purospher® RP18e column (5 μm, 10 x 250 mm) and without a pre-column. Preparative HPLC was performed with a Nucleodur® 100-10 C18ec (10 μm, 21 x 250 mm) and with a correspondent pre-column (16 x 20 mm), both purchased from *Macherey-Nagel*.

In all cases, mixtures of bidistilled water and acetonitrile (HPLC quality, *Merck*) were used. If necessary, trifluoroacetic acid (TFA, Reagent Plus grade, *Sigma Aldrich*) was added to the eluents. The resulting retention times (t_R) [min] are not dead time corrected.

The following HPLC methods were used.

Analytical HPLC:

Flow: 1 mL/min

Eluent: A – H_2O (0.1% TFA), B – MeCN

Method A-01

t [min]	0	15	20	25	32	40
B [%]	0	1	100	100	0	0

Method A-02

t [min]	0	1	30	35	40	45	50	55
B [%]	3	10	15	20	100	100	3	3

Method A-iso-01

t [min]	0	50	55	65	70	75
B [%]	9	9	100	100	9	9

Semi-preparative HPLC:

Flow: 5 mL/min

Method SP-01

Eluent: A – H_2O (0.1% TFA), B – MeCN/H_2O 80:20 (0.1% TFA)

t [min]	0	10	15	20	22	28
B [%]	0	1	100	100	5	5

Method SP-iso-01
Eluent: A – H$_2$O (0.1% TFA), B – MeCN

t [min]	0	16	20	25	30	35
B [%]	10	10	100	100	10	10

Preparative HPLC:

Flow: 10 mL/min

Eluent: A – H$_2$O (0.1% TFA), B – MeCN

Method P-01

t [min]	0	1	30	35	40	45	50
B [%]	3	10	15	100	100	3	3

Method P-iso-01

t [min]	0	75	80	90	95	105
B [%]	9	9	100	100	9	9

7.1.6 Instrumental Analytics

Nuclear magnetic resonance spectroscopy (NMR): 300 MHz and 600 MHz ^{1}H NMR as well as 75 MHz, 76 MHz and 126 MHz ^{13}C NMR, 282 MHz ^{19}F NMR and 121 MHz ^{31}P NMR spectra were recorded on *Varian* Unity 300, Mercury 300, Inova 500 and Inova 600 (at the Institute of Organic and Biomolecular Chemistry of the Georg-August University Göttingen) or *Bruker* Avance-500 and Avance-300 instruments (at the Department of Chemistry of the University of Paderborn). All ^{13}C NMR spectra are ^{1}H-decoupled. All spectra were recorded at room temperature (Göttingen) or 30 °C (Paderborn), except for samples in DMSO-d_6 and D$_2$O (standard 30 °C or 35 °C) and where indicated otherwise, and were referenced internally to solvent reference frequencies. Chemical shifts (δ) are quoted in ppm. Coupling constants (J) are reported in Hz to the nearest 0.1 Hz. The assignment of signals was carried out using ^{1}H,^{1}H COSY, ^{1}H,^{13}C HSQC and ^{1}H,^{13}C HMBC spectra obtained on the spectrometers mentioned above. For the assignment of multiplicities, the following abbreviations were used: s (singlet), d (doublet), t (triplet), q (quartet), quin (quintet), m (multiplet) and their combinations (e.g. dd (doublet of doublet), ddd (doublet of doublet of doublet)). In the case of diastereotopic protons, upfield shifted protons are indicated with an 'a' and down-field shifted protons with a 'b'.

Mass spectrometry (MS): Low resolution ESI mass spectrometry was performed on a *Finnigan* ion-trap mass spectrometer LCQ. High resolution (HR) ESI mass spectrometry was carried out on a *Bruker* microTOF spectrometer, a *Bruker* 7 T FTICR APEX IV (at the Institute of Organic and Biomolecular Chemistry of the Georg-August University

Göttingen) or a *Waters* QTOF Synapt 2G (at the Department of Chemistry of the University of Paderborn).

Melting points (mp): Melting points were measured on a *Büchi* instrument according to Dr. Tottoli or on a B-545 *Büchi* instrument and were not corrected.

Polarimetry: Optical rotations were recorded on a *Perkin-Elmer* 241 polarimeter or a *Jasco* P-2000 with a Na source using a 10 cm cell (concentrations in g/100 mL).

Infrared spectroscopy (IR): Infrared spectroscopy was either performed on a *Bruker* Vector 22 spectrometer with liquids and oils being measured as films on NaCl plates and solids as KBr pills or on spectrometers equipped with an ATR unit (*Jasco* FT-IR 4100 with *Pike* GladiATR® or *Bruker* Vertex 70 with Platinum ATR crystal). Wavenumbers ($\tilde{v}$) are quoted in cm^{-1}.

Ultraviolet spectroscopy (UV): UV spectroscopy was carried out on a Perkin-Elmer Lambda-2 spectrometer, a *Jasco* V-630 spectrophotometer or an *Agilent/Varian* Cary 50 spectrophotometer. Wavelengths of maximum absorption (λ_{max}) are reported in nm with the corresponding logarithmic molar extinction coefficient (log ε) given in parentheses.

7.2   Syntheses

7.2.1   Synthesis of Different *Garner's* Aldehyde Derivatives

7.2.1.1 Synthesis of N-Boc-D/L-Serine Methyl Ester (*R*)-122 and (*S*)-122[159]

(2R): (R)-122, (2S): (S)-122

(***R***)-**122**: Methanol (300 mL) was treated dropwise at 0 °C with acetyl chloride (49.0 mL, 54.0 g, 69.0 mmol, 2.9 eq) under an argon atmosphere. The solution was stirred for a further 5 min, then solid D-serine (***R***)-**121** (25.3 g, 241 mmol, 1.0 eq) was added in one portion and the solution was heated to reflux. Heating was continued for 2 h, then the solution was allowed to cool to room temperature, and the solvent was removed under reduced pressure. To a solution of the resultant crude methyl serinate hydrochloride (37.5 g, 241 mmol, 1.0 eq) in saturated aqueous NaHCO₃ solution (300 mL) and THF (75 mL), di-*tert*-butyl dicarbonate (65.6 g, 301 mmol, 1.3 eq) was added. After stirring the mixture at room temperature for 19 h, water (100 mL) was added, and the aqueous layer was extracted with EtOAc (6 x 300 mL). The combined organics were washed with water

(300 mL), dried over Na_2SO_4, and the solvent was removed under reduced pressure. The resultant crude product was used without further purification.

(S)-122: The L-methyl serinate hydrochloride **(S)-122** was prepared in the same way as compound **(R)-122** with **(S)-121** (25.0 g, 238 mmol, 1.0 eq), acetyl chloride (49.0 mL, 54.0 g, 69.0 mmol, 2.9 eq) and methanol (300 mL). The crude methyl serinate hydrochloride was suspended in THF (400 mL) under an argon atmosphere, and triethylamine (70.0 mL, 51.0 g, 50.0 mmol, 2.1 eq) was added. At 0 °C, a solution of di-*tert*-butyl dicarbonate (65.3 g, 299 mmol, 1.3 eq) in THF (400 mL) was added dropwise. After 15 min of additional stirring, the suspension was allowed to warm to room temperature, stirred overnight (16 h), and warmed to 50 °C for a further 3 h. The solvent was removed under reduced pressure, and the residue was partitioned between THF (110 mL) and saturated aqueous $NaHCO_3$ solution (320 mL). The aqueous phase was extracted with Et_2O (3 x 250 mL). The combined organics were dried over Na_2SO_4 and concentrated under reduced pressure. The resultant crude product was used without further purification.

Yield (R)-122: 52.8 g (100% yield: 52.8 g) as an impure colorless oil.
Yield (S)-122: 60.4 g (100% yield: 48.8 g) as an impure colorless oil.

(R)-122: ^{1}H NMR (300 MHz, DMSO-d_6, 100 °C): δ [ppm] = 1.41 (s, 9H, *t*-Bu-CH$_3$), 3.65 (s, 3H, OCH$_3$), 3.63-3.70 (m, 2H, H-3), 4.08 (dt, J = 8.0, 5.0 Hz, 1H, H-2), 4.61 (t, J = 6.0 Hz, 1H, OH), 6.38 (s$_{br}$, 1H, NH). **^{13}C NMR** (75 MHz, DMSO-d_6, 100 °C): δ [ppm] = 27.67 (*t*-Bu-CH$_3$), 50.98 (OCH$_3$), 56.06 (C-2), 61.00 (C-3), 78.02 (*t*-Bu-C), 154.62 (Boc-C=O), 170.77 (C-1). **MS** (ESI): m/z = 242.1 [M+Na]$^+$. **HRMS** (ESI): calcd for $C_9H_{17}NO_5Na$ 242.0999, found 242.0999 [M+Na]$^+$. **[α]** $_D^{20}$ = -9.1 (c = 1.4, CHCl$_3$). **IR** (film): $\tilde{v}$ [cm^{-1}] = 3406, 2978, 2369, 1744, 1703, 1514, 1366, 1248, 1165, 1062. **TLC** (isohexane/EtOAc 1:1): R$_f$ = 0.18.

(S)-122: ^{1}H NMR (301 MHz, DMSO-d_6, 100 °C): δ [ppm] = 1.41 (s, 9H, *t*-Bu-CH$_3$), 3.65 (s, 3H, OCH$_3$), 3.63-3.70 (m, 2H, H-3), 4.05-4.13 (m, 1H, H-2), 4.60 (s$_{br}$, 1H, OH), 6.36 (s$_{br}$, 1H, NH). **^{13}C NMR** (76 MHz, DMSO-d_6, 100 °C): δ [ppm] = 27.65 (*t*-Bu-CH$_3$), 50.96 (OCH$_3$), 56.07 (C-2), 61.02 (C-3), 78.03 (*t*-Bu-C), 154.63 (Boc-C=O), 170.77 (C-1). **MS** (ESI): m/z = 218.1 [M-H]$^-$. **HRMS** (ESI): calcd for $C_9H_{16}NO_5$ 218.1034, found 218.1035 [M-H]$^-$. **[α]** $_D^{20}$ = +8.1 (c = 0.49, CHCl$_3$). **IR** (ATR): $\tilde{v}$ [cm^{-1}] = 3393, 2979, 1743, 1713, 1688, 1506, 1366, 1210, 1159, 1058, 846. **TLC** (petroleum ether/EtOAc 1:1): R$_f$ = 0.21.

7.2.1.2 Synthesis of *N*-Boc-*N*,*O*-Isopropylidine-D/L-serine Methyl Ester (*R*)-101 and (*S*)-101[159]

(2*R*): (*R*)-101, (2*S*): (*S*)-101

(*R*)-101: To a solution of *N*-Boc-D-serine methyl ester (*R*)-122 (52.8 g, unpurified material) in dry acetone (300 mL), dimethoxypropane (250 mL, 211 g, 2.03 mol, 8.4 eq) and a catalytic amount of boron trifluoride etherate (2.10 mL, 2.40 g, 17.0 mol, 0.07 eq) was added under an argon atmosphere. After stirring the mixture for 4.5 h at room temperature, triethylamine (5.1 mL) was added, and the solvent was removed under reduced pressure. The residue was partitioned between Et$_2$O (200 mL) and saturated aqueous NaHCO$_3$ solution (250 mL), and the aqueous layer was extracted with Et$_2$O (2 x 150 mL). The combined organics were dried over Na$_2$SO$_4$, and the solvent was removed under reduced pressure. The resultant crude product was purified by flash chromatography (petroleum ether/EtOAc 9:1).

(*S*)-101: The isomer was prepared in the same way as compound (*R*)-101 with (*S*)-122 (60.4 g unpurified material), dimethoxypropane (225 mL, 191 g, 1.83 mol, 7.7 eq), boron trifluoride etherate (2.11 mL, 2.36 g, 16.7 mmol, 0.07 eq) and acetone (330 mL). The resulting solution was stirred for 3 h at room temperature.

Yield (*R*)-101: 49.4 g (191 mmol, 79% over 3 steps) as a yellowish oil.
Yield (*S*)-101: 52.4 g (202 mmol, 85% over 3 steps) as a yellowish oil.

(*R*)-101: **^{1}H NMR** (300 MHz, DMSO-d$_6$, 100 °C): δ [ppm] = 1.41 (s, 9H, *t*-Bu-CH$_3$), 1.47 (s, 3H, C(CH$_3$)$_2$), 1.57 (s, 3H, C(CH$_3$)$_2$), 3.70 (s, 3H, OCH$_3$), 3.94 (dd, *J* = 9.1, 3.1 Hz, 1H, H-3a), 4.17 (dd, *J* = 9.1, 7.1 Hz, 1H, H-3b), 4.40 (dd, *J* = 7.1, 3.1 Hz, 1H, H-2). **^{13}C NMR** (75 MHz, DMSO-d$_6$, 100 °C): δ [ppm] = 23.98 (C(C̲H$_3$)$_2$), 24.68 (C(C̲H$_3$)$_2$), 27.44 (*t*-Bu-CH$_3$), 51.28 (OCH$_3$), 58.36 (C-2), 65.15 (C-3), 79.16 (*t*-Bu-C), 93.51 (C̲(CH$_3$)$_2$), 150.42 (Boc-C=O), 170.58 (C-1). **MS** (ESI): *m/z* = 282.1 [M+Na]$^+$. **HRMS** (ESI): calcd for C$_{12}$H$_{21}$NO$_5$Na 282.1312, found 282.1315 [M+Na]$^+$. [α]$_D^{20}$ = +48.6 (*c* = 0.93, CHCl$_3$). **IR** (film): $\tilde{v}$ [cm^{-1}] = 3500, 2981, 2359, 1760, 1710, 1393, 1205, 1094, 848, 770. **TLC** (petroleum ether/EtOAc 1:1): R$_f$ = 0.59.

(*S*)-101: **^{1}H NMR** (301 MHz, DMSO-d$_6$, 100 °C): δ [ppm] = 1.39 (s, 9H, *t*-Bu-CH$_3$), 1.45 (s, 3H, C(CH$_3$)$_2$), 1.55 (s, 3H, C(CH$_3$)$_2$), 3.68 (s, 3H, OCH$_3$), 3.92 (dd, *J* = 9.1, 3.1 Hz, 1H, H-3a), 4.15 (dd, *J* = 9.1, 7.2 Hz, 1H, H-3b), 4.39 (dd, *J* = 7.2, 3.1 Hz, 1H, H-2). **^{13}C NMR** (76 MHz, DMSO-d$_6$, 100 °C): δ [ppm] = 24.02 (C(C̲H$_3$)$_2$), 24.71 (C(C̲H$_3$)$_2$), 27.43

(t-Bu-CH$_3$), 51.25 (OCH$_3$), 58.35 (C-2), 65.14 (C-3), 79.17 (t-Bu-C), 93.50 ($\underline{C}$(CH$_3$)$_2$), 150.41 (Boc-C=O), 170.55 (C-1). **MS** (ESI): m/z = 282.1 [M+Na]$^+$. **HRMS** (ESI): calcd for C$_{12}$H$_{21}$NO$_5$Na 282.1312, found 282.1315 [M+Na]$^+$. $[\alpha]_D^{20}$ = -56.6 (c = 0.87, CHCl$_3$). **IR** (ATR): $\tilde{v}$ [cm^{-1}] = 2978, 1758, 1698, 1456, 1377, 1364, 1250, 1201, 1163, 1091, 1053. **TLC** (petreoleum ether/EtOAc 1:1): R$_f$ = 0.54.

7.2.1.3   Synthesis of *N*-Boc-*N,O*-Isopropylidine-D/L-serinol (*R*)-123 and (*S*)-123[159] and *N*-Boc-*N,O*-Isopropylidine-D-[1,1-²H₂]serinol 100

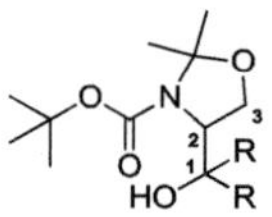

(2*R*): (*R*)-123 (R = H), (2*S*): (*S*)-123 (R = H), (2*R*): 100 (R = D)

(*R*)-123: Under an argon atmosphere, a solution of methyl ester (*R*)-101 (7.06 g, 27.2 mmol, 1.0 eq) in dry THF (40 mL) was added dropwise over a period of 20 min to a suspension of lithium aluminium hydride (1.54 g, 40.6 mmol, 1.5 eq) in dry THF (70 mL). After stirring the mixture for additional 20 min, a 10% aqueous KOH solution (20 mL) was added dropwise at 0 °C. The mixture was stirred for 1 h at room temperature. The white precipitate was removed by filtration through a Celite pad, and the pad was rinsed with Et$_2$O (3 x 20 mL). The combined filtrates were washed with aqueous phosphate buffer (c = 0.1 M, pH = 6.7, 70 mL), and the aqueous layer was extracted with Et$_2$O (3 x 25 mL). The combined organics were dried over Na$_2$SO$_4$, and the solvent was removed under reduced pressure. The resultant crude product was used without further purification.

(*S*)-123: The isomer was prepared in the same way as (*R*)-123 with (*S*)-101 (15.1 g, 58.3 mmol, 1.0 eq), lithium aluminium hydride (3.30 g, 87.0 mmol, 1.5 eq), dry THF (60 mL and 100 mL, respectively) and 10% aqueous KOH solution (36 mL).

100: The deuterated serinol was prepared in the same way as (*R*)-123 with (*R*)-101 (20.2 g, 78.0 mmol, 1.0 eq), lithium aluminium deuteride (4.89 g, 116 mmol, 1.5 eq), dry THF (90 mL and 220 mL, respectively) and 10% aqueous KOH solution (60 mL).

Yield (*R*)-123: 5.97 g (25.8 mmol, 95%) as a bright yellow oil.
Yield (*S*)-123: 10.3 g (44.6 mmol, 77%) as a bright yellow oil.
Yield 100: 15.3 g (65.5 mmol, 84%) as a yellowish oil.

(*R*)-123: **¹H NMR** (300 MHz, DMSO-d$_6$, 100 °C): δ [ppm] = 1.43 (s, 3H, C(CH$_3$)$_2$), 1.44 (s, 9H, t-Bu-CH$_3$), 1.47 (s, 3H, C(CH$_3$)$_2$), 3.22-3.32 (m, 1H, H-1a), 3.53-3.62 (m, 1H, H-1b), 3.75-3.84 (m, 1H, H-2), 3.85-3.95 (m, 2H, H-3), 4.48 (s$_{br}$, 1H, OH).

^{13}**C NMR** (75 MHz, DMSO-d$_6$, 100 °C): δ [ppm] = 23.33 (C($\underline{C}$H$_3$)$_2$), 26.33 (C($\underline{C}$H$_3$)$_2$), 27.61 (*t*-Bu-CH$_3$), 57.93 (C-2), 60.37 (C-1), 64.25 (C-3), 78.54 (*t*-Bu-C), 92.41 ($\underline{C}$(CH$_3$)$_2$), 151.00 (Boc-C=O). **MS** (ESI): *m/z* = 230.2 [M-H]$^-$. **HRMS** (ESI): calcd for C$_{11}$H$_{20}$NO$_4$ 230.1398, found 230.1396 [M-H]$^-$. [α]$_D^{20}$ = +13.9 (*c* = 1.1, CHCl$_3$). **IR** (film): $\tilde{v}$ [cm^{-1}] = 3461, 2983, 1699, 1476, 1393, 1259, 1173, 1074, 849, 771. **TLC** (petroleum ether/EtOAc 8:3): R$_f$ = 0.10.

(*S*)-123: ^{1}**H NMR** (301 MHz, DMSO-d$_6$, 100 °C): δ [ppm] = 1.43 (s, 3H, C(CH$_3$)$_2$), 1.44 (s, 9H, *t*-Bu-CH$_3$), 1.47 (s, 3H, C(CH$_3$)$_2$), 3.22-3.34 (m, 1H, H-1a), 3.53-3.64 (m, 1H, H-1b), 3.76-3.85 (m, 1H, H-2), 3.86-3.96 (m, 2H, H-3), 4.46 (t, *J* = 5.2 Hz, 1H, OH). ^{13}**C NMR** (76 MHz, DMSO-d$_6$, 100 °C): δ [ppm] = 23.32 (C($\underline{C}$H$_3$)$_2$), 26.29 (C($\underline{C}$H$_3$)$_2$), 27.58 (*t*-Bu-CH$_3$), 57.89 (C-2), 60.36 (C-1), 64.23 (C-3), 78.52 (*t*-Bu-C), 92.38 ($\underline{C}$(CH$_3$)$_2$), 150.96 (Boc-C=O). **MS** (ESI): *m/z* = 254.2 [M+Na]$^+$. **HRMS** (ESI): calcd for C$_{11}$H$_{21}$NO$_4$Na 254.1363, found 254.1367 [M+Na]$^+$. [α]$_D^{20}$ = -27.3 (*c* = 0.70, MeOH). **IR** (ATR): $\tilde{v}$ [cm^{-1}] = 3473, 2979, 1668, 1458, 1389, 1364, 1247, 1171, 1105, 1072, 846. **TLC** (petroleum ether/EtOAc 1:1): R$_f$ = 0.30.

100: ^{1}**H NMR** (300 MHz, DMSO-d$_6$, 100 °C): δ [ppm] = 1.43 (s, 3H, C(CH$_3$)$_2$), 1.44 (s, 9H, *t*-Bu-CH$_3$), 1.47 (s, 3H, C(CH$_3$)$_2$), 3.74-3.82 (m, 1H, H-2), 3.84-3.97 (m, 2H, H-3), 4.46 (s$_{br}$, 1H, OH). ^{13}**C NMR** (75 MHz, DMSO-d$_6$, 100 °C): δ [ppm] = 23.36 (C($\underline{C}$H$_3$)$_2$), 26.34 (C($\underline{C}$H$_3$)$_2$), 27.62 (*t*-Bu-CH$_3$), 57.77 (C-2), 64.20 (C-3), 64.05-64.36 (m, C-1), 78.56 (*t*-Bu-C), 92.40 ($\underline{C}$(CH$_3$)$_2$), 150.99 (Boc-C=O). **MS** (ESI): *m/z* = 256.2 [M+Na]$^+$. **HRMS** (ESI): calcd for C$_{11}$H$_{19}$NO$_4$D$_2$Na 256.1488, found 256.1493 [M+Na]$^+$. [α]$_D^{25}$ = +17.3 (*c* = 1.1, CHCl$_3$). **IR** (film): $\tilde{v}$ [cm^{-1}] = 3453, 2979, 2211, 2096, 1697, 1456, 1391, 1259, 1173, 1066, 849, 770. **TLC** (petroleum ether/EtOAc 5:1): R$_f$ = 0.15.

7.2.1.4 Synthesis of *N*-Boc-*N,O*-Isopropylidine-D/L-serinal (*R*)-98 and (*S*)-98[159] and *N*-Boc-*N,O*-Isopropylidine-D-[1-^{2}H]serinal 99 (*Garner*'s aldehyde)

(2*R*): (*R*)-98 (R = H), (2*S*): (*S*)-98 (R = H), (2*R*): 99 (R = D)

(*R*)-98: Under an argon atmosphere, dimethyl sulfoxide (5.15 mL, 5.67 g, 72.6 mmol, 3.0 eq) in dry CH$_2$Cl$_2$ (10 mL) was added dropwise at -78 °C to a solution of oxalyl chloride (3.20 mL, 4.61 g, 36.3 mmol, 1.5 eq) in dry CH$_2$Cl$_2$ (40 mL) over a period of 25 min while the temperature of the reaction mixture rised to -70 °C. At -60 °C, serinol (*R*)-123 (5.60 g, 24.2 mmol, 1.0 eq) in dry CH$_2$Cl$_2$ (40 mL) was added dropwise over a period of 50 min. After completion of the addition, the mixture had reached a temperature of -55 °C and was allowed to warm to -45 °C before N,N-diisopropylethylamine (25.2 mL, 18.1 g, 140 mmol, 2.8 eq) in dry CH_2Cl_2 (4 mL) was added over a period of 5 min. The cooling bath was removed, and the mixture was allowed to warm to 0 °C, and it was added to an ice-cold 1 M HCl solution (90 mL). The aqueous layer was extracted with CH_2Cl_2 (3 x 20 mL), the combined organics were washed with aqueous phosphate buffer ($c = 0.1$ M, pH = 6.7, 3 x 60 mL) and dried over Na_2SO_4. The solvent was removed under reduced pressure, and the obtained crude product was used without further purification.

(**S**)-**98**: The L-derivative (**S**)-**98** was obtained in the same way as compound (**R**)-**98** with oxalyl chloride (5.50 mL, 8.14 g, 64.1 mmol, 1.5 eq), dimethyl sulfoxide (9.09 mL, 10.0 g, 128 mmol, 3.0 eq), (**S**)-**123** (9.84 g, 42.6 mmol, 1.0 eq), N,N-diisopropylethylamine (42.0 mL, 31.9 g, 247 mmol, 5.8 eq) and dry CH_2Cl_2 (40 mL, 20 mL, 40 mL, 10 mL, respectively).

99: The deuterated *Garner*'s aldehyde **99** was prepared in the same way as compound (**R**)-**98** with oxalyl chloride (5.00 mL, 7.40 g, 58.3 mmol, 1.5 eq), dimethyl sulfoxide (8.20 mL, 9.02 g, 115 mmol, 3.0 eq), **100** (9.01 g, 38.6 mmol, 1.0 eq), N,N-diisopropylethylamine (38.0 mL, 28.8 g, 223 mmol, 5.8 eq) and dry CH_2Cl_2 (50 mL, 10 mL, 45 mL, 10 mL, respectively).

Yield (R)-98: 5.50 g (24.0 mmol, 99%) as a yellowish oil
Yield (S)-98: 9.20 g (40.2 mmol, 94%) as a yellowish oil.
Yield 99: 9.26 g (100% yield: 8.89 g) as an impure yellow-to-brown oil.

(**R**)-**98**: **^{1}H NMR** (300 MHz, DMSO-d_6, 100 °C): δ [ppm] = 1.43 (s, 9H, *t*-Bu-CH$_3$), 1.49 (s, 3H, C(CH$_3$)$_2$), 1.55 (s, 3H, C(CH$_3$)$_2$), 4.00-4.13 (m, 2H, H-3), 4.32-4.38 (m, 1H, H-2), 9.54 (d, $J = 2.0$ Hz, 1H, H-1). **^{13}C NMR** (75 MHz, DMSO-d_6, 100 °C): δ [ppm] = 23.69 (C($\underline{C}$H$_3$)$_2$), 25.28 (C($\underline{C}$H$_3$)$_2$), 27.43 (*t*-Bu-CH$_3$), 62.50 (C-3), 63.98 (C-2), 79.56 (*t*-Bu-C), 93.48 ($\underline{C}$(CH$_3$)$_2$), 150.98 (Boc-C=O), 198.55 (C-1). **MS** (ESI): $m/z = 252.1$ [M+Na]$^+$. **HRMS** (ESI): calcd for $C_{11}H_{19}NO_4Na$ 252.1206, found 252.1208 [M+Na]$^+$. $[\alpha]_D^{20} = +63.6$ ($c = 1.1$, CHCl$_3$). **IR** (Film): $\tilde{v}$ [cm^{-1}] = 2983, 2362, 1704, 1393, 1260, 1172, 1096, 1001, 850, 769. **TLC** (petroleum ether/EtOAc 3:1): R$_f$ = 0.28.

(**S**)-**98**: **^{1}H NMR** (301 MHz, DMSO-d_6, 100 °C): δ [ppm] = 1.43 (s, 9H, *t*-Bu-CH$_3$), 1.49 (s, 3H, C(CH$_3$)$_2$), 1.55 (s, 3H, C(CH$_3$)$_2$), 4.00-4.13 (m, 2H, H-3), 4.37 (ddd, $J = 6.9, 3.7, 2.0$ Hz, 1H, H-2), 9.54 (d, $J = 2.0$ Hz, 1H, H-1). **^{13}C NMR** (76 MHz, DMSO-d_6, 100 °C): δ [ppm] = 23.66 (C($\underline{C}$H$_3$)$_2$), 25.29 (C($\underline{C}$H$_3$)$_2$), 27.43 (*t*-Bu-CH$_3$), 62.50 (C-3), 63.97 (C-2), 79.56 (*t*-Bu-C), 93.48 ($\underline{C}$(CH$_3$)$_2$), 150.78 (Boc-C=O), 198.52 (C-1). **MS** (ESI): $m/z = 252.1$ [M+Na]$^+$. **HRMS** (ESI): calcd for $C_{11}H_{19}NO_4Na$ 252.1206, found 252.1210 [M+Na]$^+$. $[\alpha]_D^{20} = -80.2$ ($c = 0.59$, CHCl$_3$). **IR** (ATR): $\tilde{v}$ [cm^{-1}] = 2979, 1738,

1690, 1478, 1457, 1364, 1248, 1162, 1092, 1059, 849. **TLC** (cyclohexane/Et$_2$O 2:1): R$_f$ = 0.16.

99: ^{1}**H NMR** (301 MHz, DMSO-d$_6$, 100 °C): δ [ppm] = 1.43 (s, 9H, *t*-Bu-CH$_3$), 1.49 (s, 3H, C(CH$_3$)$_2$), 1.55 (s, 3H, C(CH$_3$)$_2$), 3.00-4.13 (m, 2H, H-3), 4.35 (dd, *J* = 7.0, 3.7 Hz, 1H, H-2). ^{13}**C NMR** (76 MHz, DMSO-d$_6$, 100 °C): δ [ppm] = 23.68 (C(C̲H$_3$)$_2$), 25.29 (C(C̲H$_3$)$_2$), 27.43 (*t*-Bu-CH$_3$), 62.49 (C-3), 63.82-63.92 (m, C-2), 79.56 (*t*-Bu-C), 93.48 (C̲(CH$_3$)$_2$), 150.75 (Boc-C=O), 197.89-198.62 (m, C-1). **MS** (ESI): *m/z* = 253.1 [M+Na]$^+$. **HRMS** (ESI): calcd for C$_{11}$H$_{18}$NO$_4$DNa 253.1269, found 253.1270 [M+Na]$^+$. [α]$_D^{20}$ = +74.4 (*c* = 0.48, CHCl$_3$). **IR** (ATR): ṽ [cm^{-1}] = 2979, 1690, 1375, 1364, 1248, 1165, 1103, 1084, 1057, 850, 768. **TLC** (petroleum ether/EtOAc 3:2): R$_f$ = 0.43.

7.2.2 Synthesis of 1'-Deuterium Labeled and Unlabeled TBDMS-Protected Homoallylic Alcohols (*R*)-88, (*S*)-88, (*R*)-38, and (*S*)-38

7.2.2.1 Synthesis of *N*-Boc-(4*R*,1'*RS*)-4-(1'-Hydroxy-3'-butenyl)-2,2-dimethyl-1,3-oxazolidine 39 [136] and *N*-Boc-(4*R*,1'*RS*)-4-([1'-^{2}H]-1'-Hydroxy-3'-butenyl)-2,2-dimethyl-1,3-oxazolidine 124

(1'*RS*): **39** (R = H), (1'*RS*): **124** (R = D)

39: To a solution of *Garner's* aldehyde (*R*)-98 (9.59 g, 41.9 mmol, 1.0 eq) in dry THF (200 mL), allylmagnesium chloride (2 M solution in THF, 47.0 mL, 94.0 mmol, 2.2 eq) was added dropwise at -80 °C under an argon atmosphere. After the mixture had been allowed to warm to room temperature, saturated aqueous NH$_4$Cl solution (60 mL) was added, and the aqueous layer was extracted with Et$_2$O (3 x 150 mL). The combined organics were washed with saturated aqueous NaHCO$_3$ solution (2 x 150 mL) and brine (1 x 150 mL) and dried over Na$_2$SO$_4$. After removing the solvent under reduced pressure, the resultant crude product was purified by flash chromatography (petroleum ether/EtOAc 5:1).

124: The deuterated homoallylic alcohol derivative **124** was prepared in the same way as compound **39** with **99** (13.7 g, 59.4 mmol, 1.0 eq), allylmagnesium chloride (2 M solution in THF, 65.3 mL, 130 mmol, 2.2 eq) and dry THF (160 mL).

Yield 39: 9.50 g (35.0 mmol, 84%) as a bright yellow oil as a mixture of two diastereomers in a 2:1 ratio.

Yield 124: 14.5 g (53.2 mmol, 90%) as a colorless oil as a mixture of two diastereomers in a 2:1 ratio.

39: ^{1}**H NMR** (300 MHz, DMSO-d$_6$, 100 °C): δ [ppm] = 1.42 (s, 2 x 3H, C(CH$_3$)$_2$), 1.44 (s, 2 x 9H, t-Bu-CH$_3$), 1.53 (s, 2 x 3H, C(CH$_3$)$_2$), 1.98-2.26 (m, 2 x 2H, H-2'), 3.68-4.04 (m, 2 x 4H, H-4, H-5, H-1'), 4.40 (s$_{br}$, 2 x 1H, OH), 4.96-5.11 (m, 2 x 2H, H-4'), 5.78-5.97 (m, 2 x 1H, H-3'). ^{13}**C NMR** (75 MHz, DMSO-d$_6$, 100 °C): δ [ppm] = 22.92 (C($\underline{C}$H$_3$)$_2$), 23.65 (C($\underline{C}$H$_3$)$_2$), 25.80 (C($\underline{C}$H$_3$)$_2$), 26.08 (C($\underline{C}$H$_3$)$_2$), 27.58 (t-Bu-CH$_3$), 27.62 (t-Bu-CH$_3$), 35.15 (C-2'), 38.19 (C-2'), 60.26 (C-4), 60.63 (C-4), 62.77 (C-5), 63.00 (C-5), 69.06 (C-1'), 69.59 (C-1'), 78.62 (t-Bu-C), 78.75 (t-Bu-C), 92.78 (C-2), 93.06 (C-2), 115.17 (C-4'), 115.37 (C-4'), 135.59 (C-3'), 136.05 (C-3'), 151.44 (Boc-C=O), 151.67 (Boc-C=O). **MS** (ESI): m/z = 294.2 [M+Na]$^{+}$. **HRMS** (ESI): calcd for C$_{14}$H$_{25}$NO$_4$Na 294.1676, found 294.1691 [M+Na]$^{+}$. **TLC** (petroleum ether/EtOAc 3:1): R$_f$ = 0.29.

124: ^{1}**H NMR** (301 MHz, DMSO-d$_6$, 100 °C): δ [ppm] = 1.42 (s, 2 x 3H, C(CH$_3$)$_2$), 1.44 (s, 1 x 9H, 1 x 3H, t-Bu-CH$_3$, C(CH$_3$)$_2$), 1.49 (s, 1 x 9H, t-Bu-CH$_3$), 1.53 (s, 1 x 3H, C(CH$_3$)$_2$), 2.04 (dddd, J = 14.6, 6.4, 1.8, 1.8 Hz, 1 x 1H, H-2'a), 2.10-2.16 (m, 1 x 2H, H-2'), 2.21 (dddd, J = 14.6, 7.0, 1.8, 1.8 Hz, 1 x 1H, H-2'b), 3.76 (dd, J = 6.4, 1.8 Hz, 1 x 1H, H-4a), 3.80-4.05 (m, 1 x 3H, 1 x 2H, H-4, H-4b, H-5), 4.36 (s$_{br}$, 1 x 1H, OH), 4.39 (s$_{br}$, 1 x 1H, OH), 4.96-5.09 (m, 2 x 2H, H-4'), 5.79-5.98 (m, 2 x 1H, H-3'). ^{13}**C NMR** (76 MHz, DMSO-d$_6$, 100 °C): δ [ppm] = 22.93 (C($\underline{C}$H$_3$)$_2$), 23.66 (C($\underline{C}$H$_3$)$_2$), 25.80 (C($\underline{C}$H$_3$)$_2$), 26.06 (C($\underline{C}$H$_3$)$_2$), 27.59 (t-Bu-CH$_3$), 27.62 (t-Bu-CH$_3$), 35.03 (C-2'), 38.08 (C-2'), 60.17 (C-4), 60.54 (C-4), 62.76 (C-5), 62.97 (C-5), 68.35-69.40 (m, 2 x C-1'), 78.64 (t-Bu-C), 78.77 (t-Bu-C), 92.78 ($\underline{C}$(CH$_3$)$_2$), 93.06 ($\underline{C}$(CH$_3$)$_2$), 115.19 (C-4'), 115.39 (C-4'), 135.57 (C-3'), 136.03 (C-3'), 151.43 (Boc-C=O), 151.67 (Boc-C=O). **MS** (ESI): m/z = 295.2 [M+Na]$^{+}$. **HRMS** (ESI): calcd for C$_{14}$H$_{24}$DNO$_4$Na 295.1739, found 295.1741 [M+Na]$^{+}$. **TLC** (petroleum ether/EtOAc 3:1): R$_f$ = 0.30.

7.2.2.2 Synthesis of *N*-Boc-(4*R*,1'*R*)- and (4*R*,1'*S*)-4-(1'-(*tert*-Butyldimethyl-silyloxy)3'-butenyl)-2,2-dimethyl-1,3-oxazolidine (*R*)-38 and (*S*)-38 [136] and *N*-Boc-(4*R*,1'*R*)- and (4*R*,1'*S*)-4-([1'-²H]-1'-(*tert*-Butyldimethyl-silyloxy)-3'-butenyl)-2,2-dimethyl-1,3-oxazolidine (*R*)-88 and (*S*)-88

(1'*R*): (*R*)-38 (R = H), (1'*S*): (*S*)-38 (R = H), (1'*R*): (*R*)-88 (R = D), (1'*S*): (*S*)-88 (R = D)

38: To a solution of homoallylic alcohol **39** (9.49 g, 35.0 mmol, 1.0 eq) in dry DMF (20 mL), imidazole (7.15 g, 105 mmol, 3.0 eq) followed by *tert*-butyldimethylsilyl chloride (11.0 g, 73.0 mmol, 2.0 eq) were added under an argon atmosphere. The mixture was stirred at room temperature for 24 h. Water (100 mL) was added, and the aqueous layer was extracted with Et₂O (3 x 200 mL). The combined organics were washed with saturated aqueous NaHCO₃ solution (3 x 200 mL) and brine (1 x 200 mL) and dried over Na₂SO₄. After removing the solvent under reduced pressure, the diastereomers were separated by flash chromatography (petroleum ether/Et₂O 40:1).

88: The deuterated diastereomers were prepared in the same way as compounds (*R*)-38 and (*S*)-38 with **124** (14.3 g, 52.5 mmol, 1.0 eq), imidazole (16.1 g, 236 mmol, 4.5 eq), *tert*-butyldimethylsilyl chloride (23.8 g, 158 mmol, 3.0 eq) and dry DMF (50 mL). The reaction mixture was stirred for 48 h at room temperature.

Yield (*R*)-38: 7.81 g (20.3 mmol, 58%) as a colorless oil.
Yield (*S*)-38: 2.98 g (7.73 mmol, 22%) as a colorless oil.
Yield (*R*)-88: 9.97 g (25.8 mmol, 49%) as a colorless oil.
Yield (*S*)-88: 6.99 g (18.1 mmol, 34%) as a colorless oil.

(*R*)-38: **¹H NMR** (300 MHz, DMSO-d₆, 100 °C): δ [ppm] = 0.06 (s, 3H, Si(CH₃)₂), 0.08 (s, 3H, Si(CH₃)₂), 0.88 (s, 9H, SiC(CH₃)₃), 1.42 (s, 3H, C(CH₃)₂), 1.46 (s, 9H, *t*-Bu-CH₃), 1.53 (s, 3H, C(CH₃)₂), 2.02-2.14 (m, 1H, H-2'a), 2.20-2.31 (m, 1H, H-2'b), 3.84-3.96 (m, 2H, H-4, H-5a), 3.98-4.08 (m, 1H, H-5b), 4.12-4.21 (m, 1H, H-1'), 4.97-5.08 (m, 2H, H-4'), 5.73-5.89 (m, 1H, H-3'). **¹³C NMR** (75 MHz, DMSO-d₆, 100 °C): δ [ppm] = -5.09 (Si(CH₃)₂), -5.00 (Si(CH₃)₂), 17.09 (Si*C*(CH₃)₃), 22.64 (C(*C*H₃)₂), 25.15 (SiC(*C*H₃)₃), 25.57 (C(*C*H₃)₂), 27.64 (*t*-Bu-CH₃), 34.89 (C-2'), 59.61 (C-4), 62.13 (C-5), 70.30 (C-1'), 78.96 (*t*-Bu-C), 93.53 (C-2), 115.83 (C-4'), 135.62 (C-3'), 151.32 (Boc-C=O). **MS** (ESI): *m/z* = 408.3 [M+Na]⁺. **HRMS** (ESI): calcd for C₂₀H₃₉NO₄SiNa 408.2541, found 408.2540 [M+Na]⁺. [α]$_D^{20}$ = +22.2 (*c* = 1.0, CHCl₃). **IR** (film): ṽ [cm⁻¹] = 3412, 2934,

2359, 1706, 1390, 1257, 1176, 1092, 837, 776. **TLC** (petroleum ether/EtOAc 9:1): $R_f = 0.36$.

(**S**)-**38**: ^{1}**H NMR** (300 MHz, DMSO-d_6, 100 °C): δ [ppm] = 0.05 (s, 3H, Si(CH$_3$)$_2$), 0.08 (s, 3H, Si(CH$_3$)$_2$), 0.90 (s, 9H, SiC(CH$_3$)$_3$), 1.43 (s, 3H, C(CH$_3$)$_2$), 1.45 (s, 9H, t-Bu-CH$_3$), 1.49 (s, 3H, C(CH$_3$)$_2$), 2.10-2.30 (m, 2H, H-2'), 3.79-3.86 (m, 1H, H-4), 3.90 (dd, J = 8.3, 7.1 Hz, 1H, H-5a), 4.00 (dd, J = 8.3, 4.4 Hz, 1H, H-5b), 4.26 (ddd, J = 7.6, 5.0, 2.7 Hz, 1H, H-1'), 5.00-5.14 (m, 2H, H-4'), 5.72-5.89 (m, 1H, H-3'). ^{13}**C NMR** (75 MHz, DMSO-d_6, 100 °C): δ [ppm] = -4.96 (Si(CH$_3$)$_2$), -4.92 (Si(CH$_3$)$_2$), 17.16 (Si$\underline{C}$(CH$_3$)$_3$), 23.86 (C($\underline{C}$H$_3$)$_2$), 25.29 (SiC($\underline{C}$H$_3$)$_3$), 25.75 (C($\underline{C}$H$_3$)$_2$), 27.61 (t-Bu-CH$_3$), 39.16 (C-2'), 59.72 (C-4), 61.81 (C-5), 69.65 (C-1'), 78.74 (t-Bu-C), 93.13 (C-2), 116.46 (C-4'), 133.95 (C-3'), 151.55 (Boc-C=O). **MS** (ESI): m/z = 408.3 [M+Na]$^+$. **HRMS** (ESI): calcd for C$_{20}$H$_{39}$NO$_4$SiNa 408.2541, found 408.2536 [M+Na]$^+$. [α]$_D^{20}$ = +39.2 (c = 1.0, CHCl$_3$). **IR** (film): $\tilde{\nu}$ [cm^{-1}] = 3379, 2934, 2357, 1695, 1366, 1254, 1171, 1096, 1072, 837, 775. **TLC** (petroleum ether/EtOAc 9:1): $R_f = 0.26$.

(**R**)-**88**: ^{1}**H NMR** (300 MHz, DMSO-d_6, 100 °C): δ [ppm] = 0.06 (s, 3H, Si(CH$_3$)$_2$), 0.07 (s, 3H, Si(CH$_3$)$_2$), 0.88 (s, 9H, SiC(CH$_3$)$_3$), 1.42 (s, 3H, C(CH$_3$)$_2$), 1.46 (s, 9H, t-Bu-CH$_3$), 1.53 (s, 3H, C(CH$_3$)$_2$), 2.03 (dd, J = 14.4, 7.2 Hz, 1H, H-2'a), 2.25 (ddt, J = 14.4, 6.8, 1.4 Hz, 1H, H-2'b), 3.85-3.96 (m, 2H, H-4, H-5a), 3.97-4.08 (m, 1H, H-5b), 4.99-5.09 (m, 2H, H-4'), 5.71-5.88 (m, 1H, H-3'). ^{13}**C NMR** (75 MHz, DMSO-d_6, 100 °C): δ [ppm] = -5.09 (Si(CH$_3$)$_2$), -5.01 (Si(CH$_3$)$_2$), 17.08 (Si$\underline{C}$(CH$_3$)$_3$), 22.62 (C($\underline{C}$H$_3$)$_2$), 25.14 (SiC($\underline{C}$H$_3$)$_3$), 25.54 (C($\underline{C}$H$_3$)$_2$), 27.64 (t-Bu-CH$_3$), 34.78 (C-2'), 59.54 (C-4), 62.13 (C-5), 69.83-69.94 (m, C-1'), 78.96 (t-Bu-C), 93.53 (C-2), 115.81 (C-4'), 135.60 (C-3'), 151.32 (Boc-C=O). **MS** (ESI): m/z = 409.3 [M+Na]$^+$. **HRMS** (ESI): calcd for C$_{20}$H$_{38}$DNO$_4$SiNa 409.2603 found 409.2607 [M+Na]$^+$. [α]$_D^{20}$ = +25.0 (c = 1.2, CHCl$_3$). **IR** (ATR): $\tilde{\nu}$ [cm^{-1}] = 2930, 1701, 1374, 1363, 1253, 1172, 1093, 1066, 1001, 835, 808, 774. **TLC** (petroleum ether/EtOAc 9:1): $R_f = 0.41$.

(**S**)-**88**: ^{1}**H NMR** (300 MHz, DMSO-d_6, 100 °C): δ [ppm] = 0.05 (s, 3H, Si(CH$_3$)$_2$), 0.08 (s, 3H, Si(CH$_3$)$_2$), 0.90 (s, 9H, SiC(CH$_3$)$_3$), 1.42 (s, 3H, C(CH$_3$)$_2$), 1.45 (s, 9H, t-Bu-CH$_3$), 1.49 (s, 3H, C(CH$_3$)$_2$), 2.08-2.30 (m, 2H, H-2'), 3.82 (dd, J = 7.0, 4.5 Hz, 1H, H-4), 3.90 (dd, J = 8.4, 7.0 Hz, 1H, H-5a), 3.99 (dd, J = 8.4, 4.5 Hz, 1H, H-5b), 4.99-5.14 (m, 2H, H-4'), 5.72-5.89 (m, 1H, H-3'). ^{13}**C NMR** (75 MHz, DMSO-d_6, 100 °C): δ [ppm] = -4.97 (Si(CH$_3$)$_2$), -4.89 (Si(CH$_3$)$_2$), 17.18 (Si$\underline{C}$(CH$_3$)$_3$), 23.89 (C($\underline{C}$H$_3$)$_2$), 25.31 (SiC($\underline{C}$H$_3$)$_3$), 25.76 (C($\underline{C}$H$_3$)$_2$), 27.63 (t-Bu-CH$_3$), 39.08 (C-2'), 59.66 (C-4), 61.80 (C-5), 69.48-69.64 (m, C-1'), 78.78 (t-Bu-C), 93.14 (C-2), 116.52 (C-4'), 133.96 (C-3'), 151.57 (Boc-C=O). **MS** (ESI): m/z = 409.3 [M+Na]$^+$. **HRMS** (ESI): calcd for C$_{20}$H$_{38}$DNO$_4$SiNa 409.2603 found 409.2606 [M+Na]$^+$. [α]$_D^{20}$ = +58.0 (c = 1.2, CHCl$_3$). **IR** (ATR): $\tilde{\nu}$ [cm^{-1}] = 2929, 1691, 1383, 1362, 1250, 1171, 1093, 1070, 1003, 833, 808, 771. **TLC** (petroleum ether/EtOAc 9:1): $R_f = 0.32$.

7.2.3 Synthesis of (3*R*)- and (3*S*)-3-Hydroxy-L-[5-²H]arginine (*R*)-87 and (*S*)-87 via Different Protecting Group Strategies

7.2.3.1 Synthesis of *N*-Boc-(4*R*,1'*R*)- and (4*R*,1'*S*)-4-(1'-(*tert*-Butyldiphenyl-silyloxy)-3'-butenyl)-2,2-dimethyl-1,3-oxazolidine (*R*)-147 and (*S*)-147

(1'*R*): (*R*)-147, (1'*S*): (*S*)-147

To a solution of homoallylic alcohol **39** (2.17 g, 8.00 mmol, 1.0 eq) in dry DMF (7 mL), imidazole (2.18 g, 32.0 mmol, 4.0 eq) and *tert*-butyldiphenylsilyl chloride (6.40 mL, 24.7 mmol, 3.0 eq) were added under an argon atmosphere. The mixture was stirred at room temperature for 3 d. Water (30 mL) was added, and the aqueous layer was extracted with Et$_2$O (3 x 70 mL). The combined organics were washed with saturated aqueous NaHCO$_3$ solution (3 x 70 mL) and brine (1 x 70 mL) and dried over Na$_2$SO$_4$. After removing the solvent under reduced pressure, the diastereomers were partially separated by flash chromatography (isohexane/Et$_2$O 30:1).

Yield (*R*)-147: 1.88 g (3.69 mmol, 46%) as a colorless oil.

Yield (*S*)-147: 163 mg (0.320 mmol, 4%) as a colorless oil.

Yield diastereomeric mixture: 1.17 g (2.30 mmol, 29%) as a colorless oil in a ratio (*R*):(*S*) = 5:4.

(*R*)-147: **¹H NMR** (500 MHz, DMSO-d$_6$, 100 °C): δ [ppm] = 1.07 (s, 9H, SiC(CH$_3$)$_3$), 1.31 (s, 9H, *t*-Bu-CH$_3$), 1.38 (s, 3H, C(CH$_3$)$_2$), 1.52 (s, 3H, C(CH$_3$)$_2$), 2.20 (ddd, *J* = 14.7, 7.1, 1.3 Hz, 1H, H-2'a), 2.29-2.36 (m, 1H, H-2'b), 3.90-3.96 (m, 2H, H-4, H-5a), 4.21 (dd, *J* = 8.6, 1.8 Hz, 1H, H-5b), 4.41-4.47 (m, 1H, H-1'), 4.82-4.90 (m, 2H, H-4'), 5.57-5.68 (m, 1H, H-3'), 7.39-7.49 (m, 6H, Ph), 7.61-7.68 (m, 4H, Ph). **¹³C NMR** (126 MHz, DMSO-d$_6$, 100 °C): δ [ppm] = 19.42 (Si<u>C</u>(CH$_3$)$_3$), 23.64 (C(<u>C</u>H$_3$)$_2$), 26.45 (C(<u>C</u>H$_3$)$_2$), 27.44 (SiC(<u>C</u>H$_3$)$_3$), 28.48 (*t*-Bu-CH$_3$), 36.23 (C-2'), 60.21 (C-4), 63.47 (C-5), 72.45 (C-1'), 79.93 (*t*-Bu-C), 94.71 (C-2), 116.76 (C-4'), 127.94 (CH-Ph), 128.02 (CH-Ph), 130.12 (CH-Ph), 130.19 (CH-Ph), 134.19 (C-Ph), 134.33 (C-Ph), 135.89 (CH-Ph), 135.95 (CH-Ph), 135.99 (C-3'), 152.27 (Boc-C=O). **HRMS** (ESI): calcd for C$_{30}$H$_{43}$NO$_4$SiNa 532.2854, found 532.2864 [M+Na]$^+$. [α]$_D^{20}$ = +1.1 (*c* = 1.3, CHCl$_3$). **IR** (ATR): $\tilde{v}$ [cm^{-1}] = 2932, 1699, 1391, 1374, 1364, 1262, 1174, 1106, 1070, 914, 853, 826, 741, 700, 612. **UV** (MeCN): λ$_{max}$ [nm] (log ε) = 219 (4.40), 266 (3.00). **TLC** (isohexane/EtOAc 12:1): R$_f$ = 0.26.

(**S**)-**147**: ^{1}**H NMR** (500 MHz, DMSO-d_6, 100 °C): δ [ppm] = 1.08 (s, 9H, SiC(CH$_3$)$_3$), 1.41 (s, 9H, t-Bu-CH$_3$), 1.45 (s, 3H, C(CH$_3$)$_2$), 1.53 (s, 3H, C(CH$_3$)$_2$), 2.00-2.06 (m, 2H, H-2'), 3.91 (ddd, J = 7.1, 3.8, 2.9 Hz, 1H, H-4), 3.97 (dd, J = 8.4, 7.1 Hz, 1H, H-5a), 4.17 (dd, J = 8.4, 3.8 Hz, 1H, H-5b), 4.43 (ddd, J = 7.8, 4.8, 2.9 Hz, 1H, H-1'), 4.73-4.76 (m, 1H, H-4'a), 4.76-4.80 (m, 1H, H-4'b), 5.46-5.56 (m, 1H, H-3'), 7.37-7.49 (m, 6H, Ph), 7.64-7.70 (m, 4H, Ph). ^{13}**C NMR** (126 MHz, DMSO-d_6, 100 °C): δ [ppm] = 19.44 (Si$\underline{C}$(CH$_3$)$_3$), 26.47 (C($\underline{C}$H$_3$)$_2$), 26.83 (C($\underline{C}$H$_3$)$_2$), 27.46 (SiC($\underline{C}$H$_3$)$_3$), 28.60 (t-Bu-CH$_3$), 39.75 (C-2'), 60.64 (C-4), 63.20 (C-5), 72.70 (C-1'), 79.87 (t-Bu-C), 94.44 (C-2), 117.46 (C-4'), 127.97 (CH-Ph), 128.00 (CH-Ph), 130.09 (CH-Ph), 130.12 (CH-Ph), 134.24 (C-3'), 134.91 (C-Ph), 135.00 (C-Ph), 135.82 (CH-Ph), 135.85 (CH-Ph), 152.97 (Boc-C=O). **HRMS** (ESI): calcd for C$_{30}$H$_{43}$NO$_4$SiNa 532.2854, found 532.2865 [M+Na]$^+$. [α]$_D^{20}$ = +38.0 (c = 1.1, CHCl$_3$). **IR** (ATR): $\tilde{\nu}$ [cm^{-1}] = 2935, 1697, 1389, 1365, 1255, 1174, 1092, 1075, 1011, 854, 824, 743, 703, 616. **UV** (MeCN): λ$_{max}$ [nm] (log ε) = 219 (4.63), 266 (3.22). **TLC** (isohexane/EtOAc 12:1): R$_f$ = 0.22.

7.2.3.2 Synthesis of *N*-Boc-(4*R*,1'*R*)- and (4*R*,1'*S*)-4-(1'-Hydroxy-3'-butenyl)-2,2-dimethyl-1,3-oxazolidine (*R*)-39 and (*S*)-39[136]

(1'*R*): (**R**)-**39**, (1'*S*): (**S**)-**39**

Variant 1

(**R**)-**39**: To a solution of TBDMS-protected alcohol (**R**)-**38** (7.79 g, 20.2 mmol, 1.0 eq) in dry THF (50 mL), a solution of tetra-*n*-butylammonium fluoride trihydrate (16.0 g, 50.7 mmol, 2.5 eq) in dry THF (60 mL) was added at room temperature under an argon atmosphere. After stirring the resultant mixture at room temperature for 4 h, it was partitioned between water (200 mL) and Et$_2$O (350 mL). The organic layer was washed with saturated aqueous NaHCO$_3$ solution (3 x 150 mL) and brine (1 x 150 mL), dried over Na$_2$SO$_4$, and the solvent was removed under reduced pressure. The resultant crude product was purified by flash chromatography (petroleum ether/EtOAc 3:1).

(**S**)-**39**: The isomer (**S**)-**39** was prepared in the same way as compound (**R**)-**39** with (**S**)-**38** (3.56 g, 9.23 mmol, 1.0 eq), tetra-*n*-butylammonium fluoride trihydrate (7.30 g, 23.1 mmol, 2.5 eq) and dry THF (25 mL and 30 mL, respectively). The reaction mixture was stirred for 2 h.

Yield (*R*)-39: 5.36 g (19.8 mmol, 98%) as a colorless oil.

Yield (*S*)-39: 2.34 g (8.62 mmol, 93%) as a white solid.

Variant 2

(*R*)-39: To a solution of TBDPS-protected alcohol **(*R*)-147** (56 mg, 0.11 mmol, 1.0 eq) in dry THF (3 mL), solid tetra-*n*-butylammonium fluoride trihydrate (90 mg, 0.29 mmol, 2.6 eq) was added at room temperature under an argon atmosphere. After stirring the mixture at room temperature for 3 h, it was partitioned between water (10 mL) and Et_2O (15 mL). The organic layer was washed with saturated aqueous $NaHCO_3$ solution (2 x 10 mL) and brine (1 x 10 mL), dried over Na_2SO_4, and the solvent was removed under reduced pressure. The resultant crude product was purified by flash chromatography (petroleum ether/EtOAc 3:1).

(*S*)-39: The isomer **(*S*)-39** was prepared in the same way as compound **(*R*)-39** with **(*S*)-147** (79 mg, 0.16 mmol, 1.0 eq), tetra-*n*-butylammonium fluoride trihydrate (123 mg, 0.390 mmol, 2.5 eq) and THF p.a. (3 mL).

Yield (*R*)-39: 30 mg (0.11 mmol, 100%) as a colorless oil.

Yield (*S*)-39: 35 mg (0.13 mmol, 81%) as a white solid.

(*R*)-39: **^{1}H NMR** (300 MHz, DMSO-d_6, 100 °C): δ [ppm] = 1.42 (s, 3H, C(CH$_3$)$_2$), 1.44 (s, 9H, *t*-Bu-CH$_3$), 1.53 (s, 3H, C(CH$_3$)$_2$), 1.97-2.14 (m, 1H, H-2'a), 2.15-2.30 (m, 1H, H-2'b), 3.83-4.05 (m, 4H, H-4, H-5, H-1'), 4.41 (d, *J* = 3.5 Hz, 1H, OH), 4.97-5.12 (m, 2H, H-4'), 5.81-5.98 (m, 1H, H-3'). **^{13}C NMR** (75 MHz, DMSO-d_6, 100 °C): δ [ppm] = 22.93 (C(<u>C</u>H$_3$)$_2$), 25.80 (C(<u>C</u>H$_3$)$_2$), 27.58 (*t*-Bu-CH$_3$), 35.13 (C-2'), 60.24 (C-4), 62.76 (C-5), 69.05 (C-1'), 78.76 (*t*-Bu-C), 93.05 (C-2), 115.19 (C-4'), 136.03 (C-3'), 151.66 (Boc-C=O). **MS** (ESI): *m/z* = 294.2 [M+Na]$^+$. **HRMS** (ESI): calcd for C$_{14}$H$_{25}$NO$_4$Na 294.1676, found 294.1678 [M+Na]$^+$. [α]$_D^{20}$ = +25.7 (*c* = 1.1, CHCl$_3$). **IR** (film): $\tilde{v}$ [cm^{-1}] = 3403, 2983, 2359, 1699, 1390, 1367, 1253, 1168, 1059, 850, 765. **TLC** (petroleum ether/EtOAc 3:1): R$_f$ = 0.23.

(*S*)-39: **^{1}H NMR** (300 MHz, DMSO-d_6, 100 °C): δ [ppm] = 1.45 (s, 3H, C(CH$_3$)$_2$), 1.45 (s, 9H, *t*-Bu-CH$_3$), 1.49 (s, 3H, C(CH$_3$)$_2$), 2.03-2.24 (m, 2H, H-2'), 3.65-3.79 (m, 2H, H-4, H-1'), 3.84 (dd, *J* = 8.5, 6.3 Hz, 1H, H-5a), 4.00 (dd, *J* = 8.5, 2.2 Hz, 1H, H-5b), 4.41 (d, *J* = 5.8 Hz, 1H, OH), 4.96-5.09 (m, 2H, H-4'), 5.78-5.94 (m, 1H, H-3'). **^{13}C NMR** (75 MHz, DMSO-d_6, 100 °C): δ [ppm] = 23.64 (C(<u>C</u>H$_3$)$_2$), 26.09 (C(<u>C</u>H$_3$)$_2$), 27.63 (*t*-Bu-CH$_3$), 38.20 (C-2'), 60.61 (C-4), 63.00 (C-5), 69.56 (C-1'), 78.64 (*t*-Bu-C), 92.77 (C-2), 115.41 (C-4'), 135.60 (C-3'), 151.43 (Boc-C=O). **MS** (ESI): *m/z* = 294.2 [M+Na]$^+$. **HRMS** (ESI): calcd for C$_{14}$H$_{25}$NO$_4$Na 294.1676, found 294.1676 [M+Na]$^+$. [α]$_D^{20}$ = +15.7 (*c* = 0.95, CHCl$_3$). **Mp** = 37 °C. **IR** (KBr): $\tilde{v}$ [cm^{-1}] = 3442, 2981, 2375, 1699, 1394, 1256, 1174, 1098, 913, 843, 768. **TLC** (petroleum ether/EtOAc 3:1): R$_f$ = 0.23.

7.2.3.3 Synthesis of *N*-Boc-(4*R*,1'*R*)- and (4*R*,1'*S*)-4-(1'-Benzyloxy-3'-butenyl)-2,2-dimethyl-1,3-oxazolidine (*R*)-40 and (*S*)-40[135]

(1'*R*): (*R*)-40, (1'*S*): (*S*)-40

(*R*)-40: To a suspension of sodium hydride (60 % dispersion in mineral oil, 1.03 g, 25.8 mmol, 1.3 eq) in dry THF (40 mL), a solution of **(*R*)-39** (5.38 g, 19.8 mmol, 1.0 eq) in dry THF (60 mL) was added at 0°C under an argon atmosphere, and the mixture was stirred at 0°C for 30 min. After the addition of tetra-*n*-butylammonium iodide (1.46 g, 3.97 mmol, 0.2 eq) and benzyl bromide (7.1 mL, 10 g, 59 mmol, 3.0 eq), the reaction mixture was heated under reflux for 12 h, and then it was quenched by adding saturated aqueous NH_4Cl solution (50 mL). The aqueous layer was extracted with Et_2O (3 x 50 mL). The combined organics were washed with brine (2 x 50 mL), dried over Na_2SO_4, and the solvent was removed under reduced pressure. The resultant crude product was purified by flash chromatography (petroleum ether/EtOAc 20:1).

(*S*)-40: The isomer **(*S*)-40** was prepared in the same way as compound **(*R*)-40** with **(*S*)-39** (3.50 g, 12.9 mmol, 1.0 eq), sodium hydride (60 % dispersion in mineral oil, 671 mg, 16.8 mmol, 1.3 eq), tetra-*n*-butylammonium iodide (953 mg, 2.58 mmol, 0.2 eq), benzyl bromide (4.60 mL, 6.60 g, 39.0 mmol, 3.0 eq) and dry THF (30 mL and 50 mL, respectively).

Yield (*R*)-40: 6.36 g (17.6 mmol, 89%) as a yellowish oil.
Yield (*S*)-40: 4.61 g (12.8 mmol, 99%) as a yellowish oil.

(*R*)-40: **^{1}H NMR** (300 MHz, DMSO-d$_6$, 100 °C): δ [ppm] = 1.43 (s, 9H, *t*-Bu-CH$_3$), 1.43 (s, 3H, C(CH$_3$)$_2$), 1.54 (s, 3H, C(CH$_3$)$_2$), 2.15-2.28 (m, 1H, H-2'), 2.29-2.40 (m, 1H, H-2'), 3.86 (ddd, *J* = 8.6, 4.5, 4.0 Hz, 1H, H-1'), 3.93 (dd, *J* = 9.4, 6.7 Hz, 1H, H-5a), 4.00 (dd, *J* = 9.4, 2.3 Hz, 1H, H-5b), 4.15 (ddd, *J* = 6.7, 4.5, 2.3 Hz, 1H, H-4), 4.51-4.67 (m, 2H, CH$_2$-Ph), 4.99-5.14 (m, 2H, H-4'), 5.81-5.97 (m, 1H, H-3'), 7.21-7.40 (m, 5H, Ph). **^{13}C NMR** (75 MHz, DMSO-d$_6$, 100 °C): δ [ppm] = 22.74 (C(<u>C</u>H$_3$)$_2$), 25.76 (C(<u>C</u>H$_3$)$_2$), 27.54 (*t*-Bu-CH$_3$), 33.11 (C-2'), 57.19 (C-4), 62.76 (C-5), 71.06 (CH$_2$-Ph), 77.66 (C-1'), 78.87 (*t*-Bu-C), 93.18 (C-2), 115.61 (C-4'), 126.71 (CH-Ph), 126.80 (CH-Ph), 127.51 (CH-Ph), 135.42 (C-3'), 138.26 (C-Ph), 151.30 (Boc-C=O). **MS** (ESI): *m/z* = 384.2 [M+Na]$^+$. **HRMS** (ESI): calcd for $C_{21}H_{31}NO_4Na$ 384.2145, found 384.2145 [M+Na]$^+$. $[\alpha]_D^{20}$ = +17.7 (*c* = 1.1, CHCl$_3$). **IR** (film): $\tilde{\nu}$ [cm^{-1}] = 3395, 2980, 2345, 1703, 1455, 1387,

1256, 1171, 1090, 854, 737, 698. **UV** (MeCN): λ_{max} [nm] (log ε) = 204 (5.04), 258 (3.42). **TLC** (petroleum ether/EtOAc 5:1): R_f = 0.47.

(S)-40: **^{1}H NMR** (300 MHz, DMSO-d$_6$, 100 °C): δ [ppm] = 1.45 (s, 12H, t-Bu-CH$_3$, C(CH$_3$)$_2$), 1.47 (s, 3H, C(CH$_3$)$_2$), 2.15-2.40 (m, 2H, H-2'), 3.87-4.05 (m, 4H, H-4, H-5, H-1'), 4.53 (d, J = 11.7 Hz, 1H, CH$_2$-Ph), 4.60 (d, J = 11.7 Hz, 1H, CH$_2$-Ph), 5.00-5.16 (m, 2H, H-4'), 5.76-5.92 (m, 1H, H-3'), 7.21-7.40 (m, 5H, Ph). **^{13}C NMR** (75 MHz, DMSO-d$_6$, 100 °C): δ [ppm] = 23.87 (C(<u>C</u>H$_3$)$_2$), 25.56 (C(<u>C</u>H$_3$)$_2$), 27.61 (t-Bu-CH$_3$), 35.70 (C-2'), 59.13 (C-4), 62.50 (C-5), 71.43 (CH$_2$-Ph), 76.92 (C-1'), 78.81 (t-Bu-C), 93.03 (C-2), 116.24 (C-4'), 126.69 (CH-Ph), 126.89 (CH-Ph), 127.48 (CH-Ph), 134.48 (C-3'), 138.20 (C-Ph), 151.26 (Boc-C=O). **MS** (ESI): m/z = 384.2 [M+Na]$^+$. **HRMS** (ESI): calcd for C$_{21}$H$_{31}$NO$_4$Na 384.2145, found 384.2151 [M+Na]$^+$. $[\alpha]_D^{20}$ = +43.4 (c = 1.0, CHCl$_3$). **IR** (film): $\tilde{\nu}$ [cm^{-1}] = 3483, 2979, 2361, 1698, 1454, 1366, 1257, 1096, 858, 763, 698. **UV** (MeCN): λ_{max} [nm] (log ε) = 204 (4.99), 258 (3.29). **TLC** (petroleum ether/EtOAc 5:1): R_f = 0.47.

7.2.3.4 Synthesis of *N*-Boc-(2*R*,3*R*)- and (2*R*,3*S*)-2-Amino-3-(benzyloxy)-hex-5-en-1,3-diol (*R*)-41 and (*S*)-41[135] and *N*-Boc-(2*R*,3*R*)- and (2*R*,3*S*)-2-Amino-3-(*tert*-butyldiphenylsilyloxy)-hex-5-en-1,3-diol (*R*)-148 and (*S*)-148

(3*R*): (*R*)-41, (3*S*): (*S*)-41, (3*R*): (*R*)-148, (3*S*): (*S*)-148

(R)-41: A solution of **(R)-40** (6.09 g, 16.9 mmol, 1.0 eq) in acetic acid (125 mL) and H$_2$O (25 mL) was stirred at room temperature for 2 d. The solvent was removed under reduced pressure, and the resultant crude product was purified by flash chromatography (petroleum ether/EtOAc 2:1).

(S)-41: The isomer **(S)-41** was prepared in the same way as compound **(R)-41** with **(S)-40** (2.95 g, 8.17 mmol, 1.0 eq), acetic acid (60 mL) and H$_2$O (12 mL). The mixture was stirred for 3 d.

(R)-148: The *O*-TBDPS-protected derivative **(R)-148** was prepared in the same way as compound **(R)-41** with **(R)-147** (8.44 g, 16.6 mmol, 1.0 eq), acetic acid (125 mL) and H$_2$O (25 mL). The mixture was stirred for 4 d. The resultant crude product was purified by flash chromatography (isohexane/EtOAc 3:1).

(*S*)-148: The isomer **(*S*)-148** was prepared in the same way as compound **(*R*)-41** with **(*S*)-147** (1.01 g, 1.98 mmol, 1.0 eq), acetic acid (15 mL) and H_2O (3 mL). The mixture was stirred for 2 d.

Yield (*R*)-41: 5.33 g (16.6 mmol, 98%) as a colorless oil.

Yield (*S*)-41: 2.46 g (7.65 mmol, 94%) as a colorless oil.

Yield (*R*)-148: 6.77 g (14.4 mmol, 87%) as a colorless oil.

Yield (*S*)-148: 797 mg (1.70 mmol, 86%) as a colorless oil.

(*R*)-41: ^{1}**H NMR** (300 MHz, C_6D_6): δ [ppm] = 1.40 (s, 9H, *t*-Bu-CH_3), 2.19-2.39 (m, 2H, H-4), 3.47-3.66 (m, 3H, H-1, H-3), 3.86-3.99 (m, 1H, H-2), 4.17 (d, *J* = 11.4 Hz, 1H, CH_2-Ph), 4.35 (d, *J* = 11.4 Hz, 1H, CH_2-Ph), 4.95-5.16 (m, 3H, H-6, NH), 5.70-5.87 (m, 1H, H-5), 7.03-7.21 (m, 5H, Ph). ^{13}**C NMR** (75 MHz, C_6D_6): δ [ppm] = 28.37 (*t*-Bu-CH_3), 35.90 (C-4), 54.44 (C-2), 63.40 (C-1), 72.43 (CH_2-Ph), 77.80 (C-3), 79.12 (*t*-Bu-C), 117.84 (C-6), 127.86 (CH-Ph), 128.10 (CH-Ph), 128.56 (CH-Ph), 134.58 (C-5), 138.68 (C-Ph), 156.52 (Boc-C=O). **MS** (ESI): *m/z* = 344.2 $[M+Na]^{+}$. **HRMS** (ESI): calcd for $C_{18}H_{27}NO_4Na$ 344.1832, found 344.1839 $[M+Na]^{+}$. $[\alpha]_D^{20}$ = -4.5 (*c* = 1.1, $CHCl_3$). **IR** (film): $\tilde{\nu}$ [cm^{-1}] = 3417, 2978, 2359, 1697, 1499, 1367, 1057, 917, 748, 699. **UV** (MeCN): λ$_{max}$ [nm] (log ε) = 204 (4.93), 258 (3.48). **TLC** (petroleum ether/EtOAc 2:1): R$_f$ = 0.25.

(*S*)-41: ^{1}**H NMR** (300 MHz, C_6D_6): δ [ppm] = 1.42 (s, 9H, *t*-Bu-CH_3), 2.04-2.36 (m, 2H, H-4), 3.47-3.66 (m, 2H, H-1a, H-3), 3.73-3.88 (m, 2H, H-1b, H-2), 4.16 (d, *J* = 11.6 Hz, 1H, CH_2-Ph), 4.32 (d, *J* = 11.6 Hz, 1H, CH_2-Ph), 4.94-5.08 (m, 2H, H-6), 5.19 (d, *J* = 8.2 Hz, 1H, NH), 5.66-5.84 (m, 1H, H-5), 7.02-7.27 (m, 5H, Ph). ^{13}**C NMR** (75 MHz, C_6D_6): δ [ppm] = 28.41 (*t*-Bu-CH_3), 35.86 (C-4), 53.98 (C-2), 62.06 (C-1), 72.42 (CH_2-Ph), 79.08 (*t*-Bu-C), 80.35 (C-3), 117.55 (C-6), 127.88 (CH-Ph), 127.95 (CH-Ph), 128.62 (CH-Ph), 134.56 (C-5), 138.65 (C-Ph), 156.08 (Boc-C=O). **MS** (ESI): *m/z* = 344.2 $[M+Na]^{+}$. **HRMS** (ESI): calcd for $C_{18}H_{27}NO_4Na$ 344.1832, found 344.1832 $[M+Na]^{+}$. $[\alpha]_D^{20}$ = +35.7 (*c* = 1.1, $CHCl_3$). **IR** (film): $\tilde{\nu}$ [cm^{-1}] = 3439, 2972, 2439, 1693, 1497, 1367, 1170, 1058, 913, 738, 698. **UV** (MeCN): λ$_{max}$ [nm] (log ε) = 204 (4.94), 258 (3.24). **TLC** (petroleum ether/EtOAc 2:1): R$_f$ = 0.25.

(*R*)-148: ^{1}**H NMR** (500 MHz, C_6D_6): δ [ppm] = 1.07 (s, 9H, SiC(CH_3)$_3$), 1.44 (s, 9H, *t*-Bu-CH_3), 2.15-2.24 (m, 1H, H-4a), 2.39-2.48 (m, 1H, H-4b), 3.47-3.61 (m, 1H, H-1a), 3.61-3.71 (m, 1H, H-1b), 3.94-4.02 (m, 1H, H-2), 4.05-4.12 (m, 1H, H-3), 4.83-4.92 (m, 2H, H-6), 5.12 (d, *J* = 8.7 Hz, 1H, NH), 5.52-5.66 (m, 1H, H-5), 7.12-7.23 (m, 6H, Ph), 7.67-7.76 (m, 4H, Ph). ^{13}**C NMR** (126 MHz, C_6D_6): δ [ppm] = 19.28 (Si$\underline{C}$(CH_3)$_3$), 26.90 (SiC($\underline{C}H_3$)$_3$), 28.09 (*t*-Bu-CH_3), 38.81 (C-4), 54.41 (C-2), 63.26 (C-1), 72.34 (C-3), 78.87 (*t*-Bu-C), 117.90 (C-6), 127.86 (CH-Ph), 127.95 (CH-Ph), 129.64 (CH-Ph), 129.98 (CH-Ph), 133.40 (C-Ph), 133.42 (C-5), 133.46 (C-Ph), 135.87 (CH-Ph), 135.93 (CH-Ph),

156.09 (Boc-C=O). **HRMS** (ESI): calcd for $C_{27}H_{39}NO_4SiNa$ 492.2541, found 492.2545 $[M+Na]^+$. $[\alpha]_D^{20}$ = -22.1 (c = 0.96, CHCl$_3$). **IR** (ATR): $\tilde{\nu}$ [cm^{-1}] = 2935, 1697, 1493, 1476, 1429, 1365, 1249, 1168, 1109, 1087, 1051, 1011, 918, 824, 743, 703. **UV** (MeCN): λ_{max} [nm] (log ε) = 219 (4.23), 265 (2.83). **TLC** (isohexane/EtOAc 2:1): R_f = 0.54.

(*S*)-148: **^{1}H NMR** (500 MHz, C$_6$D$_6$): δ [ppm] = 1.12 (s, 9H, SiC(CH$_3$)$_3$), 1.41 (s, 9H, *t*-Bu-CH$_3$), 2.11-2.22 (m, 2H, H-4), 3.61 (dd, J = 11.5, 3.7 Hz, 1H, H-1a), 3.79-3.86 (m, 1H, H-2), 3.86-3.93 (m, 1H, H-1b), 4.12-4.19 (m, 1H, H-3), 4.74-4.84 (m, 2H, H-6), 5.22 (d, J = 7.8 Hz, 1H, NH), 5.43-5.55 (m, 1H, H-5), 7.17-7.23 (m, 6H, Ph), 7.68-7.78 (m, 4H, Ph). **^{13}C NMR** (126 MHz, C$_6$D$_6$): δ [ppm] = 19.14 (Si$\underline{C}$(CH$_3$)$_3$), 26.89 (SiC($\underline{C}$H$_3$)$_3$), 28.00 (*t*-Bu-CH$_3$), 38.75 (C-4), 54.20 (C-2), 61.68 (C-1), 74.99 (C-3), 78.50 (*t*-Bu-C), 117.30 (C-6), 127.81 (CH-Ph), 127.95 (CH-Ph), 129.80 (CH-Ph), 129.84 (CH-Ph), 133.12 (C-Ph), 133.46 (C-5), 133.49 (C-Ph), 135.96 (CH-Ph), 156.63 (Boc-C=O). **HRMS** (ESI): calcd for $C_{27}H_{39}NO_4SiNa$ 492.2541, found 492.2545 $[M+Na]^+$. $[\alpha]_D^{20}$ = +7.7 (c = 0.96, CHCl$_3$). **IR** (ATR): $\tilde{\nu}$ [cm^{-1}] = 2935, 1697, 1505, 1429, 1365, 1243, 1168, 1109, 1051, 1034, 999, 918, 824, 743, 697. **UV** (MeCN): λ_{max} [nm] (log ε) = 219 (4.67), 266 (3.31). **TLC** (isohexane/EtOAc 2:1): R_f = 0.44.

7.2.3.5 Synthesis of *N*-Boc-(2*S*,3*R*)- and (2*S*,3*S*)-2-Amino-3-(benzyloxy)-5-hexenoic Acid (*R*)-126 and (*S*)-126[135] and *N*-Boc-(2*S*,3*R*)- and (2*S*,3*S*)-2-Amino-3-(*tert*-butyldiphenylsilyloxy)-hexenoic Acid (*R*)-149 and (*S*)-149

126: R = (benzyl group)   149: R = (*tert*-butyldiphenylsilyl group)

(3*R*): (**R**)-126, (3*S*): (**S**)-126, (3*R*): (**R**)-149, (3*S*): (**S**)-149

(*R*)-126: A mixture of (**R**)-41 (5.17 g, 16.1 mmol, 1.0 eq), bis-(acetoxy)-iodobenzene (BAIB, 11.4 g, 35.4 mmol, 2.2 eq) and 2,2,6,6-tetramethylpiperidine-1-oxyl (TEMPO, 755 mg, 4.83 mmol, 0.3 eq) in MeCN (50 mL) and H$_2$O (50 mL) was stirred at room temperature for 3 h, and then Et$_2$O (200 mL) was added. The reaction mixture was extracted with saturated aqueous NaHCO$_3$ solution, and the aqueous layer was acidified with 2 M HCl (200 mL) and extracted with EtOAc (4 x 200 mL). The combined organics were dried over Na$_2$SO$_4$, and the solvent was removed under reduced pressure. The resultant crude product was used without further purification.

(S)-126: The isomer **(S)-126** was prepared in the same way as compound **(R)-126** with **(S)-41** (3.55 g, 11.1 mmol, 1.0 eq), BAIB (7.83 g, 24.3 mmol, 2.2 eq), TEMPO (518 mg, 3.31 mmol, 0.3 eq), MeCN (30 mL) and H_2O (30 mL).

(R)-149: A mixture of **(R)-148** (6.56 g, 14.0 mmol, 1.0 eq), BAIB (9.90 g, 30.7 mmol, 2.2 eq) and TEMPO (655 mg, 4.19 mmol, 0.3 eq) in MeCN (40 mL) and H_2O (40 mL) was stirred at room temperature for 3 h. The organic layer was washed with 1 M HCl (50 mL), and the aqueous layer was extracted with EtOAc (3 x 50 mL). The combined organics were dried over Na_2SO_4, and the solvent was removed under reduced pressure. The resultant crude product was purified by flash chromatography (CH_2Cl_2/MeOH 95:5).

(S)-149: The isomer **(S)-149** was prepared in the same way as compound **(R)-149** with **(S)-148** (715 mg, 1.52 mmol, 1.0 eq), BAIB (1.08 g, 3.35 mmol, 2.2 eq), TEMPO (72 mg, 0.46 mmol, 0.3 eq), MeCN (7 mL) and H_2O (7 mL). The resultant crude product was purified by flash chromatography (CH_2Cl_2/MeOH 98:2→95:5).

Yield (R)-126: 5.25 g (15.7 mmol, 97%) as a brown oil.

Yield (S)-126: 3.90 g (100% yield: 3.71 g) as an impure brown oil.

Yield (R)-149: 6.31 g (13.0 mmol, 93%) as white crystals.

Yield (S)-149: 650 mg (1.34 mmol, 88%) as white crystals.

(R)-126: ^{1}**H NMR** (300 MHz, C_6D_6): δ [ppm] = 1.39 (s, 9H, *t*-Bu-CH$_3$), 2.03-2.32 (m, 2H, H-4), 3.98 (ddd, *J* = 7.9, 6.0, 1.9 Hz, 1H, H-3), 4.17 (d, *J* = 11.5 Hz, 1H, CH$_2$-Ph), 4.27 (d, *J* = 11.5 Hz, 1H, CH$_2$-Ph), 4.71 (dd, *J* = 9.8, 1.9 Hz, 1H, H-2), 4.87-5.08 (m, 2H, H-6), 5.41 (d, *J* = 9.8 Hz, 1H, NH), 5.51-5.59 (m, 1H, H-5), 6.99-7.19 (m, 5H, Ph), 9.46 (s$_{br}$, 1H, COOH). ^{13}**C NMR** (75 MHz, C_6D_6): δ [ppm] = 28.27 (*t*-Bu-CH$_3$), 35.82 (C-4), 56.38 (C-2), 72.42 (CH$_2$-Ph), 78.98 (C-3), 79.87 (*t*-Bu-C), 118.60 (C-6), 127.67 (CH-Ph), 128.06 (CH-Ph), 128.55 (CH-Ph), 133.62 (C-5), 138.62 (C-Ph), 156.45 (Boc-C=O), 177.24 (C-1). **MS** (ESI): *m/z* = 334.2 [M-H]$^-$. **HRMS** (ESI): calcd for $C_{18}H_{25}NO_5Na$ 358.1625, found 358.1622 [M+Na]$^+$. $[\alpha]_D^{20}$ = +26.5 (*c* = 1.1, CHCl$_3$). **IR** (film): $\tilde{v}$ [cm^{-1}] = 2984, 2368, 1720, 1499, 1366, 1163, 1073, 1026, 920, 745, 698. **UV** (MeCN): λ$_{max}$ [nm] (log ε) = 204 (4.97), 258 (3.53). **TLC** (CH_2Cl_2/MeOH 95:5): R$_f$ = 0.11 (broad spot).

(S)-126: ^{1}**H NMR** (300 MHz, C_6D_6): δ [ppm] = 1.39 (s, 9H, *t*-Bu-CH$_3$), 2.23-2.64 (m, 2H, H-4), 3.78 (ddd, *J* = 6.5, 6.5, 3.5 Hz, 1H, H-3), 4.30 (d, *J* = 11.5 Hz, 1H, CH$_2$-Ph), 4.40 (d, *J* = 11.5 Hz, 1H, CH$_2$-Ph), 4.92 (dd, *J* = 8.9, 3.5 Hz, 1H, H-2), 4.95-5.17 (m, 2H, H-6), 5.39 (d, *J* = 8.9 Hz, 1H, NH), 5.70-5.86 (m, 1H, H-5), 7.00-7.40 (m, 5H, Ph), 10.35 (s$_{br}$, 1H, COOH). ^{13}**C NMR** (75 MHz, C_6D_6): δ [ppm] = 28.28 (*t*-Bu-CH$_3$), 35.39 (C-4), 55.94 (C-2), 71.96 (CH$_2$-Ph), 79.59 (C-3), 79.81 (*t*-Bu-C), 118.05 (C-6), 127.77 (CH-Ph), 128.09 (CH-Ph), 128.51 (CH-Ph), 134.31 (C-5), 138.48 (C-Ph), 155.68 (Boc-C=O), 177.61 (C-1). **MS** (ESI): *m/z* = 334.0 [M-H]$^-$. **HRMS** (ESI): calcd for $C_{18}H_{26}NO_5$ 334.1660, found 334.1657 [M+H]$^+$. $[\alpha]_D^{20}$ = +18.6 (c = 1.2, CHCl$_3$). **IR** (film): $\tilde{v}$ [cm^{-1}] = 3357, 2341, 1715, 1497, 1393, 1369, 1163, 1059, 916, 738, 698. **UV** (MeCN): λ_{max} [nm] (log ε) = 204 (4.95), 258 (3.40). **TLC** (CH$_2$Cl$_2$/MeOH 95:5): R$_f$ = 0.11 (broad spot).

(*R*)-149: **^{1}H NMR** (500 MHz, C$_6$D$_6$): δ [ppm] = 1.11 (s, 9H, SiC(CH$_3$)$_3$), 1.45 (s, 9H, *t*-Bu-CH$_3$), 2.13-2.23 (m, 1H, H-4a), 2.44-2.56 (m, 1H, H-4b), 4.49-4.57 (m, 1H, H-3), 4.73-4.90 (m, 3H, H-2, H-6), 5.41-5.53 (m, 1H, H-5), 5.66 (d, J = 9.9 Hz, 1H, NH), 7.12-7.37 (m, 6H, Ph), 7.66-7.79 (m, 4H, Ph). **^{13}C NMR** (126 MHz, C$_6$D$_6$): δ [ppm] = 19.17 (Si$\underline{C}$(CH$_3$)$_3$), 26.82 (SiC($\underline{C}$H$_3$)$_3$), 28.05 (*t*-Bu-CH$_3$), 38.80 (C-4), 56.59 (C-2), 74.53 (C-3), 79.60 (*t*-Bu-C), 118.75 (C-6), 127.48 (CH-Ph), 127.58 (CH-Ph), 129.73 (CH-Ph), 130.08 (CH-Ph), 132.60 (C-5), 133.83 (C-Ph), 133.88 (C-Ph), 135.90 (CH-Ph), 135.95 (CH-Ph), 156.12 (Boc-C=O), 177.18 (C-1). **MS** (ESI): m/z = 506.2 [M+Na]$^+$. **HRMS** (ESI): calcd for C$_{27}$H$_{37}$NO$_5$SiNa 506.2339, found 506.2334 [M+Na]$^+$. $[\alpha]_D^{20}$ = -11.5 (c = 1.5, CHCl$_3$). **IR** (ATR): $\tilde{v}$ [cm^{-1}] = 2929, 1714, 1661, 1499, 1429, 1400, 1365, 1168, 1109, 1087, 1064, 906, 824, 743, 703. **UV** (MeCN): λ_{max} [nm] (log ε) = 219 (4.40), 266 (3.09). **Mp** = 64 °C. **TLC** (CH$_2$Cl$_2$/MeOH 95:5): R$_f$ = 0.27.

(*S*)-149: **^{1}H NMR** (500 MHz, C$_6$D$_6$): δ [ppm] = 1.18 (s, 9H, SiC(CH$_3$)$_3$), 1.40 (s, 9H, *t*-Bu-CH$_3$), 2.23-2.34 (m, 1H, H-4a), 2.60-2.73 (m, 1H, H-4b), 4.21-4.34 (m, 1H, H-3), 4.72-4.79 (m, 1H, H-2), 4.85-5.03 (m, 2H, H-6), 5.50 (d, J = 7.9 Hz, 1H, NH), 5.56-5.69 (m, 1H, H-5), 7.17-7.33 (m, 6H, Ph), 7.69-7.88 (m, 4H, Ph). **^{13}C NMR** (126 MHz, C$_6$D$_6$): δ [ppm] = 19.27 (Si$\underline{C}$(CH$_3$)$_3$), 26.87 (SiC($\underline{C}$H$_3$)$_3$), 28.03 (*t*-Bu-CH$_3$), 38.76 (C-4), 57.34 (C-2), 74.93 (C-3), 79.11 (*t*-Bu-C), 118.15 (C-6), 127.47 (CH-Ph), 127.57 (CH-Ph), 129.78 (CH-Ph), 129.88 (CH-Ph), 133.34 (C-5), 133.63 (C-Ph), 133.64 (C-Ph), 135.99 (CH-Ph), 136.01 (CH-Ph), 156.00 (Boc-C=O), 175.04 (C-1). **HRMS** (ESI): calcd for C$_{27}$H$_{37}$NO$_5$SiNa 506.2339, found 506.2358 [M+Na]$^+$. $[\alpha]_D^{20}$ = +31.6 (c = 0.98, CHCl$_3$). **IR** (ATR): $\tilde{v}$ [cm^{-1}] = 2929, 1714, 1667, 1505, 1429, 1395, 1365, 1162, 1104, 1068, 918, 824, 743, 703, 616. **UV** (MeCN): λ_{max} [nm] (log ε) = 219 (4.30), 265 (2.92). **Mp** = 59 °C. **TLC** (CH$_2$Cl$_2$/MeOH 95:5): R$_f$ = 0.24.

7.2.3.6 Synthesis of *N*-Boc-(2*S*,3*R*)- and (2*S*,3*S*)-2-Amino-3-(benzyloxy)-5-hexenoic Acid *tert*-Butyl Ester (*R*)-42 and (*S*)-42[135] and *N*-Boc-(2*S*,3*R*)-2-Amino-3-(*tert*-butyldiphenylsilyloxy)-hexenoic Acid *tert*-Butyl Ester (*R*)-150

(3*R*): (*R*)-42, (3*S*): (*S*)-42, (3*R*): (*R*)-150

(R)-42: A solution of **(R)-126** (5.11 g, 15.3 mmol, 1.0 eq) and *tert*-butanol (19.0 mL, 15.0 g, 20.0 mmol, 13 eq) in toluene (100 mL) was heated to reflux, and then *N,N*-dimethylformamide dineopentylacetal (12.0 mL, 9.90 g, 43.0 mmol, 2.8 eq) was added dropwise under reflux over 30 min. After stirring the reaction mixture for a further 5 h under reflux, it was cooled to room temperature, washed with saturated aqueous Na_2CO_3 solution (2 x 100 mL) and H_2O (2 x 100 mL), dried over Na_2SO_4, and the solvent was removed under reduced pressure. The resultant crude product was purified by flash chromatography (petroleum ether/EtOAc 12:1).

(S)-42: The isomer **(S)-42** was prepared in the same way as compound **(R)-42** with **(S)-126** (3.67 g, 11.0 mmol, 1.0 eq), *tert*-butanol (14.0 mL, 11.1 g, 15.0 mol, 13 eq), *N,N*-dimethylformamide dineopentylacetal (8.50 mL, 7.00 g, 30.0 mmol, 2.8 eq) and toluene (80 mL).

(R)-150: The *O*-TBDPS-protected derivative **(R)-150** was prepared in the same way as compound **(R)-42** with **(R)-149** (6.18 g, 12.8 mmol, 1.0 eq), *tert*-butanol (16.0 mL, 12.0 g, 16.0 mol, 13 eq), *N,N*-dimethylformamide dineopentylacetal (11.0 mL, 9.10 g, 39.0 mmol, 2.8 eq) and toluene (100 mL). The resultant crude product was purified by flash chromatography (isohexane/EtOAc 15:1).

Yield (R)-42: 3.82 g (9.76 mmol, 64%) as a yellowish oil.
Yield (S)-42: 2.49 g (6.36 mmol, 58%) as a yellowish oil.
Yield (R)-150: 3.99 g (7.39 mmol, 58%) as a colorless oil.

(R)-42: **^{1}H NMR** (300 MHz, C_6D_6): δ [ppm] = 1.29 (s, 9H, *t*-Bu-CH$_3$), 1.41 (s, 9H, *t*-Bu-CH$_3$), 2.21-2.42 (m, 2H, H-4), 4.01 (ddd, *J* = 7.8, 6.0, 2.0 Hz, 1H, H-3), 4.31 (d, *J* = 11.4 Hz, 1H, CH$_2$-Ph), 4.38 (d, *J* = 11.4 Hz, 1H, CH$_2$-Ph), 4.69 (dd, *J* = 9.9, 2.0 Hz, 1H, H-2), 4.96-5.13 (m, 2H, H-6), 5.42 (d, *J* = 9.9 Hz, 1H, NH), 5.67-5.83 (m, 1H, H-5), 7.02-7.26 (m, 5H, Ph). **^{13}C NMR** (75 MHz, C_6D_6): δ [ppm] = 27.90 (*t*-Bu-CH$_3$), 28.34 (*t*-Bu-CH$_3$), 35.91 (C-4), 57.00 (C-2), 72.37 (CH$_2$-Ph), 79.27 (*t*-Bu-C), 79.84 (C-3), 81.29 (*t*-Bu-C), 118.34 (C-6), 127.79 (CH-Ph), 128.17 (CH-Ph), 128.51 (CH-Ph), 134.05 (C-5), 138.52 (C-Ph), 156.24 (Boc-C=O), 170.66 (C-1). **MS** (ESI): *m/z* = 414.0 [M+Na]$^+$. **HRMS** (ESI): calcd for $C_{22}H_{33}NO_5Na$ 414.2251, found 414.2252 [M+Na]$^+$. $[\alpha]_D^{20}$ = +16.2 (*c* = 1.2, CHCl$_3$). **IR** (film): $\tilde{\nu}$ [cm^{-1}] = 3451, 2980, 2357, 1722, 1497, 1368, 1154, 1074, 919, 740, 698. **UV** (MeCN): λ_{max} [nm] (log ε) = 204 (4.98), 258 (3.61). **TLC** (petroleum ether/EtOAc 12:1): R$_f$ = 0.18.

(S)-42: **^{1}H NMR** (300 MHz, C_6D_6): δ [ppm] = 1.27 (s, 9H, *t*-Bu-CH$_3$), 1.39 (s, 9H, *t*-Bu-CH$_3$), 2.29-2.46 (m, 2H, H-4), 3.78 (ddd, *J* = 6.5, 6.5, 3.3 Hz, 1H, H-3), 4.34 (d, *J* = 11.6 Hz, 1H, CH$_2$-Ph), 4.51 (d, *J* = 11.6 Hz, 1H, CH$_2$-Ph), 4.82 (dd, *J* = 8.4, 3.3 Hz, 1H, H-2), 4.99-5.15 (m, 2H, H-6), 5.45 (d, *J* = 8.4 Hz, 1H, NH), 5.81-5.97 (m, 1H, H-5),

7.03-7.32 (m, 5H, Ph). ^{13}C NMR (75 MHz, C$_6$D$_6$): δ [ppm] = 27.88 (*t*-Bu-CH$_3$), 28.33 (*t*-Bu-CH$_3$), 36.01 (C-4), 56.49 (C-2), 72.00 (CH$_2$-Ph), 79.32 (*t*-Bu-C), 80.43 (C-3), 81.66 (*t*-Bu-C), 117.60 (C-6), 127.86 (CH-Ph), 128.10 (CH-Ph), 128.41 (CH-Ph), 134.87 (C-5), 138.85 (C-Ph), 155.44 (Boc-C=O), 169.65 (C-1). **MS** (ESI): *m/z* = 414.0 [M+Na]$^+$. **HRMS** (ESI): calcd for C$_{22}$H$_{33}$NO$_5$Na 414.2251, found 414.2255 [M+Na]$^+$. [α]$_D^{20}$ = +10.1 (*c* = 0.96, CHCl$_3$). **IR** (film): $\tilde{\nu}$ [cm^{-1}] = 3379, 2972, 2345, 1717, 1498, 1367, 1155, 1100, 913, 732, 698. **UV** (MeCN): λ$_{max}$ [nm] (log ε) = 204 (4.94), 258 (3.31). **TLC** (petroleum ether/EtOAc 12:1): R$_f$ = 0.18.

(***R***)-150: ^{1}**H NMR** (500 MHz, C$_6$D$_6$): δ [ppm] = 1.10 (s, 9H, SiC(CH$_3$)$_3$), 1.30 (s, 9H, *t*-Bu-CH$_3$), 1.44 (s, 9H, *t*-Bu-CH$_3$), 2.19-2.27 (m, 1H, H-4a), 2.47-2.58 (m, 1H, H-4b), 4.49 (ddd, *J* = 10.2, 4.1, 1.7 Hz, 1H, H-3), 4.72 (dd, *J* = 9.9, 1.7 Hz, 1H, H-2), 4.79-4.95 (m, 2H, H-6), 5.47-5.58 (m, 1H, H-5), 5.63 (d, *J* = 9.9 Hz, 1H, NH), 7.13-7.20 (m, 6H, Ph), 7.65-7.80 (m, 4H, Ph). ^{13}**C NMR** (126 MHz, C$_6$D$_6$): δ [ppm] = 19.18 (Si$\underline{C}$(CH$_3$)$_3$), 26.84 (SiC($\underline{C}$H$_3$)$_3$), 27.72 (*t*-Bu-CH$_3$), 28.07 (*t*-Bu-CH$_3$), 39.03 (C-4), 56.83 (C-2), 74.56 (C-3), 79.11 (*t*-Bu-C), 81.05 (*t*-Bu-C), 118.55 (C-6), 127.47 (CH-Ph), 127.55 (CH-Ph), 129.72 (CH-Ph), 129.82 (CH-Ph), 132.95 (C-5), 133.04 (C-Ph), 134.17 (C-Ph), 135.85 (CH-Ph), 135.95 (CH-Ph), 156.03 (Boc-C=O), 170.56 (C-1). **MS** (ESI): *m/z* = 562.3 [M+Na]$^+$. **HRMS** (ESI): calcd for C$_{31}$H$_{45}$NO$_5$SiNa 562.2959, found 562.2958 [M+Na]$^+$. [α]$_D^{20}$ = -20.2 (*c* = 0.84, CHCl$_3$). **IR** (ATR): $\tilde{\nu}$ [cm^{-1}] = 2975, 2929, 1720, 1493, 1429, 1365, 1325, 1151, 1109, 1087, 906, 824, 743, 703, 616. **UV** (MeCN): λ$_{max}$ [nm] (log ε) = 219 (4.68), 266 (3.30). **TLC** (isohexane/EtOAc 12:1): R$_f$ = 0.31.

7.2.3.7 Synthesis of *N*-Boc-(2*S*,3*R*)- and (2*S*,3*S*)-2-Amino-3-(benzyloxy)-[5-^{2}H]-5-hydroxy-valeric Acid *tert*-Butyl Ester (*R*)-95 and (*S*)-95 and *N*-Boc-(2*S*,3*R*)-2-Amino-3-(*tert*-butyldiphenylsilyloxy)-[5-^{2}H]-5-hydroxy-valeric Acid *tert*-Butyl Ester (*R*)-151

(3*R*): (***R***)-95, (3*S*): (***S***)-95, (3*R*): (***R***)-151

(***R***)-95: A solution of (***R***)-42 (1.87 g, 4.78 mmol, 1.0 eq) in MeOH (60 mL), CH$_2$Cl$_2$ (6.9 mL) and pyridine (1.50 mL, 1.50 g, 19.0 mmol, 4.0 eq) was cooled to -78°C, and ozone was bubbled through this solution at -78°C for 30 min. After the addition of dimethyl sulfide (4.00 mL, 3.40 g, 55.0 mmol, 12 eq), the reaction was allowed to warm to room temperature overnight. The solvent was removed under reduced pressure, and the resultant crude aldehyde was immediately dissolved in CD$_3$OD (50 mL) under an argon atmosphere. Sodium borodeuteride (2.40 g, 57.3 mmol, 12 eq) was added portionwise to this solution at 0°C, and the reaction mixture was stirred at room temperature for 22 h while being monitored by TLC (petroleum ether/EtOAc 2:1). The reaction was quenched by the addition of saturated aqueous NH$_4$Cl solution (10 mL), and the aqueous layer was extracted with EtOAc (4 x 50 mL). The combined organics were washed with H$_2$O (2 x 50 mL), dried over Na$_2$SO$_4$, and the solvent was removed under reduced pressure. The resultant crude product was purified by flash chromatography (petroleum ether/EtOAc 3:1→2:1).

(S)-95: The isomer **(S)-95** was prepared in the same way as compound **(R)-95** with **(S)-42** (1.64 g, 4.19 mmol, 1.0 eq), MeOH (60 mL), CH$_2$Cl$_2$ (6.9 mL), pyridine (1.30 mL, 1.30 g, 16.0 mmol, 3.8 eq), dimethyl sulfide (4.00 mL, 3.40 g, 55.0 mmol, 13 eq), sodium borodeuteride (3.22 g, 76.9 mmol, 18 eq) and CD$_3$OD (31 mL). The mixture was stirred for 64 h at room temperature until TLC indicated a nearly complete conversion.

(R)-151: The *O*-TBDPS-protected derivative **(R)-151** was prepared in the same way as compound **(R)-95** with **(R)-150** (1.78 g, 3.30 mmol, 1.0 eq), MeOH (60 mL), CH$_2$Cl$_2$ (6.9 mL), pyridine (1.10 mL, 1.10 g, 14.0 mmol, 4.0 eq), dimethyl sulfide (4.00 mL, 3.40 g, 55.0 mmol, 17 eq), sodium borodeuteride (2.56 g, 61.2 mmol, 19 eq) and CD$_3$OD (40 mL). The mixture was stirred for 49 h at room temperature until TLC indicated a nearly complete conversion.

Yield (R)-95: 1.68 g (4.24 mmol, 89%) as a colorless oil.
Yield (S)-95: 1.40 g (3.53 mmol, 84%) as a colorless oil.
Yield (R)-151: 1.27 g (2.33 mmol, 71%) as a colorless oil.

(R)-95: **^{1}H NMR** (301 MHz, C$_6$D$_6$): δ [ppm] = 1.30 (s, 9H, *t*-Bu-CH$_3$), 1.38 (s, 9H, *t*-Bu-CH$_3$), 1.62-1.89 (m, 2H, H-4), 2.12 (s$_{br}$, 1H, OH), 3.48-3.59 (m, 1H, H-5), 4.22-4.31 (m, 1H, H-3), 4.44 (s, 2H, CH$_2$-Ph), 4.70 (dd, *J* = 9.5, 2.1 Hz, 1H, H-2), 5.55 (d, *J* = 9.5 Hz, 1H, NH), 7.02-7.30 (m, 5H, Ph). **^{13}C NMR** (126 MHz, C$_6$D$_6$): δ [ppm] = 27.91 (*t*-Bu-CH$_3$), 28.33 (*t*-Bu-CH$_3$), 34.05 (C-4), 57.63 (C-2), 58.65-59.24 (m, C-5), 72.58 (CH$_2$-Ph), 78.17 (C-3), 79.53 (*t*-Bu-C), 81.55 (*t*-Bu-C), 127.78 (CH-Ph), 127.91 (CH-Ph), 128.53 (CH-Ph), 138.67 (C-Ph), 156.56 (Boc-C=O), 170.60 (C-1). **MS** (ESI): *m/z* = 419.2 [M+Na]$^+$. **HRMS** (ESI): calcd for C$_{21}$H$_{32}$DNO$_6$Na 419.2263, found 419.2260 [M+Na]$^+$. [α]$_D^{20}$ = +34.2 (*c* = 1.1, CHCl$_3$). **IR** (ATR): $\tilde{v}$ [cm^{-1}] = 2975, 1714, 1499, 1453, 1372, 1342, 1255, 1151, 1064, 1028, 848, 737, 703. **UV** (MeCN): λ$_{max}$ [nm] (log ε) = 205 (3.92), 257 (2.38). **TLC** (petroleum ether/EtOAc 1:1): R$_f$ = 0.36.

(S)-95: **^{1}H NMR** (301 MHz, C$_6$D$_6$): δ [ppm] = 1.27 (s, 9H, *t*-Bu-CH$_3$), 1.38 (s, 9H, *t*-Bu-CH$_3$), 1.61-1.84 (m, 2H, H-4), 1.98 (s$_{br}$, 1H, OH), 3.53-3.63 (m, 1H, H-5), 3.96-4.05 (m, 1H, H-3), 4.32 (d, *J* = 11.4 Hz, 1H, CH$_2$-Ph), 4.61 (d, *J* = 11.4 Hz, 1H, CH$_2$-Ph), 4.84

(dd, J = 7.9, 2.8 Hz, 1H, H-2), 5.58 (d, J = 7.9 Hz, 1H, NH), 7.02-7.31 (m, 5H, Ph). **^{13}C NMR** (76 MHz, C_6D_6): δ [ppm] = 27.86 (*t*-Bu-CH$_3$), 28.31 (*t*-Bu-CH$_3$), 33.89 (C-4), 56.49 (C-2), 59.68-59.88 (m, C-5), 72.08 (CH$_2$-Ph), 78.61 (C-3), 79.59 (*t*-Bu-C), 81.87 (*t*-Bu-C), 127.76 (CH-Ph), 127.99 (CH-Ph), 128.46 (CH-Ph), 138.75 (C-Ph), 155.84 (Boc-C=O), 169.57 (C-1). **MS** (ESI): m/z = 419.2 [M+Na]$^+$. **HRMS** (ESI): calcd for $C_{21}H_{32}DNO_6Na$ 419.2263, found 419.2261 [M+Na]$^+$. $[\alpha]_D^{20}$ = +1.9 (c = 0.89, CHCl$_3$). **IR** (ATR): $\tilde{v}$ [cm^{-1}] = 3435, 2978, 1708, 1498, 1366, 1249, 1150, 1096, 1056, 1027, 737, 697. **UV** (MeCN): λ_{max} [nm] (log ε) = 205 (3.97), 258 (2.45). **TLC** (petroleum ether/EtOAc 1:1): R_f = 0.43.

(*R*)-151: **^{1}H NMR** (500 MHz, C_6D_6): δ [ppm] = 1.12 (s, 9H, SiC(CH$_3$)$_3$), 1.31 (s, 9H, *t*-Bu-CH$_3$), 1.40 (s, 9H, *t*-Bu-CH$_3$), 1.66-1.74 (m, 1H, H-4a), 1.87-1.96 (m, 1H, H-4b), 3.21 (dd, J = 5.9, 5.9 Hz, 0.5H, H-5), 3.35 (dd, J = 5.4, 5.4 Hz, 0.5H, H-5), 4.73 (ddd, J = 9.5, 4.4, 2.0 Hz, 1H, H-3), 4.86 (dd, J = 9.5, 2.0 Hz, 1H, H-2), 5.75 (d, J = 9.5 Hz, 1H, NH), 7.17-7.22 (m, 6H, Ph), 7.71-7.80 (m, 4H, Ph). **^{13}C NMR** (126 MHz, C_6D_6): δ [ppm] = 19.26 (Si<u>C</u>(CH$_3$)$_3$), 26.86 (SiC(<u>C</u>H$_3$)$_3$), 27.71 (*t*-Bu-CH$_3$), 28.02 (*t*-Bu-CH$_3$), 37.02 (C-4), 58.13 (C-2), 58.21-58.89 (m, C-5), 73.26 (C-3), 79.39 (*t*-Bu-C), 81.31 (*t*-Bu-C), 127.57 (CH-Ph), 127.76 (CH-Ph), 129.65 (CH-Ph), 129.77 (CH-Ph), 133.31 (C-Ph), 134.54 (C-Ph), 135.86 (CH-Ph), 135.92 (CH-Ph), 156.37 (Boc-C=O), 170.28 (C-1). **HRMS** (ESI): calcd for $C_{30}H_{44}DNO_6SiNa$ 567.2971, found 567.2974 [M+Na]$^+$. $[\alpha]_D^{20}$ = +1.7 (c = 0.66, CHCl$_3$). **IR** (ATR): $\tilde{v}$ [cm^{-1}] = 2935, 1720, 1493, 1372, 1249, 1151, 1104, 1087, 1064, 941, 824, 790, 743, 703, 610. **UV** (MeCN): λ_{max} [nm] (log ε) = 219 (3.67), 266 (2.42). **TLC** (isohexane/EtOAc 1:1): R_f = 0.35.

7.2.3.8 Synthesis of *N*-Boc-(2*R*,3*R*)-2-Amino-3-(benzyloxy)-[1-^{2}H$_2$,5-^{2}H]pentan-1,3,5-triol (*R*)-139 as Side Product of the Reduction

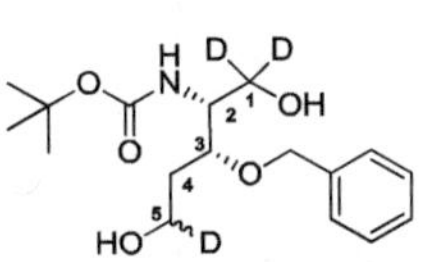

(*R*)-139

A solution of (*R*)-42 (4.59 g, 11.7 mmol, 1.0 eq) in MeOH (175 mL), CH$_2$Cl$_2$ (20 mL) and pyridine (3.80 mL, 3.73 g, 47.2 mmol, 4.0 eq) was cooled to -78°C, and ozone was bubbled through this solution at -78°C for 1 h. After the addition of dimethyl sulfide (12.0 mL, 10.1 g, 163 mmol, 14 eq), the reaction was allowed to warm to room temperature overnight. The solvent was removed under reduced pressure, and the resultant crude aldehyde was immediately dissolved in CD$_3$OD (24 mL) under an argon atmosphere. Sodium borodeuteride (1.00 g, 23.9 mmol, 2.0 eq) was added portionwise to this solution at 0°C, and the reaction mixture was stirred at room temperature for 7 d while being monitored by TLC (petroleum ether/EtOAc 2:1). Over these 7 d additional sodium borodeuteride (7.50 g, 179 mmol, 15 eq) and CD_3OD (44 mL) were added portionwise. The reaction was quenched by the addition of saturated aqueous NH_4Cl solution (60 mL), and the aqueous layer was extracted with EtOAc (4 x 150 mL). The combined organics were washed with H_2O (2 x 150 mL), dried over Na_2SO_4, and the solvent was removed under reduced pressure. The resultant crude product was purified by flash chromatography (petroleum ether/EtOAc 5:1→4:1→3:1→2:1→EtOAc).

Yield: 2.83 g (8.62 mmol, 74%) as a colorless oil.

^{1}H NMR (301 MHz, C_6D_6): δ [ppm] = 1.38 (s, 9H, t-Bu-CH_3), 1.64-1.92 (m, 2H, H-4), 3.12 (s_{br}, 1H, OH), 3.44 (s_{br}, 1H, OH), 3.61 (dd, J = 6.6, 5.1 Hz, 0.5H, H-5), 3.68 (dd, J = 5.5, 5.5 Hz, 0.5H, H-5), 3.97-4.08 (m, 2H, H-2, H-3), 4.39 (d, J = 11.4 Hz, 1H, CH_2-Ph), 4.50 (d, J = 11.4 Hz, 1H, CH_2-Ph), 5.13 (d, J = 7.6 Hz, 1H, NH), 7.02-7.29 (m, 5H, Ph). **^{13}C NMR** (76 MHz, C_6D_6): δ [ppm] = 28.37 (t-Bu-CH_3), 33.87 (C-4), 54.90 (C-2), 58.72-59.28 (m, C-5), 61.53-62.82 (m, C-1), 72.75 (CH_2-Ph), 75.75 (C-3), 79.42 (t-Bu-C), 127.83 (CH-Ph), 128.11 (CH-Ph), 128.58 (CH-Ph), 138.95 (C-Ph), 156.98 (Boc-C=O). **MS** (ESI): m/z = 351.2 [M+Na]$^+$. **HRMS** (ESI): calcd for $C_{17}H_{24}D_3NO_5Na$ 351.1970, found 351.1965 [M+Na]$^+$. **IR** (ATR): $\tilde{v}$ [cm^{-1}] = 1685, 1497, 1365, 1249, 1163, 1057, 1026, 978, 737, 697. **UV** (MeCN): λ_{max} [nm] (log ε) = 205 (3.91), 257 (2.77). **TLC** (petroleum ether/EtOAc 1:1): R_f = 0.05.

7.2.3.9 Synthesis of N^2-Boc-$N^5,N^7,N^{7'}$-Tris-Cbz-(3R)- and (3S)-3-(Benzyloxy)-L-[5-^{2}H]arginine *tert*-Butyl Ester (R)-94 and (S)-94 and N^2-Boc-$N^5,N^7,N^{7'}$-Tris-Cbz-(3R)-3-(*tert*-Butyldiphenylsilyloxy)-L-[5-^{2}H]arginine *tert*-Butyl Ester (R)-152

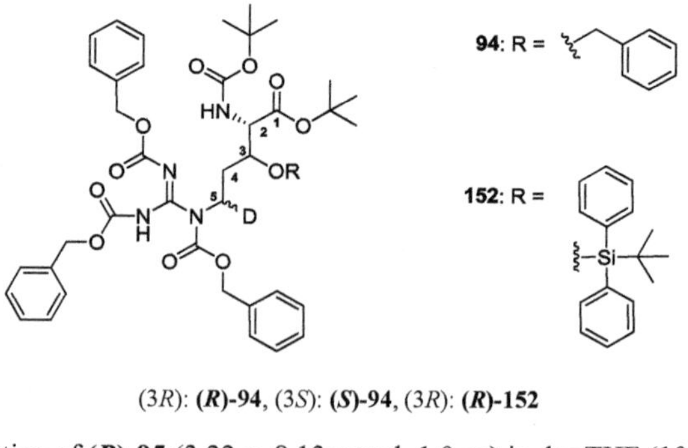

94: R =

152: R =

(3R): (R)-**94**, (3S): (S)-**94**, (3R): (R)-**152**

(R)-94: To a solution of **(R)-95** (3.22 g, 8.13 mmol, 1.0 eq) in dry THF (100 mL), tris-Cbz-guanidine **142** (11.2 g, 24.3 mmol, 3.0 eq) and triphenyl phosphine (3.20 g, 12.2 mmol,

1.5 eq) were added under an argon atmosphere at room temperature. DIAD (2.40 mL, 2.46 g, 12.2 mmol, 1.5 eq) was added dropwise at 0°C at such a rate that the reaction mixture was completely colorless before the addition of the next drop. The reaction mixture was allowed to warm to room temperature and was stirred overnight at room temperature. The reaction was quenched by the addition of H_2O (40 mL), and the solvent was removed under reduced pressure. The resultant crude product was purified by flash chromatography (petroleum ether/EtOAc 5:1→4:1→3:1).

(S)-94: The isomer **(S)-94** was prepared in the same way as compound **(R)-94** with **(S)-95** (1.24 g, 3.13 mmol, 1.0 eq), tris-Cbz-guanidine **142** (4.33 g, 9.38 mmol, 3.0 eq), triphenyl phosphine (1.23 g, 4.69 mmol, 1.5 eq), DIAD (930 μmL, 955 mg, 4.72 mmol, 1.5 eq) and dry THF (35 mL).

(R)-152: The *O*-TBDPS-protected derivative **(R)-152** was prepared in the same way as compound **(R)-94** with **(R)-151** (1.27 g, 2.33 mmol, 1.0 eq), tris-Cbz-guanidine **142** (3.23 g, 7.00 mmol, 3.0 eq), triphenyl phosphine (917 mg, 3.49 mmol, 1.5 eq), DIAD (690 μmL, 709 mg, 3.51 mmol, 1.5 eq) and dry THF (40 mL).

Yield (R)-94: 5.05 g (6.06 mmol, 74%) as a colorless viscous foam.
Yield (S)-94: 2.11 g (2.51 mmol, 80%) as a colorless viscous foam.
Yield (R)-152: 1.86 g (1.88 mmol, 81%) as a colorless oil.

(R)-94: ^{1}**H NMR** (300 MHz, C_6D_6): δ [ppm] = 1.29 (s, 9H, *t*-Bu-CH$_3$), 1.40 (s, 9H, *t*-Bu-CH$_3$), 1.95 (ddd, J = 13.8, 6.8, 6.8 Hz, 1H, H-4a), 2.14 (ddd, J = 13.8, 6.8, 6.8 Hz, 1H, H-4b), 3.92 (dd, J = 6.8, 6.8 Hz, 0.5H, H-5), 4.06 (dd, J = 6.8, 6.8 Hz, 0.5H, H-5), 4.21 (ddd, J = 6.8, 6.8, 1.9 Hz, 1H, H-3), 4.44 (d, J = 11.3 Hz, 1H, CH$_2$-Ph), 4.52 (d, 1H, J = 11.3 Hz, CH$_2$-Ph), 4.60-5.33 (m, 4H, Cbz-CH$_2$), 4.74 (dd, J = 9.6, 1.9 Hz, 1H, H-2), 4.91 (s, 2H, Cbz-CH$_2$), 5.43 (d, J = 9.6 Hz, 1H, NHBoc), 6.94-7.36 (m, 20H, Ph), 11.42 (s, 1H, NHCbz). ^{13}**C NMR** (126 MHz, C_6D_6): δ [ppm] = 27.91 (*t*-Bu-CH$_3$), 28.37 (*t*-Bu-CH$_3$), 30.47 (C-4), 44.20-45.22 (m, C-5), 57.41 (C-2), 68.07 (Cbz-CH$_2$), 68.94 (Cbz-CH$_2$), 72.15 (CH$_2$-Ph), 77.70 (C-3), 79.29 (*t*-Bu-C), 81.35 (*t*-Bu-C), 127.64 (CH-Ph), 127.81 (CH-Ph), 127.91 (CH-Ph), 128.29 (CH-Ph), 128.48 (CH-Ph), 128.57 (CH-Ph), 128.64 (CH-Ph), 128.67 (CH-Ph), 128.75 (CH-Ph), 135.43 (C-Ph), 138.74 (C-Ph), 152.30 (guanidine-C), 154.73 (Cbz-C=O), 156.33 (Boc-C=O), 170.52 (C-1). **MS** (ESI): m/z = 862.4 [M+Na]$^+$. **HRMS** (ESI): calcd for $C_{46}H_{53}DN_4O_{11}Na$ 862.3744, found 862.3759 [M+Na]$^+$. $[\alpha]_D^{20}$ = +8.6 (c = 0.78, CHCl$_3$). **IR** (ATR): $\tilde{v}$ [cm^{-1}] = 2981, 1767, 1720, 1656, 1610, 1499, 1453, 1372, 1208, 1191, 1151, 1051, 1028, 737, 697. **UV** (MeCN): λ_{max} [nm] (log ε) = 205 (4.64), 245 (4.01). **TLC** (petroleum ether/EtOAc 3:1): R_f = 0.36.

(S)-94: ^{1}**H NMR** (300 MHz, C_6D_6): δ [ppm] = 1.27 (s, 9H, *t*-Bu-CH$_3$), 1.33 (s, 9H, *t*-Bu-CH$_3$), 1.75-1.97 (m, 2H, H-4), 3.88-3.99 (m, 2H, H-3, H-5), 4.32 (d, J = 11.0 Hz, 1H,

CH$_2$-Ph), 4.67 (d, 1H, J = 11.0 Hz, CH$_2$-Ph), 4.71-5.24 (m, 4H, Cbz-CH$_2$), 4.82 (s, 2H, Cbz-CH$_2$), 4.89 (dd, J = 8.5, 3.3 Hz, 1H, H-2), 5.33 (d, J = 8.5 Hz, 1H, NHBoc), 6.94-7.40 (m, 20H, Ph), 11.34 (s, 1H, NHCbz). ^{13}C NMR (126 MHz, C$_6$D$_6$): δ [ppm] = 27.87 (*t*-Bu-CH$_3$), 28.33 (*t*-Bu-CH$_3$), 29.73 (C-4), 44.70-45.46 (m, C-5), 56.04 (C-2), 68.09 (Cbz-CH$_2$), 68.87 (Cbz-CH$_2$), 71.63 (CH$_2$-Ph), 77.74 (C-3), 79.30 (*t*-Bu-CH$_3$), 82.00 (*t*-Bu-CH$_3$), 127.54 (CH-Ph), 127.81 (CH-Ph), 127.91 (CH-Ph), 128.30 (CH-Ph), 128.49 (CH-Ph), 128.53 (CH-Ph), 128.62 (CH-Ph), 128.71 (CH-Ph), 135.38 (C-Ph), 138.96 (C-Ph), 151.71 (guanidine-C), 154.83 (Cbz-C=O), 155.67 (Boc-C=O), 169.67 (C-1). MS (ESI): m/z = 862.4 [M+Na]$^+$. HRMS (ESI): calcd for C$_{46}$H$_{53}$DN$_4$O$_{11}$Na 862.3744, found 862.3754 [M+Na]$^+$. $[\alpha]_D^{20}$ = +3.4 (c = 0.92, CHCl$_3$). IR (ATR): $\tilde{v}$ [cm^{-1}] = 2981, 1767, 1714, 1650, 1610, 1499, 1453, 1372, 1219, 1197, 1151, 1051, 1028, 737, 697. UV (MeCN): λ_{max} [nm] (log ε) = 206 (4.60), 235 (4.09). TLC (petroleum ether/EtOAc 3:1): R$_f$ = 0.28.

(*R*)-152: ^{1}H NMR (500 MHz, C$_6$D$_6$): δ [ppm] = 1.12 (s, 9H, SiC(CH$_3$)$_3$), 1.34 (s, 9H, *t*-Bu-CH$_3$), 1.42 (s, 9H, *t*-Bu-CH$_3$), 1.98-2.09 (m, 1H, H-4a), 2.10-2.21 (m, 1H, H-4b), 3.61 (dd, J = 6.9, 6.9 Hz, 0.5H, H-5), 3.98 (dd, J = 7.6, 7.6 Hz, 0.5H, H-5), 4.55-5.43 (m, 7H, H-2, Cbz-CH$_2$), 4.64-4.71 (m, 1H, H-3), 5.64 (d, J = 9.6 Hz, 1H, NHBoc), 6.50-7.46 (m, 21H, Ph), 7.74-7.83 (m, 4H, Ph), 11.20 (s$_{br}$, 1H, NHCbz). ^{13}C NMR (126 MHz, C$_6$D$_6$): δ [ppm] = 19.25 (Si<u>C</u>(CH$_3$)$_3$), 26.98 (SiC(<u>C</u>H$_3$)$_3$), 27.75 (*t*-Bu-CH$_3$), 28.07 (*t*-Bu-CH$_3$), 33.01 (C-4), 43.41-44.48 (m, C-5), 57.39 (C-2), 67.61 (Cbz-CH$_2$), 68.60 (Cbz-CH$_2$), 72.68 (C-3), 79.09 (*t*-Bu-C), 81.20 (*t*-Bu-C), 127.59 (CH-Ph), 127.75 (CH-Ph), 128.17 (CH-Ph), 128.29 (CH-Ph), 128.30 (CH-Ph), 128.35 (CH-Ph), 128.37 (CH-Ph), 129.62 (CH-Ph), 129.65 (CH-Ph), 133.52 (C-Ph), 134.04 (C-Ph), 135.14 (C-Ph), 135.94 (CH-Ph), 135.99 (CH-Ph), 150.78 (guanidine-C), 154.38 (Cbz-C=O), 156.09 (Boc-C=O), 170.32 (C-1). HRMS (ESI): calcd for C$_{55}$H$_{65}$DN$_4$O$_{11}$SiNa 1010.4452, found 1010.4476 [M+Na]$^+$. $[\alpha]_D^{20}$ = -10.4 (c = 0.99, CHCl$_3$). IR (ATR): $\tilde{v}$ [cm^{-1}] = 2935, 1767, 1714, 1661, 1610, 1499, 1372, 1214, 1191, 1151, 1109, 1051, 1028, 743, 697. UV (MeCN): λ_{max} [nm] (log ε) = 210 (5.28), 244 (4.62). TLC (isohexane/EtOAc 3:1): R$_f$ = 0.53.

7.2.3.10 Synthesis of N^2-Boc-$N^5,N^7,N^{7'}$-Tris-Alloc-(3R)-3-(*tert*-Butyldiphenyl-silyloxy)-L-[5-²H]arginine *tert*-Butyl Ester (R)-162

(R)-162

To a solution of **(R)-151** (785 mg, 1.44 mmol, 1.0 eq) in dry THF (18 mL), a solution of tris-Alloc-guanidine **159** (1.35 g, 4.34 mmol, 3.0 eq) in dry THF (12 mL) and triphenyl phosphine (570 mg, 2.17 mmol, 1.5 eq) were added under an argon atmosphere at room temperature. DIAD (425 µL, 436 mg, 2.16 mmol, 1.5 eq) was added dropwise at 0 °C at such a rate that the reaction mixture was completely colorless before the addition of the next drop. The reaction mixture was allowed to warm to room temperature and stirred overnight at room temperature. The reaction was quenched by the addition of H_2O (20 mL), and the solvent was removed under reduced pressure. The resultant crude product was purified by flash chromatography (isohexane/EtOAc 6:1).

Yield: 1.08 g (1.29 mmol, 90%) as a colorless viscous oil.

¹H NMR (500 MHz, C_6D_6): δ [ppm] = 1.14 (s, 9H, $SiC(CH_3)_3$), 1.37 (s, 9H, *t*-Bu-CH₃), 1.42 (s, 9H, *t*-Bu-CH₃), 2.01-2.11 (m, 1H, H-4a), 2.11-2.22 (m, 1H, H-4b), 3.66 (dd, J = 5.7, 5.7 Hz, 0.5H, H-5), 4.00 (dd, J = 7.4, 7.4 Hz, 0.5H, H-5), 4.22-4.73 (m, 7H, H-3, H-1'), 4.87-5.27 (m, 7H, H-2, H-3'), 5.58-5.89 (m, 4H, H-2', NHBoc), 7.19-7.34 (m, 6H, Ph), 7.73-7.86 (m, 4H, Ph), 11.17 (s_br, 1H, NHCbz). **¹³C NMR** (126 MHz, C_6D_6): δ [ppm] = 19.27 ($Si\underline{C}(CH_3)_3$), 26.98 ($SiC(\underline{C}H_3)_3$), 27.77 (*t*-Bu-CH₃), 28.07 (*t*-Bu-CH₃), 33.02 (C-4), 43.70-44.37 (m, C-5), 57.36 (C-2), 66.43 (C-1'), 67.51 (C-1'), 72.70 (C-3), 79.04 (*t*-Bu-C), 81.19 (*t*-Bu-C), 127.47 (CH-Ph), 127.58 (CH-Ph), 129.64 (CH-Ph), 131.39 (C-2'), 132.18 (C-2'), 133.48 (C-Ph), 134.16 (C-Ph), 135.98 (CH-Ph), 136.01 (CH-Ph), 150.97 (guanidine-C), 154.13 (Alloc-C=O), 156.03 (Boc-C=O), 170.30 (C-1). **HRMS** (ESI): calcd for $C_{43}H_{59}DN_4O_{11}SiNa$ 860.3983, found 860.3990 [M+Na]⁺. $[\alpha]_D^{20}$ = -7.3 (c = 0.88, CHCl₃). **IR** (ATR): $\tilde{v}$ [cm⁻¹] = 2929, 1748, 1731, 1708, 1673, 1603, 1522, 1365, 1208, 1151, 1064, 935, 767, 703, 616. **UV** (MeCN): λ_{max} [nm] (log ε) = 219 (4.76), 243 (4.37). **Mp** = 74 °C. **TLC** (isohexane/EtOAc 3:1): R_f = 0.38.

7.2.3.11 Attempt to Synthesize *N²*-Boc-*N⁵,N⁷,N⁷'*-Tris-Cbz-(3*R*)-3-hydroxy-L-[5-²H]-arginine *tert*-Butyl Ester (*R*)-154 with the Formation of Side Products 153 and 157

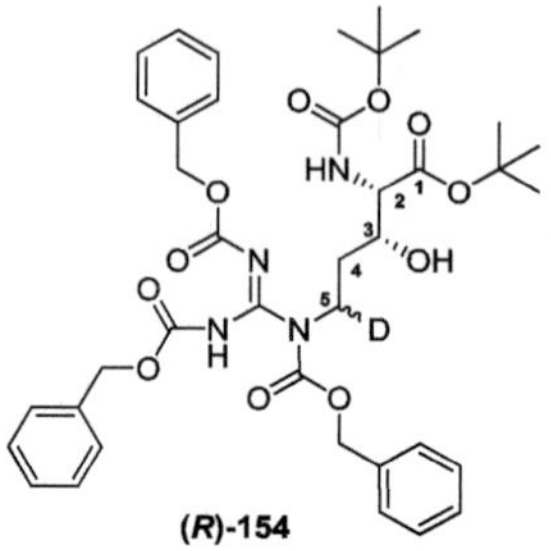

(R)-154

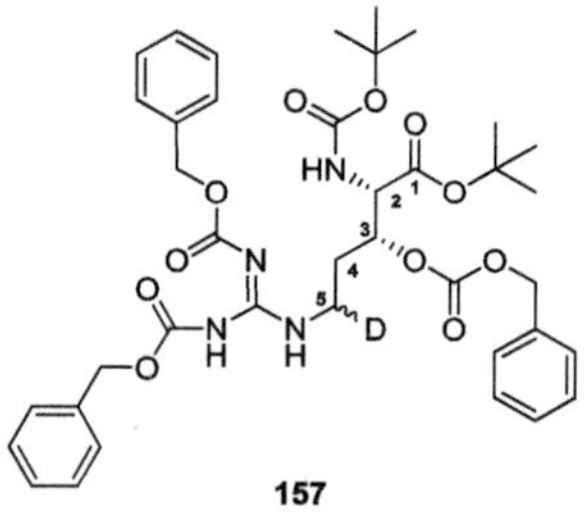

157

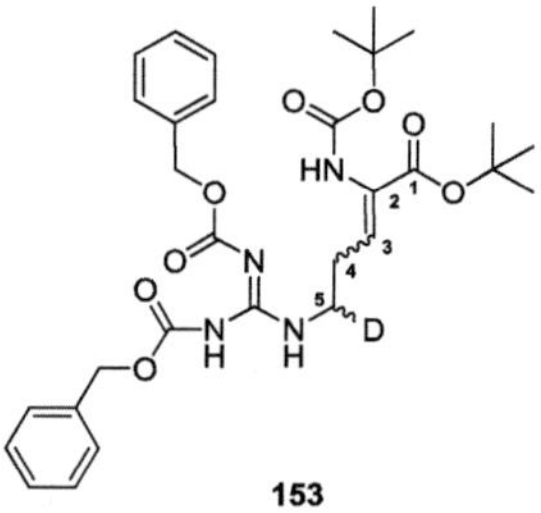

153

Variant 1 (formation of side product **157**)

To a solution of **(R)-152** (92 mg, 93 µmol, 1.0 eq) in dry THF (2 mL), acetic acid (6 µL, 6 mg, 100 µmol, 1.1 eq) and tetra-*n*-butylammonium fluoride solution (1 M in THF, 100 µL, 100 µmol, 1.1 eq) was added under an argon atmosphere. After stirring the mixture at room temperature for 3 h, it was partitioned between water (10 mL) and Et_2O (10 mL), and the organic layer was washed with saturated aqueous $NaHCO_3$ solution (1 x 5 mL) and brine (1 x 5 mL). The organic layer was dried over Na_2SO_4, and the solvent was removed under reduced pressure. The resultant crude product was purified by flash chromatography (isohexane/EtOAc 3:1).

Yield: 12 mg (16 µmol, 17%) of side product **157** as a colorless oil.

Variant 2 (formation of side product **157**)

To a solution of **(R)-152** (50 mg, 51 µmol, 1.0 eq) in dry THF (2 mL), triethylamine trihydrofluoride (0.10 mL, 99 mg, 0.51 mmol, 10 eq) was added under an argon atmosphere. The reaction was stirred at room temperature for 13 d and monitored by TLC (isohexane/EtOAc 3:1). After 6 d, additional THF (1 mL) and triethylamine trihydrofluoride (0.10 mL, 99 mg, 0.51 mmol, 10 eq) were added, and after 11 d, triethylamine (0.10 mL, 73 mg, 0.72 mmol, 14 eq) and dry THF (1 mL) were added. The mixture was partitioned between EtOAc (10 mL) and saturated aqueous $NaHCO_3$ solution (5 mL), the organic layer was washed with brine (1 x 5 mL), dried over Na_2SO_4, and the solvent was removed under reduced pressure. The resultant crude product was purified by flash chromatography (isohexane/EtOAc 4:1).

Yield: 6 mg (8 µmol, 16%) of side product **157** as a colorless oil.

Variant 3 (formation of side product **153**)

To a solution of the TBDPS-protected compound **(R)-152** (91 mg, 92 µmol, 1.0 eq) in dry THF (3 mL), solid tetra-*n*-butylammonium fluoride (73 mg, 0.23 mmol, 2.5 eq) was added at room temperature under an argon atmosphere. After stirring the mixture at room temperature for 23 h, it was partitioned between water (10 mL), Et_2O (5 mL), and EtOAc (10 mL). The organic layer was washed with saturated aqueous $NaHCO_3$ solution (1 x 10 mL) and brine (1 x 10 mL), dried over Na_2SO_4, and the solvent was removed under reduced pressure. The resultant crude product was purified by flash chromatography (isohexane/EtOAc 3:1).

Yield: 19 mg (32 µmol, 35%) of side product **153** as a colorless oil.

157: ^{1}**H NMR** (500 MHz, C_6D_6): δ [ppm] = 1.27 (s, 9H, *t*-Bu-CH$_3$), 1.37 (s, 9H, *t*-Bu-CH$_3$), 1.64-1.72 (m, 1H, H-4a), 1.73-1.83 (m, 1H, H-4b), 3.07 (ddd, J = 5.8, 5.8, 5.8 Hz, 0.5H, H-5), 3.47 (ddd, J = 5.8, 4.8, 4.8 Hz, 0.5H, H-5), 4.65-4.71 (m, 1H, H-2), 4.76 (s, 2H, Cbz-CH$_2$), 4.89 (d, J = 12.3 Hz, 1H, Cbz-CH$_2$), 4.94 (d, J = 12.3 Hz, 1H, Cbz-CH$_2$), 5.15 (d, J = 12.6 Hz, 1H, Cbz-CH$_2$), 5.17 (d, J = 12.6 Hz, 1H, Cbz-CH$_2$), 5.27 (d, J = 9.8 Hz, 1H, NHBoc), 5.41-5.47 (m, 1H, H-3), 6.98-7.40 (m, 15H, Ph), 8.30 (d, J = 5.8 Hz, 1H, N^{5}HCbz), 11.20 (s$_{br}$, 1H, N^{7}HCbz). ^{13}**C NMR** (126 MHz, C_6D_6): δ [ppm] = 27.36 (*t*-Bu-CH$_3$), 27.94 (*t*-Bu-CH$_3$), 31.02 (C-4), 36.16-37.07 (m, C-5), 56.93 (C-2), 66.81 (Cbz-CH$_2$), 67.41 (Cbz-CH$_2$), 69.53 (Cbz-CH$_2$), 76.30 (C-3), 79.33 (*t*-Bu-C), 81.83 (*t*-Bu-C), 128.16 (CH-Ph), 128.18 (CH-Ph), 128.24 (CH-Ph), 128.34 (CH-Ph), 128.38 (CH-Ph), 135.13 (CH-Ph), 135.38 (CH-Ph), 135.93 (C-Ph), 135.99 (C-Ph), 137.47 (C-Ph), 153.50 (Cbz-C=O), 154.79 (Cbz-C=O), 155.77 (guanidine-C), 156.09 (Cbz-C=O), 164.21 (Boc-C=O), 168.55 (C-1). **HRMS** (ESI): calcd for $C_{39}H_{48}DN_4O_{11}$ 750.3455, found 750.3488 [M+H]$^+$. $[\alpha]_D^{20}$ = +16.0 (c = 1.2, MeOH). **IR** (ATR): $\tilde{\nu}$ [cm^{-1}] = 2975, 1726, 1639, 1621, 1569, 1499, 1435, 1372, 1330, 1249, 1219, 1151, 1057, 743, 697. **UV** (MeCN): λ_{max} [nm] (log ε) = 207 (5.11), 236 (4.82). **TLC** (isohexane/EtOAc 3:1): R_f = 0.37.

153: ^{1}**H NMR** (500 MHz, C_6D_6): δ [ppm] = 1.36 (s, 9H, *t*-Bu-CH$_3$), 1.38 (s, 9H, *t*-Bu-CH$_3$), 2.21-2.33 (m, 2H, H-4), 3.11-3.21 (m, 1H, H-5), 4.75 (s, 2H, Cbz-CH$_2$), 5.16 (s, 2H, Cbz-CH$_2$), 6.24 (s$_{br}$, 1H, NHBoc), 6.40 (t, J = 7.4 Hz, 1H, H-3), 6.97-7.21 (m, 8H, Ph), 7.32-7.39 (m, 2H, Ph), 8.15 (d, J = 4.7 Hz, 1H, N^{5}HCbz), 12.09 (s$_{br}$, 1H, N^{7}HCbz).

^{13}**C NMR** (126 MHz, C_6D_6): δ [ppm] = 27.60 (*t*-Bu-CH$_3$), 27.88 (*t*-Bu-CH$_3$), 28.07 (C-4), 38.85-39.40 (m, C-5), 66.81 (Cbz-CH$_2$), 67.46 (Cbz-CH$_2$), 79.55 (*t*-Bu-C), 80.88 (*t*-Bu-C), 126.58 (C-2), 128.05 (CH-Ph), 128.18 (CH-Ph), 128.20 (CH-Ph), 128.24 (CH-Ph), 128.30 (CH-Ph), 128.40 (CH-Ph), 129.38 (C-3), 135.07 (C-Ph), 137.41 (C-Ph), 152.99 (guanidine-C), 153.55 (Cbz-C=O), 156.04 (Cbz-C=O), 163.56 (Boc-C=O), 164.17 (C-1). **HRMS** (ESI): calcd for $C_{31}H_{39}DN_4O_8Na$ 620.2801, found 620.2827 $[M+Na]^+$. **TLC** (isohexane/EtOAc 3:1): R$_f$ = 0.17.

7.2.3.12 Synthesis of *N²*-Boc-(3*R*)-3-(*tert*-Butyldiphenylsilyloxy)-L-[5-²H]-arginine *tert*-Butyl Ester (*R*)-158

(*R*)-158

Variant 1

To a solution of (***R***)-162 (683 mg, 0.815 mmol, 1.0 eq) in dry THF (15 mL), dimedone (572 mg, 4.08 mmol, 5.0 eq) and tetrakis(triphenylphosphine)palladium (47 mg, 41 μmol, 0.05 eq) were added under an argon atmosphere, and the resultant mixture was stirred overnight in the dark. The mixture was partitioned between Et$_2$O (30 mL) and water (50 mL), and the organic layer was washed with 1 M HCl (3 x 20 mL). The aqueous layers were combined and extracted with Et$_2$O (4 x 100 mL). The combined organics were dried over Na$_2$SO$_4$, and the solvent was removed under reduced pressure. The resultant crude product was purified by flash chromatography (CH$_2$Cl$_2$/MeOH 8:1→5:1).

Yield: 586 mg (100% yield: 477 mg) as an impure yellowish foam.

Variant 2

To a solution of (***R***)-152 (213 mg, 0.215 mmol, 1.0 eq) in EtOAc (4 mL) and EtOH (4 mL), trifluoroacetic acid (30 μL, 44 mg, 0.39 mmol, 1.8 eq) and 10% Pd/C (50 mg, 47 μmol, 0.2 eq) were added, and the reaction mixture was stirred under an atmosphere of hydrogen (1 bar) at room temperature for 80 min. It was subsequently filtered through a syringe filter. The syringe filter was washed thoroughly with EtOH, and the solvent of the filtrate was removed under reduced pressure. The resultant crude product was used without further purification.

Yield: 141 mg (100% yield: 126 mg) as an impure white foam.

^{1}H NMR (500 MHz, CD$_3$OD): δ [ppm] = 1.05 (s, 9H, SiC(CH$_3$)$_3$), 1.45 (s, 9H, *t*-Bu-CH$_3$), 1.51 (s, 9H, *t*-Bu-CH$_3$), 1.63-1.72 (m, 1H, H-4a), 1.74-1.85 (m, 1H, H-4b), 2.89 (dd, *J* = 7.8, 7.8 Hz, 0.5H, H-5), 3.07 (dd, *J* = 6.9, 6.9 Hz, 0.5H, H-5), 4.21-4.27 (m, 1H, H-2), 4.29-4.36 (m, 1H, H-3), 7.41-7.54 (m, 6H, Ph), 7.69-7.76 (m, 4H, Ph). **^{13}C NMR** (126 MHz, CD$_3$OD): δ [ppm] = 18.26 (Si$\underline{C}$(CH$_3$)$_3$), 26.10 (SiC($\underline{C}$H$_3$)$_3$), 27.03 (*t*-Bu-CH$_3$), 27.32 (*t*-Bu-CH$_3$), 33.17 (C-4), 36.78-37.44 (m, C-5), 57.19 (C-2), 71.94 (C-3), 79.98 (*t*-Bu-C), 82.34 (*t*-Bu-C), 127.43 (CH-Ph), 127.57 (CH-Ph), 129.79 (CH-Ph), 129.91 (CH-Ph), 132.50 (C-Ph), 133.56 (C-Ph), 136.60 (CH-Ph), 156.91 (guanidine-C), 156.95 (Boc-C=O), 169.86 (C-1). **HRMS** (ESI): calcd for C$_{31}$H$_{48}$DN$_4$O$_5$Si 586.3530, found 586.3524 [M+H]$^+$. **[α]$_D^{20}$** = -24.9 (*c* = 1.1, MeOH). **IR** (ATR): $\tilde{v}$ [cm^{-1}] = 2935, 1703, 1673, 1505, 1476, 1372, 1330, 1255, 1151, 1109, 1087, 824, 743, 703, 610. **UV** (MeCN): λ$_{max}$ [nm] (log ε) = 219 (4.35), 266 (3.18). **Mp** = 90 °C. **TLC** (CH$_2$Cl$_2$/MeOH 5:1): R$_f$ = 0.30.

7.2.3.13 Synthesis of *N*^{2}-Boc-(3*R*)- and (3*S*)-3-Hydroxy-L-[5-^{2}H]arginine *tert*-Butyl Ester (*R*)-143 and (*S*)-143

(3*R*): (*R*)-143, (3*S*): (*S*)-143

(*R*)-143: To a solution of **(*R*)-94** (1.05 g, 1.25 mmol, 1.0 eq) in EtOAc (20 mL) and EtOH (20 mL), trifluoroacetic acid (180 µL, 266 mg, 2.33 mmol, 1.9 eq) and 10% Pd/C (530 mg, 0.498 mmol, 0.4 eq) were added, and the reaction mixture was stirred under an atmosphere of hydrogen (1 bar) at room temperature for 22 h. It was subsequently filtered through Celite. The Celite was washed thoroughly with EtOAc and EtOH, and the solvent of the filtrate was removed under reduced pressure. The resultant crude product was purified by flash chromatography (RP silica gel 90 C$_{18}$, H$_2$O/MeCN 4:1→2:1).

(*S*)-143: The isomer **(*S*)-143** was prepared in the same way as compound **(*R*)-143** with **(*S*)-94** (2.00 g, 2.38 mmol, 1.0 eq), trifluoroacetic acid (330 µL, 488 mg, 4.28 mmol, 1.8 eq), 10% Pd/C (1.01 g, 0.952 mmol, 0.4 eq), EtOAc (40 mL) and EtOH (40 mL). After 23 h, additional 10% Pd/C (240 mg, 0.226 mmol, 0.1 eq) was added, and the reaction was stirred for a total of 60 h.

Yield (*R*)-143: 429 mg (1.23 mmol, 98%) as a white solid.

Yield (*S*)-143: 838 mg (100% yield: 827 mg) as a slightly impure white solid.

(***R***)-143: ^{1}**H NMR** (500 MHz, CD$_3$OD): δ [ppm] = 1.48 (s, 9H, *t*-Bu-CH$_3$), 1.51 (s, 9H, *t*-Bu-CH$_3$), 1.70-1.83 (m, 2H, H-4), 3.28-3.36 (m, 1H, H-5), 4.06-4.17 (m, 2H, H-2, H-3). ^{13}**C NMR** (126 MHz, CD$_3$OD): δ [ppm] = 26.89 (*t*-Bu-CH$_3$), 27.27 (*t*-Bu-CH$_3$), 32.81 (C-4), 37.42-38.17 (m, C-5), 58.81 (C-2), 68.72 (C-3), 79.56 (*t*-Bu-C), 81.85 (*t*-Bu-C), 156.94 (guanidine-C), 157.39 (Boc-C=O), 170.08 (C-1). **HRMS** (ESI): calcd for C$_{15}$H$_{30}$DN$_4$O$_5$ 348.2357, found 348.2355 [M+H]$^+$. $[\alpha]_D^{20}$ = +1.1 (*c* = 1.1, MeOH). **IR** (ATR): $\tilde{v}$ [cm^{-1}] = 3353, 1667, 1621, 1510, 1372, 1255, 1202, 1179, 1138, 1064, 843, 801, 726. **Mp** = 56 °C. **TLC** (2-propanol/H$_2$O/AcOH 5:2:1 as saturated NaCl solution): R$_f$ = 0.67-0.76. **TLC** (CH$_2$Cl$_2$/MeOH 5:1): R$_f$ = 0.00-0.12.

(***S***)-143: ^{1}**H NMR** (500 MHz, CD$_3$OD): δ [ppm] = 1.47 (s, 9H, *t*-Bu-CH$_3$), 1.51 (s, 9H, *t*-Bu-CH$_3$), 1.74-1.88 (m, 2H, H-4), 3.30-3.36 (m, 1H, H-5), 3.91 (ddd, *J* = 9.1, 4.8, 4.8 Hz, 1H, H-3), 4.09 (d, *J* = 4.8 Hz, 1H, H-2). ^{13}**C NMR** (126 MHz, CD$_3$OD): δ [ppm] = 26.93 (*t*-Bu-CH$_3$), 27.28 (*t*-Bu-CH$_3$), 31.94 (C-4), 37.43-38.32 (m, C-5), 59.32 (C-2), 69.29 (C-3), 79.49 (*t*-Bu-C), 81.84 (*t*-Bu-C), 156.51 (guanidine-C), 157.40 (Boc-C=O), 169.81 (C-1). **HRMS** (ESI): calcd for C$_{15}$H$_{30}$DN$_4$O$_5$ 348.2357, found 348.2355 [M+H]$^+$. $[\alpha]_D^{20}$ = -10.0 (*c* = 1.2, MeOH). **IR** (ATR): $\tilde{v}$ [cm^{-1}] = 3353, 1667, 1627, 1510, 1372, 1255, 1197, 1179, 1138, 1064, 1017, 843, 801, 720. **Mp** = 56 °C. **TLC** (2-propanol/H$_2$O/AcOH 5:2:1 as saturated NaCl solution): R$_f$ = 0.63-0.75. **TLC** (CH$_2$Cl$_2$/MeOH 5:1): R$_f$ = 0.00-0.20.

7.2.3.14 Synthesis of (3*R*)- and (3*S*)-3-Hydroxy-L-[5-^{2}H]arginine Dihydrochloride (*R*)-87 and (*S*)-87

(3*R*): (*R*)-87, (3*S*): (*S*)-87

Variant 1

(***R***)-87: A solution of (***R***)-143 (273 mg, 0.786 mmol, 1.0 eq) in 6 M aqueous HCl (8 mL) was stirred at room temperature for 3 h. The solvent was subsequently removed under reduced pressure. The resultant crude product was purified by flash chromatography (RP silica gel 90 C$_{18}$, H$_2$O).

(*S*)-87: The isomer **(*S*)-87** was prepared in the same way as compound **(*R*)-87** with **(*S*)-143** (739 mg, 2.13 mmol, 1.0 eq) and 6 M aqueous HCl (20 mL).

Yield (*R*)-87: 164 mg (0.621 mmol, 79%) as a white-to-yellowish hygroscopic solid.
Yield (*S*)-87: 464 mg (1.76 mmol, 83%) as a white-to-yellowish hygroscopic solid.

Variant 2

(*R*)-87: A suspension of **(*R*)-158** (298 mg impure material from the Alloc-deprotection, see chapter 7.3.3.11) in 6 M aqueous HCl (8 mL) was stirred at room temperature. In a 2 h interval, the mixture was treated for 15 min in an ultrasonic bath for the first 10 h. After 24 h, the solvent was removed under reduced pressure, and the resultant crude product was purified by flash chromatography (RP silica gel 90 C_{18}, H_2O).

Yield (*R*)-87: 105 mg (0.398 mmol, 90% over two steps) as a white-to-yellowish hygroscopic solid.

(*R*)-87: ^{1}**H NMR** (500 MHz, CD_3OD): δ [ppm] = 1.82 (ddd, *J* = 14.1, 10.5, 6.1 Hz, 1H, H-4a), 1.97 (ddd, *J* = 14.1, 7.7, 3.1 Hz, 1H, H-4b), 3.31-3.39 (m, 1H, H-5), 4.04 (d, *J* = 4.1 Hz, 1H, H-2), 4.28 (ddd, *J* = 10.5, 4.1, 3.1 Hz, 1H, H-3). ^{13}**C NMR** (126 MHz, CD_3OD): δ [ppm] = 32.04 (C-4), 37.36-37.82 (m, C-5), 57.89 (C-2), 66.52 (C-3), 156.93 (guanidine-C), 170.34 (C-1). **HRMS** (ESI): calcd for $C_6H_{14}DN_4O_3$ 192.1207, found 192.1200 [M+H]$^+$. $[\alpha]_D^{20}$ = +23.5 (*c* = 2.2, H_2O). **IR** (ATR): $\tilde{\nu}$ [cm^{-1}] = 3341, 3173, 2918, 1737, 1661, 1621, 1516, 1429, 1278, 1226, 1081, 1040. **TLC** (2-propanol/H_2O/AcOH 5:2:1 as saturated NaCl solution): R_f = 0.15.

(*S*)-87: ^{1}**H NMR** (500 MHz, CD_3OD): δ [ppm] = 1.84-1.98 (m, 2H, H-4), 3.28-3.37 (m, 1H, H-5), 4.13-4.23 (m, 2H, H-2, H-3). ^{13}**C NMR** (126 MHz, CD_3OD): δ [ppm] = 30.95 (C-4), 37.49-37.99 (m, C-5), 57.70 (C-2), 66.96 (C-3), 156.94 (guanidine-C), 169.38 (C-1). **HRMS** (ESI): calcd for $C_6H_{14}DN_4O_3$ 192.1207, found 192.1205 [M+H]$^+$. $[\alpha]_D^{20}$ = +0.9 (*c* = 2.0, H_2O). **IR** (ATR): $\tilde{\nu}$ [cm^{-1}] = 3330, 3179, 2899, 1731, 1650, 1616, 1499, 1429, 1214, 1161, 1057. **TLC** (2-propanol/H_2O/AcOH 5:2:1 as saturated NaCl solution): R_f = 0.12.

7.2.4   Synthesis of Different Guanidine Derivatives for the *Mitsunobu* Reaction

7.2.4.1 Synthesis of *N,N'*-Bis-Cbz-Guanidine 141[185]

141

To a solution of guanidine hydrochloride **140** (20.0 g, 0.209 mol, 1.0 eq) and sodium hydroxide (42.0 g, 1.05 mol, 5.0 eq) in water (200 mL), CH_2Cl_2 (400 mL) was added, and the resulting mixture was cooled to 0 °C. Benzyloxycarbonyl chloride (90.0 mL, 0.630 mol, 3.0 eq) was added dropwise, and the mixture was stirred for 20 h at 0 °C. The mixture was diluted with CH_2Cl_2 (350 mL), and the aqueous layer was extracted with CH_2Cl_2 (3 x 50 mL). The combined organics were washed with water (1 x 350 mL), dried over Na_2SO_4, and the solvent was removed under reduced pressure. The resultant crude product was recrystallized from methanol.

Yield: 50.1 g (0.153 mol, 73%) as colorless crystals.

^{1}H NMR (300 MHz, DMSO-d_6): δ [ppm] = 5.12 (s, 4H, Cbz-CH$_2$), 7.27-7.43 (m, 10H, Ph), 8.66 (s$_{br}$, 2H, NHCbz), 10.84 (s$_{br}$, 1H, NH). **^{13}C NMR** (75 MHz, DMSO-d_6): δ [ppm] = 66.16 (Cbz-CH$_2$), 127.62 (CH-Ph), 127.82 (CH-Ph), 128.28 (CH-Ph), 136.32 (C-Ph), 158.74 (guanidine-C, Cbz-C=O). **MS** (ESI): m/z = 350.2 [M+Na]$^+$. **HRMS** (ESI): calcd for $C_{17}H_{17}N_3O_4Na$ 350.1111, found 350.1120 [M+Na]$^+$. **Mp** = 145-146 °C. **IR** (KBr): $\tilde{v}$ [cm^{-1}] = 3401, 3240, 1735, 1655, 1622, 1564, 1383, 1292, 1225, 1123, 752, 669. **UV** (MeCN): λ_{max} [nm] (log ε) = 205 (4.40), 230 (4.32).

7.2.4.2 Synthesis of *N,N',N''*-Tris-Cbz-Guanidine 142[185]

142

To a suspension of *N,N'*-Bis-Cbz-guanidine **141** (12.0 g, 36.7 mmol, 1.0 eq) in dry THF (140 mL), sodium hydride (60% dispersion in mineral oil, 2.93 g, 73.3 mmol, 2.0 eq) was added in small portions at -45 °C under an argon atmosphere. After stirring the reaction mixture for 1 h at -45 °C, benzyloxycarbonyl chloride (7.90 mL, 55.0 mmol, 1.5 eq) was added. The reaction mixture was allowed to warm to room temperature, and it was stirred overnight. The solvent was removed under reduced pressure, and the residue was partitioned between CH_2Cl_2 (340 mL) and water (170 mL). The aqueous layer was extracted with CH_2Cl_2 (2 x 220 mL). The combined organics were washed with 1 M HCl (200 mL) and water (200 mL), dried over Na_2SO_4, and the solvent was removed under reduced pressure. The resultant crude product was recrystallized from EtOAc/petroleum ether.

Yield: 12.6 g (27.2 mmol, 74%) as white crystals.

^{1}H NMR (300 MHz, DMSO-d_6): δ [ppm] = 5.10 (s, 6H, Cbz-CH_2), 7.27-7.50 (m, 15H, Ph), 10.53 (s, 2H, NHCbz). **^{13}C NMR** (75 MHz, DMSO-d_6): δ [ppm] = 66.88 (Cbz-CH_2), 127.96 (CH-Ph), 128.05 (CH-Ph), 128.31 (CH-Ph), 135.69 (C-Ph), 145.27 (guanidine-C, Cbz-C=O). **MS** (ESI): m/z = 484.1 $[M+Na]^+$. **HRMS** (ESI): calcd for $C_{25}H_{23}N_3O_6Na$ 484.1479, found 484.1478 $[M+Na]^+$. **Mp** = 101 °C. **IR** (KBr): $\tilde{\nu}$ $[cm^{-1}]$ = 3423, 2357, 1792, 1730, 1640, 1552, 1455, 1310, 1278, 1184, 1055, 762, 695. **UV** (MeCN): λ_{max} [nm] (log ε) = 206 (4.58), 239 (4.28). **TLC** (CH_2Cl_2/Et_2O 98:2): R_f = 0.41.

7.2.4.3 Synthesis of *N,N',N''*-Tris-Alloc-Guanidine 159

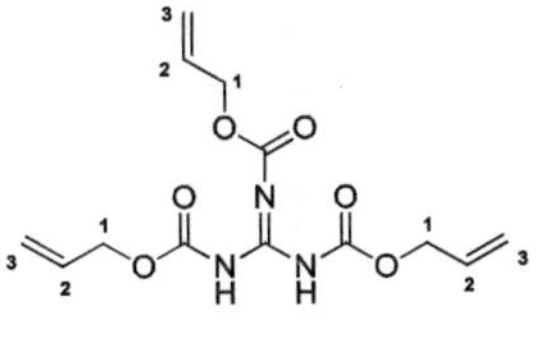

159

To a solution of guanidine hydrochloride **140** (2.01 g, 21.0 mmol, 1.0 eq) in 6 M aqueous NaOH (14 mL), benzyltriethylammonium chloride (107 mg, 0.470 mmol, 0.02 eq) and CH_2Cl_2 (40 mL) were added. After stirring the mixture at 0 °C for 15 min, allyloxycarbonyl chloride (13.5 mL, 15.3 g, 127 mmol, 6.0 eq) was added, and it was stirred for additional 8 h at 0 °C. The mixture was partitioned between water (40 mL) and CH_2Cl_2 (40 mL), and the aqueous layer was extracted with CH_2Cl_2 (2 x 40 mL). The organic extracts were combined, dried over Na_2SO_4, and the solvent was removed. The resultant crude product was purified by flash chromatography (CH_2Cl_2/Et_2O 98:2→97:3→95:5→90:10).

Yield: 3.05 g (9.80 mmol, 47%) as a colorless oil.

^{1}H NMR (500 MHz, CDCl$_3$): δ [ppm] = 4.65 (dt, J = 5.9, 1.4 Hz, 6H, H-1), 5.26 (ddt, J = 10.3, 1.4, 1.4 Hz, 3H, H-3a), 5.35 (ddt, J = 17.2, 1.4, 1.4 Hz, 3H, H-3b), 5.92 (ddt, J = 17.2, 10.3, 5.9 Hz, 3H, H-2), 10.99 (s$_{br}$, 2H, NH). **^{13}C NMR** (126 MHz, CDCl$_3$): δ [ppm] = 67.27 (C-1), 119.27 (C-3), 131.30 (C-2), 147.91 (guanidine-C), 152.17 (C=O). **MS** (ESI): m/z = 312.1 [M+H]$^+$. **HRMS** (ESI): calcd for C$_{13}$H$_{18}$N$_3$O$_6$ 312.1190, found 312.1198 [M+H]$^+$. **IR** (ATR): $\tilde{v}$ [cm^{-1}] = 3249, 1784, 1731, 1639, 1546, 1446, 1418, 1365, 1278, 1214, 1174, 1132, 1057, 994, 935, 807, 767. **UV** (MeCN): λ$_{max}$ [nm] (log ε) = 205 (4.48), 238 (4.50). **TLC** (CH$_2$Cl$_2$/MeOH 20:1): R$_f$ = 0.40.

7.2.5 Attempts and Test Reactions for the Introduction of a Second Deuterium Label at the C5-Position

7.2.5.1 Synthesis of *N*-Boc-L-Pyroglutamic Acid *tert*-Butyl Ester 129

129

To a solution of **128**^{2} (82 mg, 0.29 mmol, 1.0 eq) in MeCN (3 mL), potassium dihydrogen phosphate (10 mg, 74 µmol, 0.3 eq) in water (1.5 mL) and 30% hydrogen peroxide solution (500 µL) were added. At 0 °C, a solution of sodium chlorite (80%, 50 mg, 0.44 mmol, 1.5 eq) in water (1.5 mL) was added. The reaction was stirred at 0° C and monitored by TLC (CH$_2$Cl$_2$/EtOAc 4:1). After 1 h at 0 °C, a solution of potassium dihydrogen phosphate (120 mg, 0.882 mol, 3.1 eq) and sodium chlorite (80%, 530 mg, 4.69 mmol, 16 eq) in water (2 mL) were added at 0 °C, and the reaction was continued to stir at room temperature. After 4 d, further potassium dihydrogen phosphate (140 mg, 1.03 mol, 3.5 eq) and sodium chlorite (80%, 810 mg, 7.16 mmol, 25 eq) in water (2 mL) were added. After 5 d, the reaction was quenched with sodium sulfite (400 mg), and saturated aqueous NaHCO$_3$ solution (2 mL) and Et$_2$O (5 mL) were added. The aqueous layer was acidified with 1 M HCl and extracted with EtOAc (3 x 10 mL). The combined organics were washed with brine (20 mL), dried over Na$_2$SO$_4$, and the solvent was removed under reduced pressure. The resultant crude product was purified by flash chromatography (CH$_2$Cl$_2$/MeOH 6:1).

Yield: 63 mg (0.22 mmol, 76%) as a colorless oil.

---

^{2} Preparation of **128** see O. Ries, PhD thesis, Georg-August-University Göttingen, **2012**, p 182.[177]

^{1}H NMR (300 MHz, CDCl$_3$): δ [ppm] = 1.45 (s, 9H, *t*-Bu-CH$_3$), 1.47 (s, 9H, *t*-Bu-CH$_3$), 1.96 (dddd, *J* = 13.2, 9.4, 3.3, 2.6 Hz, 1H, H-3a), 2.25 (dddd, *J* = 13.2, 10.5, 9.4, 9.4 Hz, 1H, H-3b), 2.43 (ddd, *J* = 17.3, 9.4, 3.3 Hz, 1H, H-4a), 2.58 (ddd, *J* = 17.3, 10.5, 9.4 Hz, 1H, H-4b), 4.44 (dd, *J* = 9.4, 2.6 Hz, 1H, H-2). **^{13}C NMR** (76 MHz, CDCl$_3$): δ [ppm] = 21.61 (C-3), 27.87 (C(CH$_3$)$_3$), 27.88 (C(CH$_3$)$_3$), 31.09 (C-4), 59.55 (C-2), 82.22 (C(CH$_3$)$_3$), 83.26 (C(CH$_3$)$_3$), 149.27 (Boc-C=O), 170.33 (C-1), 173.52 (C-5). **HRMS** (ESI): calcd for C$_{14}$H$_{23}$NO$_5$Na 308.1468, found 308.1471 [M+Na]$^+$. **TLC** (petroleum ether/EtOAc 1:1): R$_f$ = 0.43.[3]

7.2.5.2 Synthesis of *N*-Boc-5-*O*-Methyl-L-glutamic Acid *tert*-Butyl Ester 135[181]

135

To a solution of *N*-Boc-glutamic acid *tert*-butylester **130** (935 mg, 3.08 mmol, 1.0 eq) in dry CH$_2$Cl$_2$ (10 mL), *N*,*N*'-dicyclohexylcarbodiimde (830 mg, 4.02 mmol, 1.3 eq) was added at 0 °C under an argon atmosphere. After 5 min, dry MeOH (250 µL, 198 mg, 6.16 mmol, 2.0 eq), triethylamine (560 µL, 409 mg, 4.04 mmol, 1.3 eq) and 4-dimethylaminopyridine (380 mg, 3.11 mmol, 1.0 eq) were added. After stirring the mixture for 1.5 h at 0 °C and for 5 h at room temperature, the white precipitant was removed by filtration, and the filtrate was concentrated under reduced pressure. The residue was redissolved in EtOAc (100 mL). The solution was washed with 1 M HCl (20 mL), saturated aqueous NaHCO$_3$ solution (20 mL) and brine (20 mL), and then dried over Na$_2$SO$_4$. After evaporation of the solvent under reduced pressure, the resultant crude product was purified by flash chromatography (petroleum ether/EtOAc 8:1→6:1) and then recrystallized from petroleum ether.

Yield: 582 mg (1.83 mmol, 59%) as a white solid.

^{1}H NMR (300 MHz, CDCl$_3$): δ [ppm] = 1.43 (s, 9H, *t*-Bu-CH$_3$), 1.46 (s, 9H, *t*-Bu-CH$_3$), 1.82-2.00 (m, 1H, H-3a), 2.06-2.23 (m, 1H, H-3b), 2.28-2.50 (m, 2H, H-4), 3.67 (s, 3H, OCH$_3$), 4.12-4.17 (m, 1H, H-2), 5.07 (d, *J* = 8.1 Hz, 1H, NH). **^{13}C NMR** (76 MHz, CDCl$_3$): δ [ppm] = 27.98 (C(CH$_3$)$_3$), 28.12 (C-3), 28.30 (C(CH$_3$)$_3$), 30.11 (C-4), 51.70 (OCH$_3$), 53.42 (C-2), 79.75 (C(CH$_3$)$_3$), 82.17 (C(CH$_3$)$_3$), 155.34 (Boc-C=O), 171.28 (C-1),

---

[3] For additional analytical data, see O. Ries, PhD thesis, Georg-August-University Göttingen, **2012**, p 182.[177]

173.27 (C-5). **HRMS** (ESI): calcd for $C_{15}H_{27}NO_6Na$ 340.1731, found 340.1747 [M+Na]$^+$. $[\alpha]_D^{20}$ = +8.6 (c = 1.1, CHCl$_3$). **IR** (ATR): $\tilde{v}$ [cm^{-1}] = 3345, 2923, 1727, 1697, 1527, 1453, 1366, 1351, 1257, 1155, 1057, 1049, 1031, 994, 852, 610. **Mp** = 62 °C. **TLC** (petroleum ether/EtOAc 3:1): R$_f$ = 0.35.

7.2.5.3 Synthesis of *N,N*-Diboc-5-*O*-Methyl-L-glutamic Acid *tert*-Butylester 136

136

To a solution of **135** (578 mg, 1.82 mmol, 1.0 eq) in dry CH_2Cl_2 (6 mL) and dry triethylamine (9 mL), di-*tert*-butyl dicarbonate (1.10 g, 5.04 mmol, 2.8 eq) and 4-dimethylaminopyridine (24 mg, 0.20 mmol, 0.1 eq) were added. The resulting mixture was heated under reflux for 4 h before the solvent was removed under reduced pressure. The residue was redissolved in EtOAc (60 mL). The solution was washed with 1 M aqueous KHSO$_4$ solution (20 mL), saturated aqueous NaHCO$_3$ solution (20 mL) and brine (20 mL), and then dried over Na$_2$SO$_4$. The solvent was removed under reduced pressure, and the resultant crude product was purified by flash chromatography (petroleum ether/EtOAc 10:1).

Yield: 615 mg (1.47 mmol, 81%) as a colorless oil.

^{1}H NMR (300 MHz, C$_6$D$_6$): δ [ppm] = 1.35 (s, 9H, *t*-Bu-CH$_3$), 1.39 (s, 18H, *t*-Bu-CH$_3$), 2.21-2.38 (m, 3H, H-3a, H-4), 2.56-2.71 (m, 1H, H-3b), 3.31 (s, 3H, OCH$_3$), 5.12 (dd, J = 9.7, 5.1 Hz, 1H, H-2). **^{13}C NMR** (75 MHz, C$_6$D$_6$): δ [ppm] = 25.27 (C-3), 27.88 (*t*-Bu-CH$_3$), 27.94 (*t*-Bu-CH$_3$), 30.88 (C-4), 51.06 (OCH$_3$), 58.68 (C-2), 80.79 (*t*-Bu-C), 82.28 (*t*-Bu-C), 152.96 (Boc-C=O), 169.20 (C-1), 172.67 (C-5). **MS** (ESI): m/z = 440.2 [M+Na]$^+$. **HRMS** (ESI): calcd for $C_{20}H_{35}NO_8Na$ 440.2255, found 440.2255 [M+Na]$^+$. $[\alpha]_D^{20}$ = -2.7 (c = 1.1, CHCl$_3$). **IR** (ATR): $\tilde{v}$ [cm^{-1}] = 2979, 1736, 1699, 1365, 1273, 1252, 1232, 1139, 1116, 850. **TLC** (petroleum ether/EtOAc 3:1): R$_f$ = 0.49.

7.2.5.4 Synthesis of *N,N*-Diboc-L-Glutamic Acid *tert*-Butyl Ester 137

137

To a solution of **136** (615 mg, 1.47 mmol, 1.0 eq) in THF (8 mL), a solution of lithium hydroxide (230 mg, 5.48 mmol, 3.7 eq) in water (8 mL) was added dropwise at 0 °C. The resulting mixture was allowed to warm to room temperature and stirred at this temperature for 23 h. The solvent was removed under reduced pressure, and the remaining aqueous layer was acidified with 2 M aqueous HCl and extracted with Et_2O (3 x 30 mL). The combined organics were washed with water (20 mL), dried over Na_2SO_4, and the solvent was removed under reduced pressure. The resultant crude product was used without further purification.

Yield: 533 mg (1.32 mmol, 90%) as a white solid.

^{1}H NMR (300 MHz, C_6D_6): δ [ppm] = 1.35 (s, 9H, *t*-Bu-CH_3), 1.40 (s, 18H, *t*-Bu-CH_3), 2.11-2.62 (m, 4H, H-3, H-4), 5.08 (dd, J = 9.3, 5.0 Hz, 1H, H-2). **^{13}C NMR** (75 MHz, C_6D_6): δ [ppm] = 24.78 (C-3), 27.87 (*t*-Bu-CH_3), 27.92 (*t*-Bu-CH_3), 30.90 (C-4), 58.64 (C-2), 80.87 (*t*-Bu-C), 82.48 (*t*-Bu-C), 152.92 (Boc-C=O), 169.13 (C-1), 179.49 (C-5). **MS** (ESI): m/z = 426.2 [M+Na]$^+$. **HRMS** (ESI): calcd for $C_{19}H_{33}NO_8Na$ 426.2098, found 426.2099 [M+Na]$^+$. $[\alpha]_D^{20}$ = -1.2 (c = 1.0, $CHCl_3$). **IR** (ATR): $\tilde{v}$ [cm^{-1}] = 2979, 1738, 1695, 1364, 1282, 1254, 1239, 1141, 1119, 1088, 1012, 847, 799, 769. **Mp** = 125 °C. **TLC** (CH_2Cl_2/MeOH 95:5): R_f = 0.31.

7.2.5.5 Synthesis of *N,N*-Diboc-(*S*)-2-Amino-5-hydroxy-valeric Acid *tert*-Butyl Ester 138

138

Variant 1

To a solution of **137** (83 mg, 0.21 mmol, 1.0 eq) in dry THF (3 mL), *N*-methylmorpholine (33 µL, 30 mg, 0.30 mmol, 1.4 eq), followed by isobutyl chloroformate (39 µL, 41 mg, 0.30 mmol, 1.4 eq) was added at -20 °C under an argon atmosphere. The resulting mixture was stirred at -20 °C for 1.5 h and then allowed to reach a temperature of 0 °C before sodium borohydride (26 mg, 0.69 mmol, 3.3 eq) and dry methanol (1.5 mL) were added. The mixture was allowed to warm to room temperature, and it was stirred for 40 h. The reaction was quenched with water (5 mL) and concentrated under reduced pressure. The residue was partitioned between EtOAc (20 mL) and saturated aqueous NH_4Cl solution (7 mL). The organic layer was washed with saturated aqueous $NaHCO_3$ solution (5 mL) and brine (5 mL), dried over Na_2SO_4, and the solvent was removed. The resultant crude product was purified by flash chromatography (petroleum ether/EtOAc 8:1→3:1→2:1).

Yield: 23 mg (59 µmol, 28%) as a colorless oil.

Variant 2

To a solution of cyanuric chloride (40 mg, 0.22 mmol, 1.1 eq) in DME (1 mL) *N*-methylmorpholine (24 µL, 22 mg, 0.22 mmol, 1.1 eq) was added at room temperature. A white suspension was formed, and a solution of **137** (80 mg, 0.20 mmol, 1.0 eq) in DME (1 mL) was added to this mixture. After 3 h at room temperature, the mixture was filtered and the filtrate was cooled to 0 °C. At this temperature, a solution of sodium borohydride (60 mg, 1.6 mmol, 8 eq) in water (2 mL) was added. The mixture was stirred for additional 30 min at 0 °C before Et_2O (15 mL) was added. The solution was acidified with saturated aqueous NH_4Cl solution (5 mL). The organic layer was washed with 10% aqueous Na_2O_3 solution (10 mL) and brine (10 mL), dried over Na_2SO_4, and the solvent was removed. The resultant crude product was purified by flash chromatography (petroleum ether/EtOAc 3:1).

Yield: 15 mg (39 µmol, 19%) as a colorless oil.

^{1}H NMR (300 MHz, C_6D_6): δ [ppm] = 1.35 (s, 9H, *t*-Bu-CH$_3$), 1.37 (s, 18H, *t*-Bu-CH$_3$), 1.46-1.60 (m, 2H, H-4), 2.06-2.24 (m, 1H, H-3a), 2.24-2.39 (m, 1H, H-3b), 3.27-3-37 (m, 2H, H-5), 5.04 (dd, J = 9.6, 5.4 Hz, 1H, H-2). **MS** (ESI): m/z = 412.2 [M+Na]$^+$. **HRMS** (ESI): calcd for $C_{19}H_{35}NO_7Na$ 412.2306, found 412.2301 [M+Na]$^+$. **TLC** (petroleum ether/EtOAc 3:1): R_f = 0.11.

7.2.5.6 Synthesis of *N*-Boc-L-Glutamic Acid *tert*-Butyl Ester 130 via Selective Boc-Deprotection

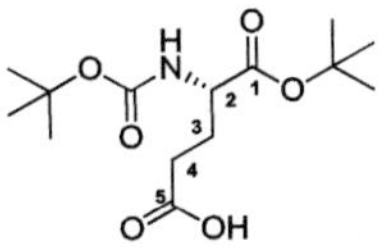

130

Variant 1

To a solution of **137** (82 mg, 0.20 mmol, 0.1 eq) in CH_2Cl_2, trifluoroacetic acid (24 µL, 36 mg, 0.32 mmol, 1.6 eq) was added, and the mixture was stirred for 4.5 d at room temperature. Et_2O (10 mL) was added, and the organic layer was washed with saturated aqueous $NaHCO_3$ solution (5 mL) and saturated aqueous NH_4Cl solution (5 mL). The organic layer was dried over Na_2SO_4, and the solvent was removed under reduced pressure. The resultant crude product was purified by flash chromatography (CH_2Cl_2/MeOH 95:5).

Yield: 20 mg (66 µmol, 33%) as a white solid.

Variant 2

To a solution of **137** (75 mg, 0.18 mmol, 1.0 eq) in toluene (3 mL), montmorillonite (200 mg) was added. The reaction mixture was heated to 65 °C, and it was stirred at this temperature for 50 min. After filtration and washing with EtOAc, the solvent was removed under reduced pressure. The resultant crude product was purified by flash chromatography (CH_2Cl_2/MeOH 97:3→95:5).

Yield: 28 mg (92 µmol, 46%) as a white solid.

^{1}H NMR (300 MHz, C_6D_6): δ [ppm] = 1.22 (s, 9H, *t*-Bu-CH_3), 1.35 (s, 9H, *t*-Bu-CH_3), 1.56-1.83 (m, 1H, H-3a), 1.93-2.25 (m, 3H, H-3b, H-4), 4.28-4.47 (m, 1H, H-2), 5.06 (d, $J = 9.7$ Hz, 1H, NH). **MS** (ESI): $m/z = 326.2$ [M+Na]$^+$. **HRMS** (ESI): calcd for $C_{14}H_{25}NO_6Na$ 326.1574, found 326.1571 [M+Na]$^+$. **TLC** (CH_2Cl_2/MeOH 95:5): $R_f = 0.19$.

7.2.5.7 Synthesis of *N,N*-Diboc-(2*S*,3*R*)-2-Amino-3-(benzyloxy)-5-hexenoic Acid *tert*-Butyl Ester 131

131

To a solution of **(*R*)-42** (51 mg, 0.13 mmol, 1.0 eq) in dry CH_2Cl_2 (2 mL) and dry triethylamine (3 mL), di-*tert*-butyl dicarbonate (120 mg, 0.550 mmol, 4.2 eq) and 4-dimethylaminopyridine (10 mg, 82 μmol, 0.6 eq) were added. The resulting mixture was heated at 75 °C for 3.5 h, concentrated under reduced pressure, and the residue was redissolved in EtOAc (10 mL). The organic layer was washed with 1 M aqueous $KHSO_4$ solution (10 mL), saturated aqueous $NaHCO_3$ solution (10 mL) and brine (10 mL), dried over Na_2SO_4, and the solvent was removed under reduced pressure. The obtained crude product was purified by flash chromatography (petroleum ether/EtOAc 20:1).

Yield: 56 mg (0.11 mmol, 85%) as white crystals.

^{1}H NMR (300 MHz, C_6D_6): δ [ppm] = 1.33 (s, 18H, *t*-Bu-CH_3), 1.36 (s, 9H, *t*-Bu-CH_3), 2.60-2.71 (m, 1H, H-4a), 2.75-2.85 (m, 1H, H-4b), 4.44-4.56 (m, 1H, H-3), 4.50 (s, 2H, CH_2-Ph), 5.11-5.20 (m, 2H, H-6), 5.22 (d, *J* = 8.6 Hz, 1H, H-2), 5.96-6.12 (m, 1H, H-5), 7.01-7.19 (m, 3H, Ph), 7.39-7.31 (m, 2H, Ph). **^{13}C NMR** (75 MHz, C_6D_6): δ [ppm] = 27.92 (*t*-Bu-CH_3), 28.00 (*t*-Bu-CH_3), 38.49 (C-4), 61.37 (C-2), 72.96 (CH_2-Ph), 76.63 (C-3), 81.13 (*t*-Bu-C), 82.01 (*t*-Bu-C), 117.69 (C-6), 127.45 (CH-Ph), 127.87 (CH-Ph), 128.10 (CH-Ph), 128.18 (CH-Ph), 135.00 (C-5), 139.43 (C-Ph), 153.38 (Boc-C=O), 168.52 (C-1). **MS** (ESI): *m/z* = 514.3 [M+Na]$^+$. **HRMS** (ESI): calcd for $C_{27}H_{41}NO_7Na$ 514.2775, found 514.2770 [M+Na]$^+$. $[\alpha]_D^{20}$ = -44.5 (*c* = 1.1, $CHCl_3$). **IR** (ATR): $\tilde{\nu}$ [cm^{-1}] = 2981, 1807, 1731, 1697, 1389, 1365, 1261, 1214, 1151, 1121, 1068, 1034, 848, 796, 703. **UV** (MeCN): λ_{max} [nm] (log ε) = 257 (2.62). **Mp** = 59 °C. **TLC** (petroleum ether/EtOAc 20:1): R_f = 0.15.

7.2.5.8 Synthesis of *N,N*-Diboc-(2*S*,3*R*)-3-(Benzyloxy)-L-glutamic Acid *tert*-Butyl Ester 133

133

A solution of **131** (100 mg, 0.203 mmol, 1.0 eq) in MeOH (26 mL), CH_2Cl_2 (3 mL) and pyridine (66 μL, 65 mg, 0.82 mmol, 4.0 eq) was cooled to -78°C, and ozone was bubbled through this solution at -78°C for 15 min. After the addition of dimethyl sulfide (300 μL, 252 mg, 4.06 mmol, 20 eq), the reaction was allowed to warm to room temperature overnight. The solution was tested with potassium iodide starch paper for remaining oxidizing species before the solvent was removed under reduced pressure. The resultant crude aldehyde **132** (85 mg) was dissolved in MeCN (2 mL), and a solution of sodium chlorite (80%, 100 mg, 0.885 mmol, 4.4 eq) in water (2 mL) was added dropwise at 0 °C. The resulting solution was treated with a solution of potassium dihydrogen phosphate (58 mg, 0.43 mmol, 2.1 eq) in water (1 mL) and hydrogen peroxide (30%, 50 μL, 56 mg, 1.6 mmol, 7.9 eq). During this addition, the temperature stayed under 10 °C. At this temperature, the mixture was stirred for 45 min, allowed to warm to room temperature and then stirred overnight. The reaction was quenched by the addition of solid Na_2SO_3. The resulting mixture was acidified with 1 M aqueous HCl solution, partitioned between brine (10 mL) and CH_2Cl_2 (20 mL), and the aqueous layer was extracted with CH_2Cl_2 (3 x 5 mL). The combined organics were washed with brine (15 mL), dried over Na_2SO_4, and the solvent was removed under reduced pressure. The resultant crude product was purified by flash chromatography (CH_2Cl_2/MeOH 20:1).

Yield: 69 mg (0.14 mmol, 69%) as a colorless oil.

¹H NMR (300 MHz, C_6D_6): δ [ppm] = 1.35 (s, 18H, *t*-Bu-CH₃), 1.36 (s, 9H, *t*-Bu-CH₃), 2.82 (dd, *J* = 16.1, 7.2 Hz, 1H, H-4a), 3.08 (dd, *J* = 16.1, 3.6 Hz, 1H, H-4b), 4.67 (d, *J* = 11.0 Hz, 1H, CH₂-Ph), 4.79 (d, *J* = 11.0 Hz, 1H, CH₂-Ph), 4.88-4.98 (m, 1H, H-3), 5.33 (d, *J* = 8.5 Hz, 1H, H-2), 7.03-7.21 (m, 3H, Ph), 7.36-7.43 (m, 2H, Ph). **¹³C NMR** (76 MHz, C_6D_6): δ [ppm] = 27.87 (*t*-Bu-CH₃), 27.95 (*t*-Bu-CH₃), 40.11 (C-4), 61.78 (C-2), 73.92 (CH₂-Ph), 74.58 (C-3), 81.58 (*t*-Bu-C), 82.30 (*t*-Bu-C), 127.51 (CH-Ph), 127.86 (CH-Ph), 128.20 (CH-Ph), 139.16 (C-Ph), 153.30 (Boc-C=O), 168.21 (C-1), 176.71 (C-5). **MS** (ESI): *m/z* = 532.3 [M+Na]⁺. **HRMS** (ESI): calcd for $C_{26}H_{39}NO_9Na$ 532.2517, found 532.2520 [M+Na]⁺. **TLC** (CH_2Cl_2/MeOH 20:1): R_f = 0.28.

7.2.5.9 Synthesis of *N*,*N*-Diboc-(2*S*,3*R*)-2-Amino-3-(benzyloxy)-5-hydroxy-valeric Acid *tert*-Butyl Ester 134

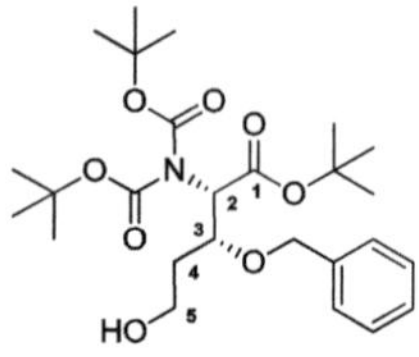

134

To a solution of **133** (42 mg, 82 µmol, 1.0 eq) in dry THF (1 mL), *N*-methylmorpholine (10 µL, 9.2 mg, 91 µmol, 1.1 eq), followed by isobutyl chloroformate (12 µL, 13 mg, 95 µmol, 1.2 eq) was added under an argon atmosphere at -15 °C. After stirring the mixture for 10 min at this temperature, sodium borohydride (20 mg, 0.53 mmol, 6.5 eq) was added, followed by a dropwise addition of dry MeOH (2 mL) over 15 min. The resulting mixture was stirred for an additional 15 min at -15 °C. The mixture was treated with 1 M aqueous HCl solution (200 µmol) and concentrated under reduced pressure before water (5 mL) and EtOAc (5 mL) were added. The aqueous layer was extracted with EtOAc (3 x 4 mL). The combined organics were washed with 1 M aqueous HCl solution (4 mL), water (2 x 3 mL), saturated aqueous NaHCO$_3$ solution (3 mL) and brine (3 mL), dried over Na$_2$SO$_4$, and the solvent was removed under reduced pressure. The resultant crude product was purified by flash chromatography (petroleum ether/EtOAc 3:1).

Yield: 15 mg (30 µmol, 37%) as a colorless oil.

^{1}H NMR (301 MHz, C$_6$D$_6$): δ [ppm] = 1.34 (s, 18H, *t*-Bu-CH$_3$), 1.38 (s, 9H, *t*-Bu-CH$_3$), 1.90-2.07 (m, 1H, H-4a), 2.10-2.24 (m, 1H, H-4b), 3.67-3.82 (m, 2H, H-5), 4.55-4.66 (m, 3H, H-3, CH$_2$-Ph), 5.28 (d, *J* = 8.2 Hz, 1H, H-2), 7.03-7.20 (m, 3H, Ph), 7.33-7.44 (m, 2H, Ph). **^{13}C NMR** (126 MHz, C$_6$D$_6$): δ [ppm] = 27.93 (*t*-Bu-CH$_3$), 27.98 (*t*-Bu-CH$_3$), 36.97 (C-4), 59.41 (C-5), 62.03 (C-2), 73.46 (CH$_2$-Ph), 75.48 (C-3), 81.30 (*t*-Bu-C), 82.23 (*t*-Bu-C), 127.61 (CH-Ph), 127.91 (CH-Ph), 128.29 (CH-Ph), 128.29 (CH-Ph), 139.20 (C-Ph), 153.47 (Boc-C=O), 168.71 (C-1). **MS** (ESI): *m/z* = 518.3 [M+Na]$^+$. **HRMS** (ESI): calcd for C$_{26}$H$_{41}$NO$_8$Na 518.2724, found 518.2723 [M+Na]$^+$. [α]$_D^{20}$ = -20.5 (*c* = 1.0, CHCl$_3$). **IR** (ATR): $\tilde{v}$ [cm^{-1}] = 2981, 1737, 1697, 1389, 1365, 1255, 1151, 1121, 1057, 854, 801, 737, 703. **UV** (MeCN): λ$_{max}$ [nm] (log ε) = 252 (2.76). **TLC** (petroleum ether/EtOAc 3.1): R$_f$ = 0.21.

7.2.5.10 Synthesis of *N*-Boc-2-Aminoethanol 199 by Reduction of *N*-Boc-Glycine

199

To a solution of *N*-Boc-glycine (102 mg, 0.582 mmol, 1.0 eq) in dry THF (3 mL), *N*-methylmorpholine (94 µL, 86 mg, 0.85 mmol, 1.5 eq), followed by isobutyl chloroformate (111 µL, 117 mg, 0.857 mmol, 1.5 eq) was added under an argon atmosphere at -20 °C. After stirring the mixture for 75 min at this temperature, sodium borohydride (69 mg, 1.8 mmol, 3.1 eq) was added, followed by a slow dropwise addition of dry MeOH (6 mL) at 0 °C. The resulting mixture was stirred for an additional 15 min at 0 °C. As TLC analysis (CH_2Cl_2/MeOH 20:1) indicated no complete conversion of the starting material, additional sodium borohydride (43 mg, 1.1 mmol, 1.9 eq) was added at 5 °C. The reaction was allowed to warm to room temperature, and after 17 h, more sodium borohydride (100 mg, 2.64 mmol, 4.5 eq) was added at room temperature. The reaction was stirred for an additional 22 h before the reaction mixture was quenched with water (5 mL) and concentrated under reduced pressure. The residue was partitioned between EtOAc (10 mL) and saturated aqueous NH_4Cl (5 mL) solution. The aqueous layer was extracted with EtOAc (2 x 7 mL). The combined organics were washed with saturated aqueous NH_4Cl solution (2 x 7 mL), water (7 mL), 5% aqueous $NaHCO_3$ solution (7 mL) and brine (7 mL), and then dried over Na_2SO_4. After removing the solvent under reduced pressure, the obtained crude product was purified by flash chromatography (petroleum ether/EtOAc 1:1).

Yield: 66 mg (0.41 mmol, 70%) as a colorless oil.

^{1}H NMR (300 MHz, DMSO-d_6): δ [ppm] = 1.37 (s, 9H, *t*-Bu-CH_3), 2.98 (dt, *J* = 6.1, 6.1 Hz, 2H, H-2), 3.36 (dt, *J* = 6.1, 6.1 Hz, 2H, H-1), 4.51 (t, *J* = 6.1 Hz, 1H, OH), 6.60 (s_{br}, 1H, NH). **^{13}C NMR** (75 MHz, DMSO-d_6): δ [ppm] = 28.18 (*t*-Bu-CH_3), 42.62 (C-2), 60.00 (C-1), 77.34 (*t*-Bu-C), 155.40 (Boc-C=O). **MS** (ESI): *m/z* = 184.1 $[M+Na]^+$. **HRMS** (ESI): calcd for $C_7H_{15}NO_3Na$ 184.0944, found 184.0938 $[M+Na]^+$. **TLC** (CH_2Cl_2/MeOH 20:1): R_f = 0.27. **TLC** (petroleum ether/EtOAc 1:1): R_f = 0.15.

7.2.6   Synthesis of [5',5'-²H₂]Uridine 89

7.2.6.1 Synthesis of 2',3'-*O*-Isopropylidene Uridine 71[211]

71

To a solution of uridine **1** (5.00 g, 20.5 mmol, 1.0 eq) in dry acetone (300 mL), 2,2-dimethoxypropane (7.60 mL, 6.40 g, 61.0 mmol, 3.0 eq) and concentrated sulfuric acid (97%, 1.20 mL, 2.21 g, 21.9 mmol, 1.1 eq) were added at room temperature under an argon atmosphere. The resulting red solution was stirred at room temperature for 20 h, quenched by the addition of triethylamine (8 mL), and the solvent was removed under reduced pressure. The obtained crude product was purified by flash chromatography (CH₂Cl₂/MeOH 95:5).

Yield: 5.52 g (19.4 mmol, 95%) as a white solid.

¹H NMR (300 MHz, DMSO-d₆, 35 °C): δ [ppm] = 1.29 (s, 3H, C(CH₃)₂), 1.49 (s, 3H, C(CH₃)₂), 3.50-3.65 (m, 2H, H-5'), 4.03-4.10 (m, 1H, H-4'), 4.75 (dd, J = 6.4, 3.6 Hz, 1H, H-3'), 4.89 (dd, J = 6.4, 2.7 Hz, 1H, H-2'), 5.03 (t, J = 5.2 Hz, 1H, 5'-OH), 5.62 (d, J = 8.0 Hz, 1H, H-5), 5.83 (d, J = 2.7 Hz, 1H, H-1'), 7.78 (d, J = 8.0 Hz, 1H, H-6), 11.32 (s, 1H, NH). **¹³C NMR** (75 MHz, DMSO-d₆, 35 °C): δ [ppm] = 25.12 (C(CH₃)₂), 26.97 (C(CH₃)₂), 61.22 (C-5'), 80.43 (C-3'), 83.62 (C-2'), 86.43 (C-4'), 91.06 (C-1'), 101.67 (C-5), 112.91 (C(CH₃)₂), 141.78 (C-6), 150.26 (C-4), 163.06 (C-2). **MS** (ESI): m/z = 307.1 [M+Na]⁺. **HRMS** (ESI): calcd for C₁₂H₁₆N₂O₆Na 307.0901, found 307.0901 [M+Na]⁺. [α]$_D^{25}$ = +18.8 (c = 1.0, MeOH). **IR** (KBr): $\tilde{v}$ [cm⁻¹] = 3307, 3246, 2935, 2860, 2773, 1776, 1704, 1618, 1468, 1288, 1076, 768, 570. **UV** (MeOH): λ_max [nm] (log ε) = 198 (4.65), 260 (3.99), 370 (2.40), 419 (2.28). **Mp** = 157 °C. **TLC** (CH₂Cl₂/MeOH 9:1): R_f = 0.48.

7.2.6.2 Synthesis of 2',3'-*O*-Isopropylidene Uridine-5'-carboxylic Acid 105[226]

105

To a solution of 2',3'-*O*-isopropylidene uridine **71** (6.33 g, 22.3 mmol, 1.0 eq) in MeCN (25 mL) and water (25 mL), BAIB (15.8 g, 40.0 mmol, 2.2 eq) and TEMPO (696 mg, 4.46 mmol, 0.2 eq) were added, and the reaction mixture was stirred at room temperature for 4 d. After removing the solvent, the resultant crude product was recrystallized from acetone.

Yield: 4.78 g (16.1 mmol, 72%) as a white solid.

^{1}H NMR (300 MHz, DMSO-d_6, 35 °C): δ [ppm] = 1.30 (s, 3H, C(CH$_3$)$_2$), 1.45 (s, 3H, C(CH$_3$)$_2$), 4.55 (d, J = 1.8 Hz, 1H, H-4'), 5.05-5.33 (m, 2H, H-3', H-2'), 5.60 (d, J = 8.0 Hz, 1H, H-5), 5.76 (s, 1H, H-1'), 7.77 (d, J = 8.0 Hz, 1H, H-6), 11.29 (s, 1H, NH), 12.70 (s$_{br}$, 1H, COOH). **^{13}C NMR** (125 MHz, DMSO-d_6, 35 °C): δ [ppm] = 24.74 (C($\underline{C}$H$_3$)$_2$), 26.42 (C($\underline{C}$H$_3$)$_2$), 83.60 (C-3'), 83.99 (C-2'), 86.73 (C-4'), 95.58 (C-1'), 101.23 (C-5), 112.00 ($\underline{C}$(CH$_3$)$_2$), 144.52 (C-6), 150.59 (C-4), 163.22 (C-2) 170.57 (COOH). **MS** (ESI): m/z = 297.1 [M-H]⁻. **HRMS** (ESI): calcd for C$_{12}$H$_{13}$N$_2$O$_7$ 297.0728, found 297.0730 [M-H]⁻. $[\alpha]_D^{25}$ = +15.1 (c = 1.0, MeOH). **IR** (KBr): $\tilde{\nu}$ [cm⁻¹] = 3164, 3087, 2938, 2615, 1702, 1467, 1315, 1128, 767, 560. **UV** (MeOH): λ$_{max}$ [nm] (log ε) = 198 (4.69), 260 (3.97), 370 (2.65), 419 (1.71). **Mp** = 181 °C. **TLC** (CH$_2$Cl$_2$/MeOH 9:1): R$_f$ = 0.05.

7.2.6.3 Synthesis of 2',3'-*O*-Isopropylidene Uridine 71 and 2',3'-*O*-Isopropylidene [5',5'-^{2}H$_2$]Uridine 167

71: R = H, **167**: R = D

71: To a solution of 2',3'-*O*-isopropylidene uridine-5'-carboxylic acid **105** (101 mg, 0.339 mmol, 1.0 eq) in dry THF (4 mL), *N*-methylmorpholine (55 µL, 51 mg, 0.50 mmol,

1.5 eq) and isobutyl chloroformate (65 µL, 68 mg, 0.50 mmol, 1.5 eq) were added at -20 °C. The reaction was allowed to reach a temperature of 0 °C over a period of 30 min before sodium borohydride (40 mg, 1.1 mmol, 3.2 eq) and dry MeOH (2 mL) were added. The mixture was allowed to reach room temperature and stirred at room temperature for 21 h. After the addition of water (2.5 mL), the solvent was removed under reduced pressure, and the resultant crude product was purified by flash chromatography (CH_2Cl_2/MeOH 95:5).

167: The deuterated derivative **167** was prepared in the same way as compound **71** with **105** (3.00 g, 10.1 mmol, 1.0 eq), *N*-methylmorpholine (1.66 mL, 1.53 g, 15.1 mmol, 1.5 eq), isobutyl chloroformate (1.96 mL, 2.06 g, 15.1 mmol, 1.5 eq), sodium borodeuteride (2.11 g, 50.3 mmol, 5.0 eq), dry THF (120 mL) and CD_3OD (9.7 mL). Before the addition of sodium borodeuteride, the mixture was allowed to reach 0 °C over a period of 1.5 h. The obtained crude product was purified by flash chromatography (CH_2Cl_2/MeOH 98:2→97:3→95:5→90:10).

Yield 71: 77 mg (0.27 mmol, 80%) as a yellowish solid.
Yield 167: 2.13 g (7.46 mmol, 74%) as a white solid.

71: **^{1}H NMR** (300 MHz, DMSO-d_6, 35 °C): δ [ppm] = 1.29 (s, 3H, C(CH$_3$)$_2$), 1.49 (s, 3H, C(CH$_3$)$_2$), 3.50-3.65 (m, 2H, H-5'), 4.03-4.10 (m, 1H, H-4'), 4.75 (dd, J = 6.4, 3.6 Hz, 1H, H-3'), 4.89 (dd, J = 6.4, 2.7 Hz, 1H, H-2'), 5.03 (t, J = 5.2 Hz, 1H, 5'-OH), 5.62 (d, J = 8.0 Hz, 1H, H-5), 5.83 (d, J = 2.7 Hz, 1H, H-1'), 7.78 (d, J = 8.0 Hz, 1H, H-6), 11.32 (s, 1H, NH). **^{13}C NMR** (75 MHz, DMSO-d_6, 35 °C): δ [ppm] = 25.12 (C($\underline{C}$H$_3$)$_2$), 26.97 (C($\underline{C}$H$_3$)$_2$), 61.22 (C-5'), 80.43 (C-3'), 83.62 (C-2'), 86.43 (C-4'), 91.06 (C-1'), 101.67 (C-5), 112.91 ($\underline{C}$(CH$_3$)$_2$), 141.78 (C-6), 150.26 (C-4), 163.06 (C-2). **MS** (ESI): m/z = 307.1 [M+Na]$^+$. **HRMS** (ESI): calcd for $C_{12}H_{16}N_2O_6Na$ 307.0901, found 307.0901 [M+Na]$^+$. **[α]$_D^{25}$** = +18.8 (c = 1.0, MeOH). **IR** (KBr): $\tilde{\nu}$ [cm^{-1}] = 3307, 3246, 2935, 2860, 2773, 1776, 1704, 1618, 1468, 1288, 1076, 768, 570. **UV** (MeOH): λ_{max} [nm] (log ε) = 198 (4.65), 260 (3.99), 370 (2.40), 419 (2.28). **Mp** = 157 °C. **TLC** (CH_2Cl_2/MeOH 9:1): R$_f$ = 0.48.

167: **^{1}H NMR** (300 MHz, DMSO-d_6, 35 °C): δ [ppm] = 1.28 (s, 3H, C(CH$_3$)$_2$), 1.48 (s, 3H, C(CH$_3$)$_2$), 4.06 (d, J = 3.4 Hz, 1H, H-4'), 4.75 (dd, J = 6.4, 3.4 Hz, 1H, H-3'), 4.88 (dd, J = 6.4, 2.7 Hz, 1H, H-2'), 4.99 (s, 1H, 5'-OH), 5.62 (d, J = 8.0 Hz, 1H, H-5), 5.83 (d, J = 2.7 Hz, 1H, H-1'), 7.77 (d, J = 8.0 Hz, 1H, H-6), 11.30 (s, 1H, NH). **^{13}C NMR** (125 MHz, DMSO-d_6, 35 °C): δ [ppm] = 25.14 (C($\underline{C}$H$_3$)$_2$), 27.00 (C($\underline{C}$H$_3$)$_2$), 60.08-60.42 (m, C-5'), 80.46 (C-3'), 83.67 (C-2'), 86.38 (C-4'), 91.15 (C-1'), 101.70 (C-5), 112.95 ($\underline{C}$(CH$_3$)$_2$), 141.84 (C-6), 150.31 (C-4), 163.13 (C-2). **MS** (ESI): m/z = 309.1 [M+Na]$^+$. **HRMS** (ESI): calcd for $C_{12}H_{14}D_2N_2O_6Na$ 309.1026, found 309.1032 [M+Na]$^+$. **[α]$_D^{25}$** = +21.7 (c = 1.0, MeOH). **IR** (KBr): $\tilde{\nu}$ [cm^{-1}] = 3307, 3245, 2985, 2934, 2183, 2086,

1617, 1472, 1394, 1303, 1075, 773, 529. **UV** (MeOH): λ_{max} [nm] (log ε) = 198 (4.67), 262 (4.03), 377 (2.33), 419 (2.22). **Mp** = 156 °C. **TLC** (CH_2Cl_2/MeOH 9:1): R_f = 0.43.

7.2.6.4 Synthesis of Uridine 1 and [5',5'-²H₂]Uridine 89

1: R = H, **89**: R = D

1: A solution of 2',3'-isopropylidene uridine **71** (61 mg, 0.21 mmol, 1.0 eq) in 80% aqueous trifluoroacetic acid (1.5 mL) was stirred at room temperature for 19 h. The solvent was removed under reduced pressure, and the residue was coevaporated with EtOH (5 x 2 mL). The resultant crude product was purified by flash chromatography (CH_2Cl_2/MeOH 8:2→7:3).

89: The deuterated derivative **89** was prepared in the same way as compound **1** with **167** (1.86 g, 6.50 mmol, 1.0 eq) and 80% aqueous trifluoroacetic acid (20 mL). The resulting mixture was stirred at room temperature for 4 h and then diluted with toluene (10 mL) before the solvent was removed under reduced pressure. The resultant crude product was purified by flash chromatography (CH_2Cl_2/MeOH 90:10→85:15→80:20→70:30).

Yield 1: 49 mg (0.20 mmol, 94%) as a yellowish solid.
Yield 89: 1.31 g (5.33 mmol, 82%) as a white solid.

1: **¹H NMR** (300 MHz, DMSO-d₆, 35 °C): δ [ppm] = 3.42-3.74 (m, 2H, H-5'), 3.74-3.88 (m, 1H, H-4'), 3.88-4.12 (m, 2H, H-2', H-3'), 4.95-5.19 (m, 2H, 3'-OH, 5'-OH), 5.47-5.25 (m, 1H, 2'-OH), 5.62 (d, J = 8.1 Hz, 1H, H-5), 5.76 (d, J = 5.3 Hz, 1H, H-1'), 7.86 (d, J = 8.1 Hz, 1H, H-6), 11.29 (s, 1H, NH). **MS** (ESI): m/z = 267.1 [M+Na]⁺. **HRMS** (ESI): calcd for $C_9H_{12}N_2O_6Na$ 267.0588, found 267.0586 [M+Na]⁺. **TLC** (CH_2Cl_2/MeOH 8:2): R_f = 0.28.

89: **¹H NMR** (300 MHz, DMSO-d₆, 35 °C): δ [ppm] = 3.83 (d, J = 3.9 Hz, 1H, H-4'), 3.90-4.08 (m, 2H, H-2', H-3'), 4.99 (s_br, 2H, 3'-OH, 5'-OH), 5.31 (s_br, 1H, 2'-OH), 5.63 (d, J = 8.1 Hz, 1H, H-5), 5.78 (d, J = 5.4 Hz, 1H, H-1'), 7.87 (d, J = 8.1 Hz, 1H, H-6), 11.24 (s_br, 1H, NH). **¹³C NMR** (75 MHz, DMSO-d₆, 35 °C): δ [ppm] = 59.48-60.97 (m, C-5'), 69.85 (C-3'), 73.52 (C-2'), 84.73 (C-4'), 87.77 (C-1'), 101.73 (C-5), 140.73 (C-6), 150.74 (C-4), 163.09 (C-2). **MS** (ESI): m/z = 269.1 [M+Na]⁺. **HRMS** (ESI): calcd for

$C_9H_{10}D_2N_2O_6Na$ 269.0713, found 269.0710 $[M+Na]^+$. $[\alpha]_D^{25} = +2.8$ ($c = 1.0$, MeOH). **IR** (KBr): $\tilde{\nu}$ $[cm^{-1}] = 3350, 3104, 2963, 2801, 2228, 2169, 2103, 1680, 1470, 1270, 1132, 1058, 766, 570$. **UV** (MeOH): λ_{max} [nm] (log ε) = 198 (4.63), 261 (3.98), 369 (2.59), 419 (2.25). **Mp** = 152 °C. **TLC** (CH_2Cl_2/MeOH 9:1): $R_f = 0.07$.

7.2.7 Synthesis of Nucleosyl Amino Acid Derivatives for Biosynthetic Studies

7.2.7.1 Synthesis of (6'*S*)- and (6'*R*)-Nucleosyl Amino Acids (*S*)-90 and (*R*)-90

(6'*R*): (***R***)**-90**, (6'*S*): (***S***)**-90**

(***R***)**-90**: A solution of protected nucleosyl amino acid derivative (***R***)**-78** (21 mg, 33 µmol, 1.0 eq) in 80% aqueous TFA (4.9 mL) was stirred at room temperature for 24 h. The reaction mixture was diluted with water (20 mL), and the solvent was removed under reduced pressure. The resultant crude product was purified by semi-preparative HPLC (method SP-01, see chapter 7.1.5) and then lyophilized.

(***S***)**-90**: The isomer (***S***)**-90** was prepared in the same way as compound (***R***)**-90** with (***S***)**-78** (22 mg, 34 µmol, 1.0 eq) and 80% aqueous TFA (5.1 mL).

Yield (***R***)**-90**: 12 mg (20 µmol, 61%) as a white fluffy solid.
Yield (***S***)**-90**: 11 mg (19 µmol, 56%) as a white fluffy solid.

(***R***)**-90**: **^{1}H NMR** (301 MHz, D_2O, 35 °C): δ [ppm] = 2.06-2.21 (m, 2H, H-2''), 2.32 (ddd, J = 15.6, 9.4, 5.2 Hz, 1H, H-5'a), 2.50 (ddd, J = 15.6, 5.2, 3.0 Hz, 1H, H-5'b), 3.13 (dd, J = 7.7, 7.7 Hz, 2H, H-3''), 3.22 (dd, J = 9.2, 6.3 Hz, 2H, H-1''), 3.98 (dd, J = 5.2, 5.2 Hz, 1H, H-6'), 4.04-4.20 (m, 2H, H-3', H-4'), 4.46 (dd, J = 5.5, 3.6 Hz, 1H, H-2'), 5.78 (d, J = 3.6 Hz, 1H, H-1'), 5.94 (d, J = 8.1 Hz, 1H, H-5), 7.70 (d, J = 8.1 Hz, 1H, H-6). **^{13}C NMR** (126 MHz, D_2O, 35 °C): δ [ppm] = 26.30 (C-2''), 34.94 (C-5'), 39.17 (C-3''), 46.89 (C-1''), 63.14 (C-6'), 75.12 (C-2'), 75.32 (C-3'), 82.09 (C-4'), 94.70 (C-1'), 104.90 (C-5), 118.98 (q, J = 291.8 Hz, $\underline{C}F_3COO$), 145.60 (C-6), 154.08 (C-2), 165.46 (q, J = 35.6 Hz, $CF_3\underline{C}OO$), 168.74 (C-4), 174.66 (C-7'). **^{19}F NMR** (282 MHz, D_2O, 35 °C): δ [ppm] = -75.43 (CF_3). **MS** (ESI): m/z = 359.2 $[M+H]^+$. **HRMS** (ESI): calcd for $C_{14}H_{23}N_4O_7$ 359.1561, found 359.1568 $[M+H]^+$. $[\alpha]_D^{20} = +54.6$ ($c = 0.48$, H_2O).

IR (ATR): $\tilde{\nu}$ [cm⁻¹] = 1673, 1633, 1470, 1423, 1389, 1272, 1197, 1121, 1064, 837, 801, 760, 726. **UV** (H_2O): λ_{max} [nm] (log ε) = 204 (4.57), 263 (4.67). **Mp** = 165 °C. **HPLC** (analytical): t_R = 2.0 min (method A-01, injection 5 μL (c ~ 1 mg/mL, in MeOH). **HPLC** (semi-preparative): t_R = 4.3 min (method SP-01, injection volume 100 μL, concentration of injection: 15 mg in H_2O (0.8 mL)).

(S)-90: **¹H NMR** (301 MHz, D_2O, 35 °C): δ [ppm] = 2.09-2.23 (m, 2H, H-2"), 2.38 (ddd, J = 15.1, 10.0, 5.5 Hz, 1H, H-5'a), 2.53 (ddd, J = 15.1, 7.0, 2.9 Hz, 1H, H-5'b), 3.14 (dd, J = 7.7, 7.7 Hz, 2H, H-3"), 3.22 (dd, J = 7.8, 7.8 Hz, 2H, H-1"), 4.09-4.24 (m, 3H, H-3', H-4', H-6'), 4.46 (dd, J = 5.5, 3.7 Hz, 1H, H-2'), 5.80 (d, J = 3.7 Hz, 1H, H-1'), 5.93 (d, J = 8.1 Hz, 1H, H-5), 7.67 (d, J = 8.1 Hz, H-6). **¹³C NMR** (126 MHz, D_2O, 35 °C): δ [ppm] = 26.38 (C-2"), 35.15 (C-5'), 39.11 (C-3"), 46.48 (C-1"), 61.51 (C-6'), 75.25 (C-2'), 75.53 (C-3'), 81.93 (C-4'), 94.32 (C-1'), 104.84 (C-5), 118.93 (q, J = 291.7 Hz, $\underline{C}F_3COO$), 145.22 (C-6), 154.02 (C-2), 165.43 (q, J = 35.5 Hz, $CF_3\underline{C}OO$), 168.75 (C-4), 173.87 (C-7'). **¹⁹F NMR** (283 MHz, D_2O, 35 °C): δ [ppm] = -75.43 (CF_3). **MS** (ESI): m/z = 359.2 [M+H]⁺. **HRMS** (ESI): calcd for $C_{14}H_{23}N_4O_7$ 359.1561, found 359.1563 [M+H]⁺. $[\alpha]_D^{20}$ = +52.5 (c = 0.48, H_2O). **IR** (ATR): $\tilde{\nu}$ [cm⁻¹] = 1672, 1638, 1468, 1423, 1393, 1269, 1200, 1130, 1086, 837, 802, 763, 723. **UV** (H_2O): λ_{max} [nm] (log ε) = 204 (4.78), 262 (4.98). **Mp** = 158 °C. **HPLC** (analytical): t_R = 1.9 min (method A-01, injection 5 μL (c ~ 1 mg/mL, in MeOH). **HPLC** (semi-preparative): t_R = 4.5 min (method SP-01, injection volume 100 μL, concentration of injection: 13 mg in H_2O (0.8 mL)).

7.2.7.2 Synthesis of Methyl (E)-1',5',6'-Trideoxy-2',3'-O-isopropylidene-1'-(uracil-1-yl)-β-D-ribo-5'-eneheptofuranuronate 72[150]

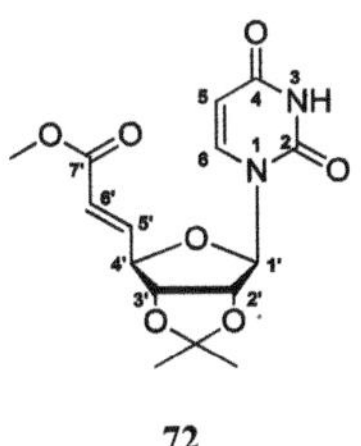

72

To a solution of 2',3'-O-isopropylidene uridine **71** (104 mg, 0.366 mmol, 1.0 eq) in dry MeCN (4 mL), 2-iodoxybenzoic acid (IBX, 260 mg, 0.929 mmol, 2.5 eq) was added under an argon atmosphere and heated at 80 °C for 1 h. The reaction mixture was cooled in an ice bath and the resulting white precipitates were removed by filtration through a Celite pad. The pad was rinsed with EtOAc, and the combined filtrates were concentrated under reduced pressure. The residue was redissolved in dry CH_2Cl_2 (3.5 mL) under an argon atmosphere and cooled to -20 °C before a solution of methyl (triphenylphosphoranylidene)acetate (150 mg, 0.449 mmol, 1.2 eq) in dry CH_2Cl_2 (1 mL) was added dropwise. After stirring the mixture for 1 h at -20 °C, it was partitioned between EtOAc (15 mL) and water (7 mL). The organic layer was washed with brine (7 mL), dried over Na_2SO_4, and the solvent was removed under reduced pressure. The resultant crude product was purified by flash chromatography (isohexane/EtOAc 3:1→11:9→1:1).

Yield: 51 mg (0.15 mmol, 41%) as a colorless viscous oil.

^{1}H NMR (500 MHz, CDCl$_3$): δ [ppm] = 1.35 (s, 3H, C(CH$_3$)$_2$), 1.58 (s, 3H, C(CH$_3$)$_2$), 3.74 (s, 3H, OCH$_3$), 4.66 (ddd, J = 5.9, 4.5, 1.6 Hz, 1H, H-4'), 4.85 (dd, J = 6.4, 4.5 Hz, 1H, H-3'), 5.08 (dd, J = 6.4, 1.7 Hz, 1H, H-2'), 5.63 (d, J = 1.7 Hz, 1H, H-1'), 5.75 (dd, J = 8.0, 1.7 Hz, 1H, H-5), 6.03 (dd, J = 15.8, 1.6 Hz, 1H, H-6'), 7.01 (dd, J = 15.8, 5.9 Hz, 1H, H-5'), 7.20 (d, J = 8.0 Hz, 1H, H-6), 9.04 (s$_{br}$, 1H, NH). **^{13}C NMR** (126 MHz, CDCl$_3$): δ [ppm] = 25.28 (C(CH$_3$)$_2$), 27.11 (C(CH$_3$)$_2$), 51.76 (OCH$_3$), 84.06 (C-3'), 84.47 (C-2'), 86.84 (C-4'), 95.21 (C-1'), 102.85 (C-5), 114.84 (C(CH$_3$)$_2$), 122.25 (C-6'), 142.58 (C-6), 143.71 (C-5'), 149.79 (C-2), 163.00 (C-4), 166.15 (C-7'). **HRMS** (ESI): calcd for $C_{15}H_{18}N_2O_7Na$ 361.1006, found 361.1004 [M+Na]$^+$. $[\alpha]_D^{20}$ = +40.8 (c = 1.2, CHCl$_3$). **IR** (ATR): $\tilde{\nu}$ [cm^{-1}] = 1684, 1633, 1459, 1435, 1382, 1266, 1214, 1174, 1162, 1068, 987, 883, 860, 819, 732. **UV** (MeCN): λ$_{max}$ [nm] (log ε) = 205 (4.53), 257 (4.22). **TLC** (isohexane/ EtOAc 1:3): R$_f$ = 0.27.

7.2.7.3 Synthesis of Amino Alcohol Derivative 73[150]

73

Preparation of tert-*butyl hypochlorite*:

Aqueous bleach solution (5%, 100 mL) was cooled to a temperature of under 10 °C. At this temperature, *tert*-butanol (8 mL) and acetic acid (4.5 mL) were added in the dark under vigorous stirring. After an additional 3 min of stirring at room temperature, the layers were separated, and the oily organic layer was washed with saturated aqueous Na_2CO_3 solution (20 mL) and water (20 mL). The resultant *tert*-butyl hypochlorite was dried over $CaCl_2$, filtered and stored under an argon atmosphere over $CaCl_2$ at 4 °C with the exclusion of light.

Sharpless asymmetric aminohydroxylation:

To a solution of benzyl carbamate (730 mg, 4.83 mmol, 3.0 eq) in 0.6 M aqueous NaOH solution (8 mL) and *n*-propanol (8 mL), *tert*-butyl hypochlorite (800 µL, 768 mg, 7.07 mmol, 4.4 eq) was added dropwise at 0 °C. After 10 min of stirring at 0 °C, the ice bath was exchanged with a water bath (T ~ 15 °C). A solution of olefin **72** (542 mg, 1.60 mmol, 1.0 eq) in *n*-propanol (8 mL), [DHQD]$_2$AQN (206 mg, 0.240 mmol, 0.15 eq) and K$_2$OsO$_2$(OH)$_4$ (88 mg, 0.24 mmol, 0.15 eq) were sequentially added. The resulting mixture was stirred at 15 °C for 6 h and then quenched by the addition of saturated aqueous Na$_2$S$_2$O$_3$ solution (10 mL). The mixture was extracted with EtOAc (3 x 20 mL). The organic layer was washed with brine (1 x 20 mL), dried over Na$_2$SO$_4$ and concentrated under reduced pressure. The residue was directly applied to a silica gel column and purified by flash chromatography (isohexane/EtOAc 1:1→1:3→1:4).

Yield: 380 mg (0.752 mmol, 47%) as a white foam.

^{1}H NMR (500 MHz, CDCl$_3$): δ [ppm] = 1.32 (s, 3H, C(CH$_3$)$_2$), 1.53 (s, 3H, C(CH$_3$)$_2$), 3.63-3.72 (m, 4H, OH, OCH$_3$), 4.11-4.33 (m, 2H, H-4', H-5'), 4.52-4.59 (m, 1H, H-6'), 4.91 (dd, *J* = 6.4, 3.7 Hz, 1H, H-3'), 4.96 (dd, *J* = 6.4, 2.9 Hz, 1H, H-2'), 5.06 (d, *J* = 11.4 Hz, 1H, Cbz-CH$_2$), 5.12 (d, *J* = 11.4 Hz, 1H, Cbz-CH$_2$), 5.46-5.54 (m, 1H, H-1'), 5.67 (d, *J* = 7.9 Hz, 1H, H-5), 5.88 (d, *J* = 8.1 Hz, 1H, NH-6'), 7.22-7.37 (m, 6H, H-6, Ph), 9.45 (s$_{br}$, 1H, NH-3). **^{13}C NMR** (126 MHz, CDCl$_3$): δ [ppm] = 25.30 (C($\underline{C}$H$_3$)$_2$), 27.22 (C($\underline{C}$H$_3$)$_2$), 52.73 (OCH$_3$), 56.64 (C-6'), 67.14 (Cbz-CH$_2$), 71.39 (C-5'), 81.44 (C-3'), 82.85 (C-2'), 86.19 (C-4'), 95.93 (C-1'), 102.78 (C-5), 114.81 ($\underline{C}$(CH$_3$)$_2$), 128.16 (CH-Ph), 128.19 (CH-Ph), 128.51 (CH-Ph), 136.23 (C-Ph), 143.03 (C-6), 150.45 (C-2), 156.34 (Cbz-C=O), 163.19 (C-4), 170.88 (C-7'). **MS** (ESI): *m/z* = 528.2 [M+Na]$^+$. **HRMS** (ESI): calcd for C$_{23}$H$_{27}$N$_3$O$_{10}$Na 528.1589, found 528.1590 [M+Na]$^+$. **IR** (ATR): $\tilde{v}$ [cm^{-1}] = 2958, 2922, 1684, 1522, 1464, 1376, 1261, 1214, 1081, 1022, 860, 801, 760, 697. **UV** (MeCN): λ$_{max}$ [nm] (log ε) = 205 (4.40), 259 (3.99). **Mp** = 96 °C. **TLC** (isohexane/EtOAc 1:3): R$_f$ = 0.15.

7.2.7.4 Synthesis of 5'-*O*-Glycosylated Nucleosyl Amino Acid Derivative β-75[150]

β-75

A mixture of **73** (46 mg, 91 µmol, 1.0 eq) in dry CH_2Cl_2 (2.5 mL) and molecular sieves (380 mg, 3 Å) was cooled to 0 °C, and a solution of azide **74**[4] (27 mg, 0.12 mmol, 1.3 eq) in dry CH_2Cl_2 (330 µL) was added under an argon atmosphere. The mixture was stirred at 0 °C for 1 h before boron trifluoride diethyl etherate (0.2 M in dry CH_2Cl_2, 190 µL, 0.124 mmol, 1.4 eq) was added. After 2.5 h of stirring at 0 °C, additional boron trifluoride diethyl etherate (0.2 M in dry CH_2Cl_2, 190 µL, 0.124 mmol, 1.4 eq) was added. After an additional hour of stirring, the reaction was quenched at 0 °C with saturated aqueous $NaHCO_3$ solution (5 mL) and allowed to warm to room temperature. The molecular sieves were removed by filtration and rinsed with CH_2Cl_2 (3 x 15 mL). The combined organics were washed with brine (15 mL), dried over Na_2SO_4, and the solvent was removed under reduced pressure. The resultant crude product was obtained as a mixture of α- and β-anomer, which could be purified and separated by flash chromatography (isohexane/EtOAc 65:35→60:40→55:45→50:50→40:60).

Yield β-75: 34 mg (47 µmol, 52%) as a white foam.
Yield α-75: 4 mg (5 µmol, 5%) as a white foam.

β-75: **^{1}H NMR** (500 MHz, $CDCl_3$): δ [ppm] = 0.76-0.99 (m, 6H, H-1'''), 1.30 (s, 3H, $C(CH_3)_2$), 1.44-1.81 (m, 4H, H-2'''), 1.50 (s, 3H, $C(CH_3)_2$), 3.34 (dd, J = 12.8, 5.8 Hz, 1H, H-5"a), 3.40 (dd, J = 12.8, 5.8 Hz, 1H, H-5"b), 3.77 (s, 3H, OCH_3), 4.22 (dd, J = 6.9, 3.9 Hz, 1H, H-4'), 4.25 (dd, J = 5.8, 5.8 Hz, 1H, H-4"), 4.46 (d, J = 6.9, 1H, H-5'), 4.58-4.62 (m, 2H, H-2", H3"), 4.66 (d, J = 9.8 Hz, 1H, H-6'), 4.77-4.84 (m, 2H, H-2', H-3'), 5.05 (d, J = 12.2 Hz, 1H, Cbz-CH_2), 5.14 (s, 1H, H-1"), 5.23 (d, J = 12.2 Hz, 1H, Cbz-CH_2), 5.66 (d, J = 2.3 Hz, 1H, H-1'), 5.70 (d, J = 8.0 Hz, 1H, H-5), 5.82 (d, J = 9.8 Hz, 1H, NH-6'), 7.28-7.39 (m, 6H, H-6, Ph), 8.80 (s$_{br}$, 1H, NH-3). **^{13}C NMR** (126 MHz, $CDCl_3$): δ [ppm] = 7.38 (C-1'''), 8.33 (C-1'''), 25.38 (C($\underline{C}H_3$)$_2$), 27.12 (C($\underline{C}H_3$)$_2$), 28.95 (C-2'''), 29.39 (C-2'''), 52.84 (OCH_3), 53.19 (C-5"), 54.48 (C-6'), 67.26 (Cbz-CH_2), 78.55 (C-5'), 80.64 (C-3'), 81.91 (C-3"), 83.69 (C-2'), 85.34 (C-4"), 86.06 (C-2"), 86.33 (C-4'), 93.55 (C-1'), 102.58 (C-5), 111.34 (C-1"), 115.04 ($\underline{C}(CH_3)_2$), 117.35 (C-3'''), 128.21 (CH-Ph), 128.25 (CH-Ph), 128.49 (CH-Ph), 136.29 (C-Ph), 142.10 (C-6), 149.87 (C-2), 156.29 (Cbz-C=O), 162.90 (C-4), 170.47 (C-7'). **HRMS** (ESI): calcd for $C_{33}H_{42}N_6O_{13}Na$ 753.2702, found 753.2700 $[M+Na]^+$. $[\alpha]_D^{20}$ = -1.9 (c = 1.6, $CHCl_3$). **IR** (ATR): $\tilde{\nu}$ [cm^{-1}] = 2929, 2103, 1691, 1516, 1459, 1383, 1261, 1214, 1168, 1068, 1040, 1011, 924, 813, 697. **UV** (MeCN): λ_{max} [nm] (log ε) = 205 (4.23), 258 (3.93). **Mp:** 79 °C. **TLC** (isohexane/EtOAc 1:3): R_f = 0.44.

α-75: **^{1}H NMR** (500 MHz, $CDCl_3$): δ [ppm] = 0.78-0.99 (m, 6H, H-1'''), 1.35 (s, 3H, $C(CH_3)_2$), 1.47-1.78 (m, 4H, H-2'''), 1.52 (s, 3H, $C(CH_3)_2$), 3.36 (dd, J = 12.8, 7.6 Hz, 1H,

---

[4] For the preparation of azide **74**, see D. Wiegmann, Master thesis, Georg-August-University Göttingen, **2012**, pp 93-96,[212] and A. Spork, PhD thesis, Georg-August-University Göttingen, **2012**, pp 427-431.[152]

H-5"a), 3.55 (dd, J = 12.8, 6.1 Hz, 1H, H-5"b), 3.78 (s, 3H, OCH$_3$), 4.08 (dd, J = 8.6, 3.2 Hz, 1H, H-4'), 4.31 (ddd, J = 7.6, 6.1, 1.2 Hz, 1H, H-4"), 4.47 (d, J = 8.6 Hz, 1H, H-5'), 4.50-4.64 (m, 3H, H-2", H3", H-6'), 5.05-5.21 (m, 5H, H-2', H-3', H-1", Cbz-CH$_2$), 5.35 (d, J = 10.2 Hz, 1H, NH-6'), 5.48 (s, 1H, H-1'), 5.71 (d, J = 8.0 Hz, 1H, H-5), 7.21 (d, J = 8.0 Hz, 1H, H-6), 7.28-7.39 (m, 5H, H-6, Ph), 8.22 (s$_{br}$, 1H, NH-3). **^{13}C NMR** (126 MHz, CDCl$_3$): δ [ppm] = 7.43 (C-1'''), 8.33 (C-1'''), 25.20 (C($\underline{C}$H$_3$)$_2$), 26.96 (C($\underline{C}$H$_3$)$_2$), 28.94 (C-2'''), 29.40 (C-2'''), 52.83 (OCH$_3$), 53.56 (C-5"), 55.09 (C-6'), 67.36 (Cbz-CH$_2$), 80.44 (C-5'), 82.14 (C-3'), 82.22 (C-3"), 84.22 (C-2'), 85.55 (C-4"), 86.10 (C-2"), 86.40 (C-4'), 97.02 (C-1'), 102.71 (C-5), 111.60 (C-1"), 114.19 ($\underline{C}$(CH$_3$)$_2$), 117.34 (C-3'''), 128.19 (CH-Ph), 128.28 (CH-Ph), 128.56 (CH-Ph), 136.03 (C-Ph), 143.40 (C-6), 149.66 (C-2), 156.29 (Cbz-C=O), 162.44 (C-4), 171.14 (C-7'). **HRMS** (ESI): calcd for C$_{33}$H$_{42}$N$_6$O$_{13}$Na 753.2702, found 753.2699 [M+Na]$^+$. [α]$_D^{20}$ = -20.0 (c = 0.52, CHCl$_3$). **IR** (ATR): $\tilde{v}$ [cm^{-1}] = 2935, 2109, 1714, 1697, 1522, 1459, 1376, 1261, 1214, 1064, 1040, 1011, 924, 813, 703. **UV** (MeCN): λ$_{max}$ [nm] (log ε) = 205 (4.59), 257 (4.30). **Mp** = 79 °C. **TLC** (isohexane/EtOAc 1:3): R$_f$ = 0.49.

7.2.7.5 Synthesis of Partially Deprotected Nucleosyl Amino Acid Derivative 169

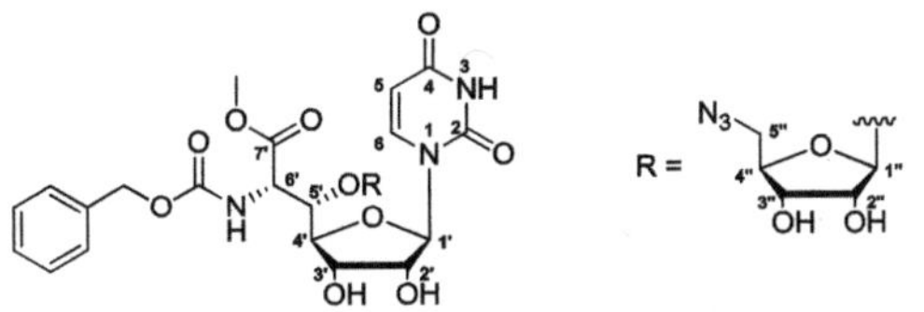

169

A solution of protected nucleosyl amino acid derivative **β-75** (96 mg, 0.13 mmol, 1.0 eq) in 80% aqueous TFA (2 mL) was stirred at room temperature for 8 h. The reaction mixture was diluted with water (15 mL), and the solvent was removed under reduced pressure. The resultant crude product was purified by flash chromatography (CH$_2$Cl$_2$/MeOH 9:1→7:1).

Yield: 68 mg (0.11 mmol, 85%) as a white solid.

^{1}H NMR (500 MHz, CD$_3$OD): δ [ppm] = 3.65 (dd, J = 13.3, 3.6 Hz, 1H, H-5"a), 3.74 (dd, J = 13.3, 3.6 Hz, 1H, H-5"b), 3.79 (s, 3H, OCH$_3$), 3.93 (d, J = 4.7 Hz, 1H, H-2"), 4.00 (ddd, J = 7.5, 3.6, 3.6 Hz, 1H, H-4"), 4.05-4.13 (m, 3H, H-3', H-2', H3"), 4.15 (dd, J = 5.6, 2.9 Hz, 1H, H-4'), 4.49 (dd, J = 2.9, 1.8 Hz, 1H, H-5'), 4.75 (d, J = 1.8 Hz, 1H, H-6'), 4.92 (d, J = 12.3 Hz, 1H, Cbz-CH$_2$), 5.00 (s, 1H, H-1"), 5.28 (d, J = 12.3 Hz, 1H, Cbz-CH$_2$), 5.62-5.66 (m, 2H, H-1', H-5), 7.24-7.40 (m, 5H, Ph), 7.82 (d, J = 8.1 Hz, 1H, H-6). **^{13}C NMR** (126 MHz, CD$_3$OD): δ [ppm] = 51.50 (C-5"), 51.79 (OCH$_3$), 55.85 (C-6'), 66.44

(Cbz-CH$_2$), 70.25 (C-3'), 70.38 (C-3"), 73.96 (C-2'), 75.06 (C-2"), 77.60 (C-5'), 80.66 (C-4"), 85.48 (C-4'), 90.14 (C-1'), 100.80 (C-5), 109.37 (C-1"), 127.79 (CH-Ph), 127.81 (CH-Ph), 127.88 (CH-Ph), 136.70 (C-Ph), 140.49 (C-6), 150.67 (C-2), 156.87 (Cbz-C=O), 164.72 (C-4), 170.92 (C-7'). **HRMS** (ESI): calcd for C$_{25}$H$_{30}$N$_6$O$_{13}$Na 645.1763, found 645.1771 [M+Na]$^+$. **[α]$_D^{20}$** = +49.8 (c = 1.2, MeOH). **IR** (ATR): $\tilde{v}$ [cm^{-1}] = 3318, 2109, 1673, 1546, 1435, 1272, 1197, 1132, 1068, 1034, 964, 801, 720. **UV** (MeOH): λ_{max} [nm] (log ε) = 206 (4.27), 264 (3.97). **Mp** = 166 °C (decomposition). **TLC** (CH$_2$Cl$_2$/MeOH 7:1): R$_f$ = 0.14.

7.2.7.6 Synthesis of Partially Deprotected Nucleosyl Amino Acid Derivative 174

174

A solution of nucleosyl amino acid derivative β-75 (85 mg, 0.12 mmol, 1.0 eq) and barium hydroxide octahydrate (37 mg, 0.12 mmol, 1.0 eq) in THF (4 mL) and water (1 mL) was stirred at room temperature for 10 h. Aqueous HCl solution (1 M, 20 mL) was added, and the aqueous layer was extracted with EtOAc (3 x 20 mL). The combined organics were washed with brine (20 mL), dried over Na$_2$SO$_4$, and the solvent was removed under reduced pressure. The resultant crude product was purified by flash chromatography (CH$_2$Cl$_2$/MeOH/AcOH 90:9.5:0.5).

Yield: 45 mg (100% yield: 16 mg) as an impure material.

^{1}H NMR (500 MHz, CDCl$_3$): δ [ppm] = 0.74-0.94 (m, 6H, H-1'''), 1.33 (s, 3H, C(CH$_3$)$_2$), 1.44-1.63 (m, 4H, H-2'''), 1.49 (s, 3H, C(CH$_3$)$_2$), 3.29 (dd, J = 11.7, 7.2 Hz, 1H, H-5"a), 3.44 (dd, J = 11.7, 4.4 Hz, 1H, H-5"b), 4.18-4.27 (m, 1H, H-4"), 4.27-4.33 (m, 1H, H-4'), 4.45-5.04 (m, 6H, H-2', H-3', H-5', H-6', H-2", H-3"), 5.08 (d, J = 12.6 Hz, 1H, Cbz-CH$_2$), 5.20 (d, J = 12.6 Hz, 1H, Cbz-CH$_2$), 5.23 (s$_{br}$, 1H, H-1"), 5.52 (s$_{br}$, 1H, H-1'), 5.71 (d, J = 8.0 Hz, 1H, H-5), 5.85 (s$_{br}$, 1H, NH-6'), 7.23-7.42 (m, 6H, H-6, Ph), 10.11 (s$_{br}$, 1H, NH-3). **^{13}C NMR** (126 MHz, CDCl$_3$): δ [ppm] = 7.34 (C-1'''), 8.47 (C-1'''), 25.28 (C(CH$_3$)$_2$), 27.08 (C(CH$_3$)$_2$), 28.88 (C-2'''), 29.69 (C-2'''), 53.04 (C-5"), 54.48 (C-6'), 67.07 (Cbz-CH$_2$), 79.75, 81.49, 82.07, 82.51, 85.34, 86.02, 86.45 (C-2', C-3', C-4', C-5', C-2", C-3", C-4"), 97.42 (C-1'), 103.02 (C-5), 110.77 (C-1"), 114.52 (C(CH$_3$)$_2$), 117.20 (C-3'''),

128.07 (CH-Ph), 128.25 (CH-Ph), 128.49 (CH-Ph), 136.19 (C-Ph), 143.13 (C-6), 150.91 (C-2), 156.53 (Cbz-C=O), 163.28 (C-4), 175.36 (C-7'). **HRMS** (ESI): calcd for $C_{32}H_{40}N_6O_{13}Na$ 739.2546, found 739.2550 $[M+Na]^+$. **TLC** (CH_2Cl_2/MeOH/AcOH 85:14:1): $R_f = 0.28$.

7.2.7.7 Synthesis of Partially Deprotected Nucleosyl Amino Acid Derivative 170

170

Variant 1

A mixture of nucleosyl amino acid derivative **169** (18 mg, 29 µmol, 1.0 eq) and barium hydroxide octahydrate (11 mg, 35 µmol, 1.2 eq) in THF (800 µL) and water (200 µL) was stirred at 0 °C for 7 h. As TLC analysis (CH_2Cl_2/MeOH/AcOH 85:14:1) indicated no complete conversion, the reaction was allowed to reach room temperature overnight. After 28 h of stirring in total, the reaction mixture was partitioned between 0.3 M aqueous HCl solution (5 mL) and EtOAc (10 mL). The aqueous layer was extracted with EtOAc (3 x 10 mL). The combined organics were dried over Na_2SO_4, and the solvent was removed under reduced pressure. The resultant crude product was purified by flash chromatography (CH_2Cl_2/MeOH/AcOH 90:9.5:0.5).

Recovered starting material 169: 1.7 mg (2.7 µmol, 10%) as a white solid.

Variant 2

A solution of protected nucleosyl amino acid derivative **174** (54 mg as impure material) in 80% aqueous TFA (3 mL) was stirred at room temperature for 6 h. The reaction mixture was diluted with water (15 mL), and the solvent was removed under reduced pressure. The resultant crude product was purified by three different sequentially performed HPLC runs.

1st Run: **HPLC** (preparative, gradient): t_R = 23.8 min (method P-01, injection volume 100 µL, concentration of injections: 65 mg in H_2O/MeCN 1:1 (0.45 mL) → 48 mg).

2nd Run: **HPLC** (preparative, isocratic): t_R = 67.5 min (method P-iso-01, injection volume 200 µL, concentration of injections: 31 mg in H_2O/MeCN 9:1 (0.3 mL) → 31 mg).

3rd Run: **HPLC** (semi-preparative, isocratic): t_R = 13.8 min (method SP-iso-01, injection volume 100 µL, concentration of injections: 31 mg in H$_2$O/MeCN 85:15 (0.6 mL)→ 31 mg).

Yield: 31 mg (51 µmol, 36% over two steps) as a white solid.

^{1}H NMR (500 MHz, CD$_3$OD): δ [ppm] = 3.66 (dd, J = 13.2, 4.3 Hz, 1H, H-5"a), 3.71 (dd, J = 13.2, 3.3 Hz, 1H, H-5"b), 3.94 (d, J = 4.5 Hz, 1H, H-2"), 4.02 (ddd, J = 7.6, 3.3, 4.3 Hz, 1H, H-4"), 4.03-4.17 (m, 3H, H-2', H-3', H3"), 4.18 (dd, J = 5.8, 3.4 Hz, 1H, H-4'), 4.51 (dd, J = 3.4, 1.9 Hz, 1H, H-5'), 4.69 (d, J = 1.9 Hz, 1H, H-6'), 4.93 (d, J = 12.3 Hz, 1H, Cbz-CH$_2$), 5.08 (s, 1H, H-1"), 5.28 (d, J = 12.3 Hz, 1H, Cbz-CH$_2$), 5.64 (d, J = 8.1 Hz, 1H, H-5), 5.67 (d, J = 3.4 Hz, 1H, H-1'), 7.23-7.42 (m, 5H, Ph), 7.83 (d, J = 8.1 Hz, 1H, H-6). **^{13}C NMR** (126 MHz, CD$_3$OD): δ [ppm] = 51.58 (C-5"), 55.51 (C-6'), 66.42 (Cbz-CH$_2$), 70.28 (C-3'), 70.61 (C-3"), 73.99 (C-2"), 75.09 (C-2'), 76.89 (C-5'), 80.72 (C-4"), 85.63 (C-4'), 90.02 (C-1'), 100.86 (C-5), 108.76 (C-1"), 127.79 (CH-Ph), 127.81 (CH-Ph), 127.88 (CH-Ph), 136.73 (C-Ph), 140.52 (C-6), 150.68 (C-2), 157.00 (Cbz-C=O), 164.77 (C-4), 172.16 (C-7'). **HRMS** (ESI): calcd for C$_{24}$H$_{27}$N$_6$O$_{13}$ 607.1642, found 607.1630 [M-H]$^-$. **[α]**$_D^{20}$ = +69.3 (c = 0.59, MeOH). **IR** (ATR): $\tilde{v}$ [cm^{-1}] = 3318, 2109, 1673, 1533, 1459, 1406, 1348, 1272, 1208, 1115, 1064, 1022, 958, 924, 813, 760, 703. **UV** (H$_2$O): λ$_{max}$ [nm] (log ε) = 206 (4.66), 264 (4.37). **Mp** = 150 °C (decomposition). **TLC** (CH$_2$Cl$_2$/MeOH/AcOH 85:14:1): R$_f$ = 0.00-0.24. **HPLC** (analytical, gradient): t_R = 26.9 min (method A-02, injection 5 µL (c ~ 1 mg/mL, in MeOH). **HPLC** (analytical, isocratic): t_R = 41.7 min (method A-iso-01, injection 5 µL (c ~ 1 mg/mL, in MeOH).

7.2.7.8 Synthesis of Nucleosyl Amino Acid 12

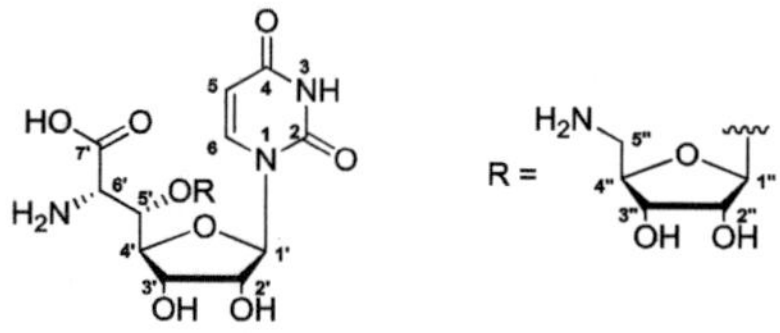

12

To a solution of the $N^{6'}$-Cbz-protected azide **170** (14 mg, 23 µmol, 1.0 eq) in degassed bidistilled water (3 mL), 10% Pd/C (5 mg, 5 µmol, 0.2 eq) and 1,4-cyclohexadiene (97%, 45 µL, 37 mg, 0.45 mmol, 20 eq) were added under an argon atmosphere, and the reaction mixture was stirred at room temperature for 5.5 h. After filtration of the mixture through a syringe filter and rinsing of the filter with bidistilled water (4 × 5 mL), the filtrate was lyophilized.

Yield: 12 mg (100% yield: 10 mg) as a white fluffy solid.

^{1}H NMR (500 MHz, D$_2$O): δ [ppm] = 3.06-3.15 (m, 1H, H-5"a), 3.25-3.38 (m, 1H, H 5"b), 3.84-3.91 (m, 1H, H-6'), 4.08-4.17 (m, 3H, H-2", H-3", H-4"), 4.18-4.26 (m, 2H, H-3', H-4'), 4.35 (dd, J = 5.0, 2.7 Hz, 1H, H-2'), 4.47-4.53 (m, 1H, H-5'), 5.17 (s, 1H, H-1"), 5.81 (d, J = 2.7 Hz, 1H, H-1'), 5.87 (d, J = 8.0 Hz, 1H, H-5), 7.74 (d, J = 8.0 Hz, 1H, H-6). **^{13}C NMR** (126 MHz, D$_2$O): δ [ppm] = 42.52 (C-5"), 57.21 (C-6'), 69.32 (C-3'), 71.96 (C-3"), 73.14 (C-2'), 74.97 (C-2"), 77.21 (C-5'), 79.05 (C-4"), 84.66 (C-4'), 91.04 (C-1'), 102.03 (C-5), 108.97 (C-1"), 141.73 (C-6), 152.26 (C-2), 167.41 (C-4, C-7'). **HRMS** (ESI): calcd for C$_{16}$H$_{25}$N$_4$O$_{11}$ 449.1514, found 449.1525 [M+H]$^+$. [α]$_D^{20}$ = +3.8 (c = 0.90, H$_2$O). **IR** (ATR): $\tilde{v}$ [cm^{-1}] = 3139, 2935, 2365, 2336, 1703, 1684, 1627, 1510, 1470, 1395, 1272, 1197, 1121, 1051, 1005, 819, 796, 720. **UV** (H$_2$O): λ$_{max}$ [nm] (log ε) = 202 (4.13), 262 (4.10). **Mp** = 172 °C (decomposition). **TLC** (2-propanol/H$_2$O/AcOH 5:2:1 as saturated NaCl solution): R$_f$ = 0.05-0.16.

7.2.7.9 Synthesis of Benzyl (*E*)-1',5',6'-Trideoxy-2',3'-*O*-isopropylidene-1'-(uracil-1-yl)-*β*-D-ribo-5'-eneheptofuranuronate 172

172

To a solution of 2',3'-*O*-isopropylidene uridine **71** (206 mg, 0.725 mmol, 1.0 eq) in dry MeCN (8 mL), 2-iodoxybenzoic acid (IBX, 510 mg, 1.82 mmol, 2.5 eq) was added under an argon atmosphere and heated at 80 °C for 1.5 h. The reaction mixture was cooled in an ice bath, and the resulting white precipitates were removed by filtration through a Celite pad. The pad was rinsed with EtOAc, and the combined filtrates were concentrated under reduced pressure. The residue was redissolved under an argon atmosphere in dry CH$_2$Cl$_2$ (7 mL) and cooled to -20 °C before a solution of benzyl (triphenylphosphoranylidene) acetate (355 mg, 0.865 mmol, 1.2 eq) in dry CH$_2$Cl$_2$ (2 mL) was added dropwise. After 1 h of stirring at -20 °C, the mixture was partitioned between EtOAc (20 mL) and water (8 mL). The aqueous layer was extracted with EtOAc (2 x 5 mL) before the combined organics were washed with brine (10 mL) and dried over Na$_2$SO$_4$. The solvent was removed under reduced pressure, and the resultant crude product was purified by flash chromatography (isohexane/EtOAc 1:2).

Yield: 281 mg (0.678 mmol, 94%) as a white foam.

^{1}H NMR (500 MHz, CDCl$_3$): δ [ppm] = 1.35 (s, 3H, C(CH$_3$)$_2$), 1.57 (s, 3H, C(CH$_3$)$_2$), 4.65 (ddd, J = 5.8, 4.6, 1.6 Hz, 1H, H-4'), 4.84 (dd, J = 6.4, 4.6 Hz, 1H, H-3'), 5.06 (dd, J = 6.4, 1.7 Hz, 1H, H-2'), 5.16 (d, J = 12.4 Hz, 1H, CH$_2$-Ph), 5.19 (d, J = 12.4 Hz, 1H, CH$_2$-Ph), 5.63 (d, J = 1.7 Hz, 1H, H-1'), 5.73 (dd, J = 8.0, 1.9 Hz, 1H, H-5), 6.08 (dd, J = 15.8, 1.6 Hz, 1H, H-6'), 7.05 (dd, J = 15.8, 5.8 Hz, 1H, H-5'), 7.18 (d, J = 8.0 Hz, 1H, H-6), 7.29-7.38 (m, 5H, Ph), 9.13 (s$_{br}$, 1H, NH). **^{13}C NMR** (126 MHz, CDCl$_3$): δ [ppm] = 25.29 (C(<u>C</u>H$_3$)$_2$), 27.12 (C(<u>C</u>H$_3$)$_2$), 66.47 (CH$_2$-Ph), 83.95 (C-3'), 84.44 (C-2'), 86.62 (C-4'), 94.94 (C-1'), 102.87 (C-5), 114.89 (<u>C</u>(CH$_3$)$_2$), 122.31 (C-6'), 128.24 (CH-Ph), 128.29 (CH-Ph), 128.57 (CH-Ph), 135.76 (C-Ph), 142.47 (C-6), 143.96 (C-5'), 149.80 (C-2), 163.01 (C-4), 165.48 (C-7'). **HRMS** (ESI): calcd for C$_{21}$H$_{21}$N$_2$O$_7$ 413.1354, found 413.1345 [M-H]$^-$. **[α]$_D^{20}$** = +46.0 (c = 1.0, CHCl$_3$). **IR** (ATR): $\tilde{v}$ [cm^{-1}] = 1708, 1691, 1633, 1459, 1376, 1266, 1214, 1168, 1064, 987, 883, 860, 807, 749, 697. **UV** (MeCN): λ$_{max}$ [nm] (log ε) = 207 (4.72), 257 (4.27). **Mp**: 77 °C. **TLC** (isohexane/EtOAc 3:1): R$_f$ = 0.27.

7.2.7.10 Synthesis of Amino Alcohol Derivative 173 as Isomeric Mixture

To a solution of benzyl carbamate (88 mg, 0.58 mmol, 3.0 eq) in 0.6 M aqueous NaOH solution (1 mL) and *n*-propanol (1 mL), *tert*-butyl hypochlorite (96 µL, 92 mg, 0.85 mmol, 4.4 eq) was added dropwise at 0 °C. After 10 min of stirring at 0 °C, the ice bath was exchanged with a water bath (T ~ 15 °C). A solution of olefin **172** (80 mg, 0.19 mmol, 1.0 eq) in *n*-propanol (1 mL), [DHQD]$_2$AQN (25 mg, 29 µmol, 0.15 eq) and K$_2$OsO$_2$(OH)$_4$ (10 mg, 0.27 µmol, 0.15 eq) were sequentially added. The resulting mixture was stirred at 15 °C for 7 h and then quenched by the addition of a saturated aqueous Na$_2$S$_2$O$_3$ solution (2 mL). The mixture was extracted with EtOAc (3 x 20 mL). The organic layer was washed with brine (1 x 30 mL), dried over Na$_2$SO$_4$, and concentrated under reduced pressure. The residue was directly applied to a silica gel column and purified by flash chromatography (isohexane/EtOAc 1:1→1:3).

Yield: 32 mg (55 µmol, 29%) as a white solid as an isomeric mixture.[5]

HRMS (ESI): calcd for $C_{29}H_{31}N_3O_{10}Na$ 604.1902, found 604.1909 $[M+Na]^+$. **TLC** (isohexane/EtOAc 1:3): $R_f = 0.13$.

7.2.8 Synthesis of Phosphonate 109

7.2.8.1 Synthesis of *N*-(Benzyloxycarbonyl)-2-methoxyglycine Methyl Ester 110[166-167]

110

A solution of glyoxylic acid monohydrate **175** (25.4 g, 276 mmol, 1.1 eq) in dry Et_2O (240 mL) was treated with benzyl carbamate (37.9 g, 251 mmol, 1.0 eq) under an argon atmosphere. The reaction mixture was stirred at room temperature for 19 h and then concentrated to approximately 100 mL under reduced pressure. The resulting suspension was filtered, and the solid was washed with ice-cooled Et_2O (2 x 50 mL). The resultant *N*-(benzyloxycarbonyl)-2-hydroxy glycine was dried in vacuo before it was redissolved in dry MeOH (550 mL) and treated with concentrated sulfuric acid (7.75 mL, 4.22 g, 145 mmol, 0.5 eq) at 0 °C under an argon atmosphere. After stirring the mixture at room temperature for 6 d, it was added to ice-cooled saturated aqueous $NaHCO_3$ solution (1000 mL). The aqueous layer was extracted with EtOAc (1 x 1200 mL, 3 x 600 mL). The combined organics were dried over Na_2SO_4, and the solvent was removed under reduced pressure. The resultant crude product was washed with petroleum ether (2 x 50 mL) and dried in vacuo.

Yield: 48.9 g (193 mmol, 70%) as a white solid.

^{1}H NMR (300 MHz, CD_3OD): δ [ppm] = 3.36 (s, 3H, H-1'), 3.74 (s, 3H, $COOCH_3$), 4.84 (s, 1H, H-2), 5.12 (s, 2H, Cbz-CH_2), 5.23 (s, 1H, NH), 7.24-7.41 (m, 5H, Ph). **^{13}C NMR** (75 MHz, CD_3OD): δ [ppm] = 53.05 (COO$\underline{C}H_3$), 55.84 (C-1'), 67.96 (Cbz-CH_2), 82.04 (C-2), 128.94 (CH-Ph), 129.14 (CH-Ph), 129.50 (CH-Ph), 137.85 (C-Ph), 158.16 (Cbz-C=O), 169.75 (C-1). **MS** (ESI): m/z = 276.1 $[M+Na]^+$. **HRMS** (ESI): calcd for $C_{12}H_{15}NO_5Na$ 276.0842, found 276.0846 $[M+Na]^+$. **UV** (MeCN): λ_{max} [nm]

---

[5] No reliable assignment of the signals in the ^{1}H or ^{13}C NMR spectra was possible.

(log ε) = 206 (3.96), 258 (2.42). **IR** (ATR): $\tilde{\nu}$ [cm^{-1}] = 3307, 1750, 1686, 1529, 1455, 1361, 1257, 1218, 1195, 1099, 978. **Mp** = 69 °C. **TLC** (CH$_2$Cl$_2$/EtOAc 1:1): R$_f$ = 0.64.

7.2.8.2 Synthesis of *N*-(Benzyloxycarbonyl)-2-(diethoxyphosphoryl)-glycine Methyl Ester 176[166]

176

A solution of *N*-(benzyloxycarbonyl)-2-methoxyglycine methyl ester **110** (24.9 g, 98.4 mmol, 1.0 eq) in dry toluene (100 mL) was heated to 80 °C under an argon atmosphere before phosphorus trichloride (9.70 mL, 15.3 g, 111 mmol, 1.1 eq) was added dropwise. After 4 h of heating to reflux, triethyl phosphite (18.0 mL, 17.5 g, 105 mmol, 1.1 eq) was added, and heating was continued for a further 2 h. The solvent was removed under reduced pressure, and the residue was redissolved in EtOAc (500 mL). The solution was washed with saturated aqueous NaHCO$_3$ solution (3 x 150 mL), and the aqueous layers were extracted with EtOAc (2 x 100 mL). The combined organics were dried over Na$_2$SO$_4$, and the solvent was removed under reduced pressure. The resultant liquid residue was treated with petroleum ether (100 mL) and stirred for 30 min at room temperature. The suspension was filtered, and the solid was washed with ice-cooled petroleum ether (3 x 50 mL) and dried in vacuo.

Yield: 28.4 g (79.1 mmol, 80%) as a white solid.

^{1}H NMR (301 MHz, CDCl$_3$): δ [ppm] = 1.25-1.38 (m, 6H, H-2'), 3.82 (s, 3H, COOCH$_3$), 4.05-4.26 (m, 4H, H-1'), 4.88 (dd, $^2J_{HP}$ = 22.3 Hz, J = 9.5 Hz, 1H, H-2), 5.06-5.20 (m, 2H, Cbz-CH$_2$), 5.60 (d, J = 9.5 Hz, 1 H, NH), 7.30-7.39 (m, 5H, Ph). **^{13}C NMR** (76 MHz, CDCl$_3$): δ [ppm] = 16.21 (d, $^3J_{CP}$ = 6.2 Hz, C-2'a), 16.23 (d, $^3J_{CP}$ = 5.9 Hz, C-2'b), 52.53 (d, $^1J_{CP}$ = 146.9 Hz, C-2), 53.15 (COOCH$_3$), 63.74 (d, $^2J_{CP}$ = 6.7 Hz, C-1'a), 63.79 (d, $^2J_{CP}$ = 6.4 Hz, C-1'b), 67.52 (Cbz-CH$_2$), 128.11 (CH-Ph), 128.27 (CH-Ph), 128.49 (CH-Ph), 135.82 (C-Ph), 155.58 (d, $^3J_{CP}$ = 8.4 Hz, Cbz-C=O), 167.39 (d, $^2J_{CP}$ = 2.2 Hz, C-1). **^{31}P NMR** (122 MHz, CDCl$_3$): δ [ppm] = 15.68. **MS** (ESI): *m/z* = 382.1 [M+Na]$^+$. **HRMS** (ESI): calcd for C$_{15}$H$_{22}$NO$_7$PNa 382.1026, found 382.1024 [M+Na]$^+$. **UV** (MeCN): λ$_{max}$ [nm] (log ε) = 207 (3.96), 258 (2.59). **IR** (ATR): $\tilde{\nu}$ [cm^{-1}] = 3221, 2980, 1748, 1707, 1537, 1327, 1264, 1230, 1206, 1039, 962. **Mp** = 73 °C. **TLC** (petroleum ether/EtOAc 1:1): R$_f$ = 0.08.

7.2.8.3 Synthesis of *N*-(Benzyloxycarbonyl)-2-(diethoxyphosphoryl)-glycine (2-Trimethylsilylethyl) Ester 109[227]

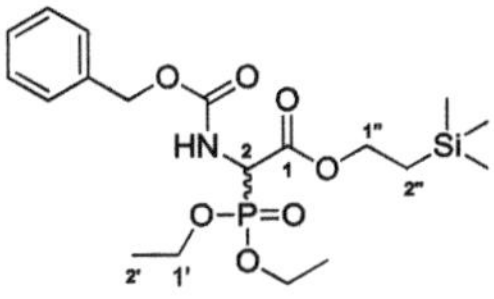

109

To a solution of methyl ester phosphonate **176** (23.6 g, 65.7 mmol, 1.0 eq) in a mixture of THF and water (4:1, 600 mL), lithium hydroxide monohydrate (5.53 g, 132 mmol, 2.0 eq) was added at 0 °C. After stirring the reaction mixture at 0 °C for 3 h, it was partitioned between EtOAc (500 mL) and water (200 mL). The aqueous layer was acidified to a pH value of 1, the layers were separated, and the aqueous layer was extracted with EtOAc (3 x 300 mL). The combined organics were washed with brine (300 mL) and dried over Na_2SO_4. After removing the solvent under reduced pressure, the resultant crude carboxylic acid was dried in vacuo and redissolved in dry CH_2Cl_2 (400 mL). Under an argon atmosphere, this solution was treated sequentially with 4-dimethylaminopyridine (814 mg, 6.66 mmol, 0.1 eq), 2-(trimethylsilyl)ethanol (10.9 mL, 8.99 g, 76.0 mmol, 1.2 eq) and 1-ethyl-3-(3-dimethylaminopropyl)carbodiimide hydrochloride (EDC) (15.8 g, 82.2 mmol, 1.3 eq). After 16 h at room temperature, EtOAc (500 mL) was added, and the organic layer was washed with 1 M aqueous HCl solution (2 x 100 mL). The aqueous layer was extracted with EtOAc (2 x 50 mL), and the combined organics were washed with brine (250 mL) and dried over Na_2SO_4. After removing the solvent under reduced pressure, the resultant crude product was purified by flash chromatography (CH_2Cl_2).

Yield: 19.6 g (44.0 mmol, 67%) as a white solid.

^{1}H NMR (301 MHz, $CDCl_3$): δ [ppm] = 0.01 (s, 9H, Si(CH$_3$)$_3$), 0.97-1.09 (m, 2H, H-2"), 1.22-1.35 (m, 6H, H-2'), 4.05-4.19 (m, 4H, H-1'), 4.20-4.33 (m, 2H, H-1"), 4.82 (dd, $^2J_{HP}$ = 22.1 Hz, J = 10.0 Hz, 1H, H-2), 5.03-5.18 (m, 2H, Cbz-CH$_2$), 5.58 (d, J = 10.0 Hz, 1H, NH), 7.27-7.36 (m, 5H, Ph). **^{13}C NMR** (126 MHz, $CDCl_3$): δ [ppm] = -1.64 (Si(CH$_3$)$_3$), 16.24 (d, $^3J_{CP}$ = 5.9 Hz, C-2'a), 16.25 (d, $^3J_{CP}$ = 5.7 Hz, C-2'b), 17.33 (C-2"), 52.75 (d, $^1J_{CP}$ = 146.8 Hz, C-2), 63.67 (d, $^2J_{CP}$ = 6.6 Hz, C-1'), 64.95 (C-1"), 67.46 (Cbz CH$_2$), 128.10 (CH-Ph), 128.24 (CH-Ph), 128.48 (CH-Ph), 135.90 (C-Ph), 155.58 (d, $^3J_{CP}$ = 7.6 Hz, Cbz-C=O), 166.91 (C-1). **^{31}P NMR** (122 MHz, $CDCl_3$): δ [ppm] = 15.98. **MS** (ESI): *m/z* = 468.2 [M+Na]$^+$. **HRMS** (ESI): calcd for $C_{19}H_{32}NO_7PSiNa$ 468.1578, found 468.1579 [M+Na]$^+$. **UV** (MeCN): λ_{max} [nm] (log ε) = 207 (3.94), 258 (2.28). **IR** (ATR): $\tilde{v}$ [cm^{-1}] = 3245, 2953, 1740, 1716, 1551, 1263, 1237, 1183, 1041, 1016, 836. **Mp** = 42 °C. **TLC** (petroleum ether/EtOAc 1:1): R_f = 0.22.

7.2.9 Synthesis of Enduracididine Precursor 108 via *Wittig* Reaction with Phosphonate 109 and (*S*)-*Garner's* Aldehyde (*S*)-98

7.2.9.1 Synthesis of Didehydroamino Acid 108

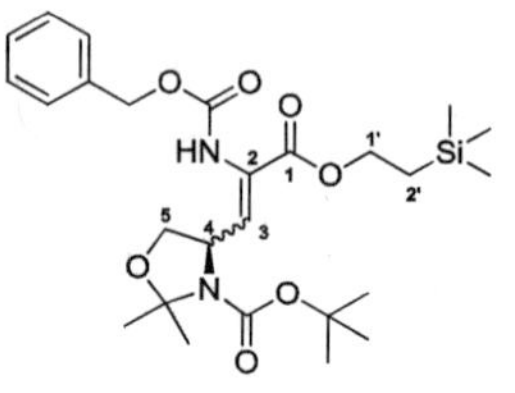

108

To a solution of the base in dry THF (2 mL/mmol), a solution of the phosphonate **109** in dry THF (3 mL/mmol) was added dropwise at -78 °C under an argon atmosphere. The solution was stirred at -78 °C for 5 min before a solution of (*S*)-*Garner's* aldehyde (*S*)-98 in dry THF (1.5 mL/mmol) was added at -78 °C. The resulting reaction mixture was stirred for 16 h and slowly warmed to room temperature during this period. The reaction was quenched at 0 °C by the addition of MeOH (5 mL). After adding EtOAc (300 mL), the organic layer was washed with water (3 x 100 mL), dried over Na_2SO_4, and the solvent was removed under reduced pressure.

Variant 1

The diastereomeric mixture **108** was prepared using potassium *tert*-butylate (26 mg, 0.23 mmol, 0.9 eq), phosphonate **109** (109 mg, 0.245 mmol, 1.0 eq) and (*S*)-*Garner's* aldehyde (*S*)-98 (56 mg, 0.24 mmol, 1.0 eq). The diastereomers **108a** and **108b** were separated by flash chromatography (cyclohexane/Et$_2$O 2:1).

Yield 108a: 40 mg (77 µmol, 31%) as a yellowish oil.
Yield 108b: 35 mg (67 µmol, 28%) as a yellowish oil.

Variant 2

The diastereomeric mixture **108** was prepared using potassium hexamethyldisilazane (KHMDS) (49 mg, 0.25 mmol, 1.1 eq), phosphonate **109** (131 mg, 0.294 mmol, 1.3 eq) and (*S*)-*Garner's* aldehyde (*S*)-98 (51 mg, 0.22 mmol, 1.0 eq). The diastereomers **108a** and **108b** were separated by flash chromatography (cyclohexane/Et$_2$O 2:1).

Yield 108a: 38 mg (73 µmol, 33%) as a yellowish oil.
Yield 108b: 13 mg (26 µmol, 12%) as a yellowish oil.

108a: **^{1}H NMR** (301 MHz, DMSO-d$_6$, 100 °C): δ [ppm] = 0.04 (s, 9H, Si(CH$_3$)$_3$), 0.83-1.02 (m, 2H, H-2'), 1.38 (s, 9H, *t*-Bu-CH$_3$), 1.46 (s, 3H, C(CH$_3$)$_2$), 1.53 (s, 3H, C(CH$_3$)$_2$), 3.69 (dd, *J* = 9.0, 3.2 Hz, 1H, H-5a), 4.10-4.17 (m, 1H, H-5b), 4.17-4.25 (m, 2H, H-1'), 4.85-4.97 (m, 1H, H-4), 5.08 (s, 2H, Cbz-CH$_2$), 5.97 (d, *J* = 8.8 Hz, 1H, H-3), 7.28-7.42 (m, 5H, Ph), 8.65 (s, 1H, NH). **^{13}C NMR** (151 MHz, DMSO-d$_6$, 100 °C): δ [ppm] = -2.12 (Si(CH$_3$)$_3$), 16.37 (C-2'), 23.78 (C($\underline{C}$H$_3$)$_2$), 26.10 (C($\underline{C}$H$_3$)$_2$), 27.52 (*t*-Bu-CH$_3$), 54.34 (C-4), 62.39 (C-1'), 65.50 (Cbz-CH$_2$), 68.15 (C-5), 78.79 (*t*-Bu-C), 92.90 ($\underline{C}$(CH$_3$)$_2$), 127.16 (CH-Ph), 127.31 (CH-Ph), 127.49 (C-2), 127.75 (CH-Ph), 129.64 (C-3), 136.15 (C-Ph), 150.74 (Boc-C=O), 153.43 (Cbz-C=O), 163.00 (C-1). **MS** (ESI): *m/z* = 543.3 [M+Na]$^+$. **HRMS** (ESI): calcd for C$_{26}$H$_{40}$N$_2$O$_7$SiNa 543.2497, found 543.2490 [M+Na]$^+$. [α]$_D^{20}$ = +6.6 (*c* = 0.63, CHCl$_3$). **UV** (MeCN): λ$_{max}$ [nm] (log ε) = 208 (3.72), 250 (3.39). **IR** (ATR): $\tilde{v}$ [cm^{-1}] = 2921, 1699, 1515, 1456, 1377, 1250, 1214, 1172, 1086, 1052, 1027. **TLC** (cyclohexane/Et$_2$O 2:1): R$_f$ = 0.40.

108b: **^{1}H NMR** (301 MHz, DMSO-d$_6$, 100 °C): δ [ppm] = 0.04 (s, 9H, Si(CH$_3$)$_3$), 0.91-1.01 (m, 2H, H-2'), 1.39 (s, 9H, *t*-Bu-CH$_3$), 1.46 (s, 3H, C(CH$_3$)$_2$), 1.53 (s, 3H, C(CH$_3$)$_2$), 3.70 (dd, *J* = 9.2, 3.2 Hz, 1H, H-5a), 4.08 (dd, *J* = 9.2, 6.4 Hz, 1H, H-5b), 4.14-4.25 (m, 2H, H-1'), 4.61 (ddd, *J* = 9.1, 6.4, 3.2 Hz, 1H, H-4), 5.07 (s, 2H, Cbz-CH$_2$), 6.14 (d, *J* = 9.1 Hz, 1H, H-3), 7.26-7.45 (m, 5H, Ph), 8.52 (s, 1H, NH). **^{13}C NMR** (126 MHz, DMSO-d$_6$, 50 °C): δ [ppm] = -1.66 (Si(CH$_3$)$_3$), 16.41 (C-2'), 23.90 (C($\underline{C}$H$_3$)$_2$), 26.08 (C($\underline{C}$H$_3$)$_2$), 27.75 (*t*-Bu-CH$_3$), 53.88 (C-4), 62.72 (C-1'), 65.87 (Cbz-CH$_2$), 67.87 (C-5), 79.18 (*t*-Bu-C), 93.11 ($\underline{C}$(CH$_3$)$_2$), 127.44 (C-2), 127.56 (CH-Ph), 127.71 (CH-Ph), 128.10 (CH-Ph), 131.76 (C-3), 136.31 (C-Ph), 151.91 (Boc-C=O), 153.87 (Cbz-C=O), 163.86 (C-1). **MS** (ESI): *m/z* = 543.3 [M+Na]$^+$. **HRMS** (ESI): calcd for C$_{26}$H$_{40}$N$_2$O$_7$SiNa 543.2497, found 543.2498 [M+Na]$^+$. [α]$_D^{20}$ = +0.45 (*c* = 0.58, CHCl$_3$). **UV** (MeCN): λ$_{max}$ [nm] (log ε) = 206 (4.12), 242 (3.02). **IR** (ATR): $\tilde{v}$ [cm^{-1}] = 3351, 2953, 1719, 1518, 1455, 1391, 1355, 1249, 1174, 1050, 834. **TLC** (cyclohexane/Et$_2$O 2:1): R$_f$ = 0.29.

7.2.9.2 Attempt to Synthesize Enduracididine Precursor 93

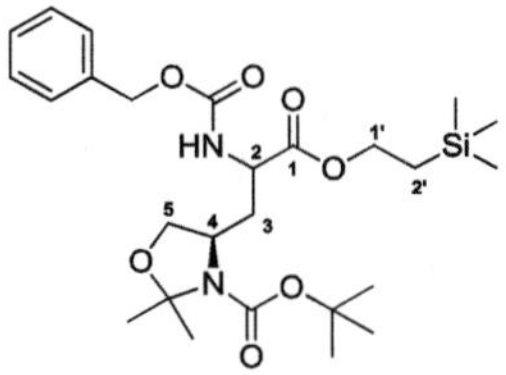

93

To a solution of didehydroamino acid **108a** (25 mg, 48 µmol, 1.0 eq) in dry MeOH (2 mL), (+)-1,2-bis[(2*S*,5*S*)-2,5-dimethylpholphospholano]benzene(cyclooctadiene)-rhodium(I)   tetra-fluoroborate ((S,S)-Me-DUPHOS-Rh) **177** (3 mg, 5 µmol, 0.1 eq) was added under strictly anaerobic conditions. The resulting reaction mixture was stirred at room temperature under an atmosphere of H_2 (1 bar) for 7 d. After 3 d, additional catalyst **177** (3 mg, 5 µmol, 0.1 eq) was added. The solvent was evaporated under reduced pressure, and the resultant crude product was purified by flash chromatography (cyclohexane/Et$_2$O 2:1).

Recovered starting material 108a: 20 mg (39 µmol, 81%) as a yellowish oil.

7.2.10 Approach for the Synthesis of Enduracididine Precursors via *Sharpless* Asymmetric Dihydroxylation

7.2.10.1 Synthesis of 4-Methyl L-Aspartate Hydrochloride 179[222]

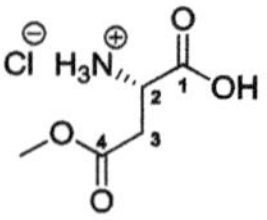

179

To a suspension of L-aspartic acid **178** (15.0 g, 113 mmol, 1.0 eq) in dry MeOH (80 mL), thionyl chloride (11.4 mL, 18.7 g, 157 mmol, 1.4 eq) was added dropwise at -10 °C under an argon atmosphere. The suspension was stirred for 30 min and warmed slowly to room temperature during this period. Et$_2$O (300 mL) was added, and the mixture was cooled in an ice bath to complete precipitation. The resulting white solid was filtered off, washed with ice-cooled Et$_2$O (2 x 50 mL) and dried in vacuo.

Yield: 15.9 g (86.6 mmol, 77%) as a white solid.

^{1}H NMR (301 MHz, CD$_3$OD): δ [ppm] = 3.00-3.07 (m, 2H, H-3), 3.76 (s, 3H, OCH$_3$), 4.32 (dd, J = 6.3, 4.9 Hz, 1H, H-2). **^{13}C NMR** (76 MHz, CD$_3$OD): δ [ppm] = 34.85 (C-3), 50.07 (C-2), 52.92 (OCH$_3$), 170.44 (C-4), 171.61 (C-1). **MS** (ESI): m/z = 170.1 [M+Na]$^+$. **HRMS** (ESI): calcd for C$_5$H$_9$NO$_4$Na 170.0424, found 170.0427 [M+Na]$^+$. **Mp** = 191 °C. **[α]$_D^{20}$** = +15.3 (c = 0.99, MeOH). **IR** (ATR): $\tilde{\nu}$ [cm^{-1}] = 2824, 1731, 1499, 1418, 1389, 1232, 1202, 1179, 1104, 854.

7.2.10.2 Synthesis of 4-Methyl *N*-(Benzyloxycarbonyl)-L-aspartate 180[228]

180

To a solution of 4-methyl L-aspartate **179** (15.4 g, 83.9 mmol, 1.0 eq) in 1 M aqueous NaHCO$_3$ solution (300 mL), benzyloxycarbonyl chloride (19 mL, 23 g, 128 mmol, 1.5 eq) was added dropwise at 0 °C. The resulting mixture was stirred at room temperature for 42 h. The aqueous layer was extracted with Et$_2$O (3 x 150 mL) before it was acidified with 2 M aqueous HCl solution to a pH value of 2-3. After the extraction with EtOAc (1 x 500 mL, 2 x 200 mL), the combined organics were dried over Na$_2$SO$_4$, and the solvent was removed under reduced pressure. The resultant crude product was used without further purification.

Yield: 20.1 g (71.5 mmol, 85%) as a colorless viscous oil.

^{1}H NMR (300 MHz, CDCl$_3$): δ (ppm) = 2.87 (dd, J = 17.4, 4.6 Hz, 1H, H-3a), 3.08 (dd, J = 17.4, 4.6 Hz, 1H, H-3b), 3.69 (s, 3H, OCH$_3$), 4.69 (ddd, J = 8.6, 4.6, 4.6 Hz, 1H, H-2), 5.12 (s, 2H, Cbz-CH$_2$), 5.83 (d, J = 8.6 Hz, 1H, NH), 7.19-7.39 (m, 5H, Ph), 9.76 (s$_{br}$, 1H, COOH). **^{13}C NMR** (126 MHz, CDCl$_3$): δ [ppm] = 36.25 (C-3), 50.15 (C-2), 52.18 (OCH$_3$), 67.31 (Cbz-CH$_2$), 127.99 (CH-Ph), 128.12 (CH-Ph), 128.40 (CH-Ph), 135.79 (C-Ph), 156.07 (Cbz-C=O), 171.31 (C-4), 175.04 (C-1). **MS** (ESI): m/z = 304.1 [M+Na]$^+$. **HRMS** (ESI): calcd for C$_{13}$H$_{15}$NO$_6$Na 304.0792, found 304.0793 [M+Na]$^+$. [α]$_D^{20}$ = +34.2 (c = 0.99, CHCl$_3$). **IR** (ATR): $\tilde{v}$ [cm^{-1}] = 3336, 2952, 1720, 1516, 1440, 1412, 1372, 1330, 1214, 1179, 1057, 1028, 994, 854, 778, 743, 697. **UV** (MeCN): λ_{max} [nm] (log ε) = 206 (3.74), 258 (2.15). **TLC** (CH$_2$Cl$_2$/MeOH 5:1): R$_f$ = 0.00-0.29.

7.2.10.3 Synthesis of 4-Methyl 1-(2'-Trimethylsilylethyl) *N*-(Benzyloxycarbonyl)-L-aspartate 181

181

A solution of protected L-aspartate derivative **180** (2.16 g, 7.68 mmol, 1.0 eq) in dry CH_2Cl_2 (40 mL) was treated sequentially under an argon atmosphere with 4-dimethylaminopyridine (1.41 g, 11.5 mmol, 1.5 eq), 2-(trimethylsilyl)ethanol (1.65 mL, 1.36 g, 11.5 mmol, 1.5 eq) and 1-ethyl-3-(3-dimethylaminopropyl)carbodiimide hydrochloride (EDC) (2.30 g, 12.0 mmol, 1.6 eq). After 19 h at room temperature, the mixture was washed with saturated aqueous $NaHCO_3$ solution (20 mL). The aqueous layer was extracted with CH_2Cl_2 (3 x 50 mL), and the combined organics were washed with brine (50 mL) and dried over Na_2SO_4. After removing the solvent under reduced pressure, the resultant crude product was purified by flash chromatography (petroleum ether/EtOAc 6:1).

Yield: 1.99 g (5.22 mmol, 68%) as a colorless oil.

^{1}H NMR (301 MHz, $CDCl_3$): δ (ppm) = 0.01 (s, 9H, $Si(CH_3)_3$), 0.90-1.02 (m, 2H, H-2'), 2.82 (dd, J = 17.0, 4.5 Hz, 1H, H-3a), 3.00 (dd, J = 17.0, 4.5 Hz, 1H, H-3b), 3.64 (s, 3H, OCH_3), 4.15-4.28 (m, 2H, H-1'), 4.58 (ddd, J = 8.6, 4.5, 4.5 Hz, 1H, H-2), 5.09 (s, 2H, Cbz-CH_2), 5.77 (d, J = 8.6 Hz, 1H, NH), 7.22-7.39 (m, 5H, Ph). **^{13}C NMR** (76 MHz, $CDCl_3$): δ [ppm] = -1.65 ($Si(\underline{C}H_3)_3$), 17.15 (C-2'), 36.38 (C-3), 50.39 (C-2), 51.88 (OCH_3), 64.25 (C-1'), 66.95 (Cbz-CH_2), 127.99 (CH-Ph), 128.08 (CH-Ph), 128.41 (CH-Ph), 136.10 (C-Ph), 155.83 (Cbz-C=O), 170.64 (C-4), 171.17 (C-1). **MS** (ESI): m/z = 404.1 [M+Na]$^+$. **HRMS** (ESI): calcd for $C_{18}H_{27}NO_6SiNa$ 404.1500, found 404.1493 [M+Na]$^+$. $[\alpha]_D^{20}$ = +16.9 (c = 0.96, $CHCl_3$). **IR** (ATR): $\tilde{v}$ [cm^{-1}] = 3353, 2952, 1726, 1510, 1440, 1336, 1249, 1208, 1174, 1051, 993, 935, 860, 837, 697. **UV** (MeCN): λ_{max} [nm] (log ε) = 207 (3.95), 258 (2.42). **TLC** (petroleum ether/EtOAc 3:1): R_f = 0.35.

7.2.10.4 Synthesis of 1-*tert*-Butyl 4-Methyl *N*-(Benzyloxycarbonyl)-L-aspartate 183

183

A solution of protected L-aspartate derivative **180** (9.29 g, 33.0 mmol, 1.0 eq) and *tert*-butanol (40 mL, 31 g, 0.42 mol, 13 eq) in toluene (100 mL) was heated to reflux, and then *N,N*-dimethylformamide dineopentylacetal (27 mL, 22 g, 95 mmol, 2.9 eq) was added dropwise under reflux over 30 min. After stirring the reaction mixture for a further 5 h under reflux, it was cooled to room temperature, washed with saturated aqueous Na_2CO_3

solution (2 x 100 mL) and H$_2$O (1 x 100 mL), dried over Na$_2$SO$_4$, and the solvent was removed under reduced pressure. The resultant crude product was purified by flash chromatography (isohexane/EtOAc 5:1).

Yield: 6.54 g (19.4 mmol, 59%) as a yellowish oil.

^{1}H NMR (500 MHz, CDCl$_3$): δ [ppm] = 1.45 (s, 9H, *t*-Bu-CH$_3$), 2.81 (dd, J = 16.7, 4.5 Hz, 1H, H-3a), 2.97 (dd, J = 16.7, 4.5 Hz, 1H, H-3b), 3.68 (s, 3H, OCH$_3$), 4.52 (ddd, J = 8.1, 4.5, 4.5 Hz, 1H, H-2), 5.12 (s, 2H, Cbz-CH$_2$), 5.69 (d, J = 8.1 Hz, 1H, NH), 7.29-7.36 (m, 5H, Ph). **^{13}C NMR** (126 MHz, CDCl$_3$): δ [ppm] = 27.98 (*t*-Bu-CH$_3$), 36.96 (C-3), 50.99 (C-2), 51.84 (OCH$_3$), 66.98 (Cbz-CH$_2$), 82.58 (*t*-Bu-C), 128.07 (CH-Ph), 128.15 (CH-Ph), 128.51 (CH-Ph), 136.30 (C-Ph), 155.93 (Cbz-C=O), 169.53 (C-1), 171.13 (C-4). **HRMS** (ESI): calcd for C$_{17}$H$_{22}$NO$_6$ 336.1453, found 336.1451 [M-H]$^-$. $[\alpha]_D^{20}$ = +17.5 (c = 1.3, CHCl$_3$). **IR** (ATR): $\tilde{\nu}$ [cm^{-1}] = 3359, 2975, 1720, 1505, 1459, 1435, 1372, 1342, 1219, 1156, 1028, 848, 749, 703. **UV** (MeCN): λ_{max} [nm] (log ε) = 208 (4.13), 258 (2.51). **TLC** (isohexane/EtOAc 4:1): R$_f$ = 0.19.

7.2.10.5 Synthesis of 1-*tert*-Butyl *N*-(Benzyloxycarbonyl)-L-aspartate 200

200

To a solution of methyl ester **183** (6.54 g, 19.4 mmol, 1.0 eq) in THF (160 mL) and water (40 mL), lithium hydroxide monohydrate (1.63 g, 38.8 mmol, 2.0 eq) was added at 0 °C. The resulting mixture was stirred at this temperature for 3 h before it was partitioned between EtOAc (140 mL) and water (60 mL). The aqueous layer was acidified with 1 M aqueous HCl solution to a pH value of 1 and extracted with EtOAc (3 x 90 mL). The combined organics were dried over Na$_2$SO$_4$, and the solvent was removed under reduced pressure. The resultant crude product was used without further purification.

Yield: 5.62 g (17.4 mmol, 90%) as a colorless oil.

^{1}H NMR (500 MHz, C$_6$D$_6$): δ [ppm] = 1.27 (s, 9H, *t*-Bu-CH$_3$), 2.62 (dd, J = 16.9, 4.7 Hz, 1H, H-3a), 2.73 (dd, J = 16.9, 4.7 Hz, 1H, H-3b), 4.55 (ddd, J = 7.6, 4.7, 4.7 Hz, 1H, H-2), 4.98 (s, 2H, Cbz-CH$_2$), 5.74 (d, J = 7.6 Hz, 1H, NH), 7.00-7.22 (m, 5H, Ph), 8.89 (s$_{br}$, 1H, COOH). **^{13}C NMR** (126 MHz, C$_6$D$_6$): δ [ppm] = 27.74 (*t*-Bu-CH$_3$), 36.33 (C-3), 50.92

(C-2), 66.96 (Cbz-CH$_2$), 81.03 (*t*-Bu-C), 127.96 (CH-Ph), 128.07 (CH-Ph), 128.26 (CH-Ph), 136.65 (C-Ph), 156.15 (Cbz-C=O), 169.15 (C-1), 175.64 (C-4). **MS** (ESI): *m/z* = 346.1 [M+Na]$^+$. **HRMS** (ESI): calcd for C$_{16}$H$_{21}$NO$_6$Na 346.1261, found 346.1262 [M+Na]$^+$. [α]$_D^{20}$ = +13.2 (*c* = 1.0, CHCl$_3$). **IR** (ATR): $\tilde{v}$ [cm^{-1}] = 2975, 1708, 1516, 1453, 1400, 1372, 1342, 1226, 1156, 1051, 1028, 848, 778, 749, 697. **UV** (MeCN): λ$_{max}$ [nm] (log ε) = 207 (4.41), 258 (2.80). **TLC** (CH$_2$Cl$_2$/MeOH 9:1): R$_f$ = 0.37.

7.2.10.6 Synthesis of *tert*-Butyl *N*^{2}-(Benzyloxycarbonyl)-*N*^{5}-methoxy-*N*^{5}-methyl-L-asparaginate 185

185

To a solution of carboxylic acid **200** (326 mg, 1.01 mmol, 1.0 eq) in dry CH$_2$Cl$_2$ (9 mL), (benzotriazol-1-yloxy)tripyrrolidinophosphonium hexafluorophosphate (PyBOP) (754 mg, 1.45 mmol, 1.4 eq), triethylamine (200 μL, 146 mg, 1.44 mmol, 1.4 eq) and *N,O*-dimethylhydroxylamine hydrochloride (177 mg, 2.90 mmol, 2.9 eq) were added under an argon atmosphere, and the reaction mixture was stirred at room temperature for 6 d. After the addition of CH$_2$Cl$_2$, the organic layer was washed with saturated aqueous NH$_4$Cl solution (10 mL), saturated aqueous NaHCO$_3$ solution (10 mL) and brine (7 mL), and dried over Na$_2$SO$_4$. The solvent was removed under reduced pressure, and the resultant crude product was purified by flash chromatography (isohexane/EtOAc 3:1).

Yield: 272 mg (0.742 mmol, 73%) as a colorless oil.

^{1}**H NMR** (500 MHz, C$_6$D$_6$): δ [ppm] = 1.44 (s, 9H, *t*-Bu-CH$_3$), 2.77 (s, 3H, NCH$_3$), 2.80-2.90 (m, 1H, H-3a), 3.01 (s, 3H, NOCH$_3$), 3.08-3.24 (m, 1H, H-3b), 4.83 (ddd, *J* = 8.3, 3.5, 3.5 Hz, 1H, H-2), 5.06 (d, *J* = 12.5 Hz, 1H, Cbz-CH$_2$), 5.09 (d, *J* = 12.5 Hz, 1H, Cbz-CH$_2$), 6.29 (d, 1H, *J* = 8.3 Hz, NH), 7.21-7.36 (m, 5H, Ph). ^{13}**C NMR** (126 MHz, C$_6$D$_6$): δ [ppm] = 27.52 (*t*-Bu-CH$_3$), 31.09 (NCH$_3$), 34.45 (C-3), 51.92 (C-2), 60.03 (NOCH$_3$), 65.87 (Cbz-CH$_2$), 81.03 (*t*-Bu-C), 127.96 (CH-Ph), 127.99 (CH-Ph), 128.18 (CH-Ph), 137.13 (C-Ph), 156.15 (Cbz-C=O), 170.03 (C-1), 171.51 (C-4). **HRMS** (ESI): calcd for C$_{18}$H$_{26}$N$_2$O$_6$Na 389.1683, found 389.1685 [M+Na]$^+$. [α]$_D^{20}$ = +9.3 (*c* = 1.3, CHCl$_3$). **IR** (ATR): $\tilde{v}$ [cm^{-1}] = 2975, 1726, 1656, 1505, 1400, 1372, 1336, 1261, 1219, 1156, 1051, 1028, 1005, 848, 749, 703. **UV** (MeCN): λ$_{max}$ [nm] (log ε) = 206 (4.21), 257 (2.70). **TLC** (isohexane/EtOAc 3:1): R$_f$ = 0.09. **TLC** (CH$_2$Cl$_2$/MeOH 9:1): R$_f$ = 0.37.

7.2.10.7 Synthesis of *tert*-Butyl *N*-(benzyloxycarbonyl)-L-aspartate Semialdehyde 184

184

Variant 1

To a solution of methyl ester **183** (84 mg, 0.25 mmol, 1.0 eq) in dry toluene (3 mL), diisobutylaluminium hydride (DIBAL-H) (1.2 M in toluene, 230 µL, 276 µmol, 1.1 eq) was added at -78 °C under an argon atmosphere. The reaction mixture was stirred at this temperature for 10 min before the reaction was quenched by the addition of acetone (0.5 mL), water (0.5 mL) and solid NaHCO$_3$ (500 mg). The mixture was allowed to reach room temperature, and it was filtered through a pad of NaHCO$_3$ on Celite. The pad was rinsed with EtOAc, and the filtrate was concentrated under reduced pressure. The resultant crude product was purified by flash chromatography (isohexane/EtOAc 4:1).

Yield: 8.8 mg (0.029 mmol, 12%) as a colorless oil.

Variant 2

To a solution of Weinreb amide **185** (91 mg, 0.25 mmol, 1.0 eq) in dry THF (2 mL), Schwartz's reagent Cp$_2$ZrHCl (64 mg, 0.25 mmol, 1.0 eq) in dry THF (1 mL) was added under an argon atmosphere. After stirring the reaction at room temperature for 3 d, it was quenched by the addition of water (200 µL). The solvent was removed under reduced pressure, and the resultant crude product was purified by flash chromatography (isohexane/EtOAc 4:1).

Yield: 38 mg (0.12 mmol, 48%) as a colorless oil.

^{1}H NMR (500 MHz, CDCl$_3$): δ [ppm] = 1.24 (s, 9H, *t*-Bu-CH$_3$), 2.36 (dd, J = 17.9, 4.9 Hz, 1H, H-3a), 2.42 (dd, J = 17.9, 4.9 Hz, 1H, H-3b), 4.44 (ddd, J = 7.0, 4.9, 4.9 Hz, 1H, H-2), 5.00 (d, J = 12.3 Hz, 1H, Cbz-CH$_2$), 5.05 (d, J = 12.3 Hz, 1H, Cbz-CH$_2$), 5.41 (d, J = 7.0 Hz, 1H, NH), 6.98-7.36 (m, 5H, Ph), 9.03 (s, 1H, CHO). **^{13}C NMR** (126 MHz, C$_6$D$_6$): [ppm] = 27.33 (*t*-Bu-CH$_3$), 45.62 (C-3), 49.72 (C-2), 66.63 (Cbz-CH$_2$), 81.67 (*t*-Bu-C), 127.96 (CH-Ph), 128.06 (CH-Ph), 128.26 (CH-Ph), 136.72 (C-Ph), 156.66 (Cbz-C=O), 169.40 (C-1), 197.85 (C-4). **MS** (ESI): m/z = 308.2 [M+H]$^+$. **HRMS** (ESI): calcd for C$_{16}$H$_{22}$NO$_5$ 308.1492, found 308.1488 [M+H]$^+$. **[α]**$_D^{20}$ = +14.1 (c = 2.0, CHCl$_3$). **IR** (ATR): $\tilde{\nu}$ [cm^{-1}] = 3341, 2975, 2939, 1720, 1510, 1464, 1400, 1372, 1336, 1255, 1219,

1151, 1057, 970, 848, 778, 749. **UV** (MeCN): λ_{max} [nm] (log ε) = 206 (3.94), 258 (2.35). **TLC** (isohexane/EtOAc 4:1): R_f = 0.19.

7.2.10.8 Synthesis of *N*-(Benzyloxycarbonyl)-L-allylglycine *tert*-Butyl Ester 186

186

To a solution of methyltriphenylphosphonium bromide (227 mg, 0.636 mmol, 2.3 eq) in dry THF (2.2 mL), potassium hexamethyldisilazane (KHMDS) (0.7 M in toluene, 830 µL, 0.581 mmol, 2.1 eq) was added slowly under an argon atmosphere. The reaction mixture was stirred at room temperature for 30 min before a solution of aldehyde **184** (85 mg, 0.28 mmol, 1.0 eq) in dry THF (2 mL) was added. After 3 h, the reaction was quenched by the addition of saturated aqueous NH$_4$Cl solution. The mixture was extracted with Et$_2$O (2 x 5 mL). The combined organics were washed with water (5 mL) and brine (5 mL), and dried over Na$_2$SO$_4$. After removing the solvent under reduced pressure, the resultant crude product was purified by flash chromatography (isohexane/EtOAc 4:1).

Yield: 54 mg (0.18 mmol, 64%) as a yellowish oil.

^{1}H NMR (500 MHz, CDCl$_3$): δ (ppm) = 1.26 (s, 9H, *t*-Bu-CH$_3$), 2.36 (ddd, J = 13.1, 6.5, 6.5 Hz, 1H, H-3a), 2.45 (ddd, J = 13.1, 6.5, 6.5 Hz, 1H, H-3b), 4.46 (ddd, J = 6.5, 6.5, 6.5 Hz, 1H, H-2), 4.89 (d, J = 12.0 Hz, 2H, H-5), 5.03 (d, J = 12.3 Hz, 1H, Cbz-CH$_2$), 5.07 (d, J = 12.3 Hz, 1H, Cbz-CH$_2$), 5.31 (d, J = 6.5 Hz, 1H, NH), 5.52-5.63 (m, 1H, H-4), 7.01-7.23 (m, 5H, Ph). **^{13}C NMR** (126 MHz, C$_6$D$_6$): δ [ppm] = 27.52 (*t*-Bu-CH$_3$), 36.68 (C-3), 53.80 (C-2), 66.53 (Cbz-CH$_2$), 81.12 (*t*-Bu-C), 118.26 (C-5), 127.96 (CH-Ph), 128.06 (CH-Ph), 128.22 (CH-Ph), 132.49 (C-4), 136.89 (C-Ph), 155.52 (Cbz-C=O), 170.55 (C-1). **MS** (ESI): *m/z* = 328.2 [M+Na]$^+$. **HRMS** (ESI): calcd for C$_{17}$H$_{23}$NO$_4$Na 328.1519, found 328.1519 [M+Na]$^+$. [α]$_D^{20}$ = +10.8 (*c* = 1.2, CHCl$_3$). **IR** (ATR): $\tilde{\nu}$ [cm^{-1}] = 2975, 1708, 1505, 1376, 1348, 1255, 1219, 1151, 1051, 1028, 987, 930, 843, 743, 697. **UV** (MeCN): λ_{max} [nm] (log ε) = 207 (4.02), 258 (2.64). **TLC** (isohexane/EtOAc 4:1): R_f = 0.43.

7.2.10.9 Synthesis of (4*R*,1'*RS*)-*N*-Boc-4-(1'-Hydroxy-2'-propenyl)-2,2-dimethyl 1,3-oxazolidine 120[173]

120

To a solution of (*R*)-*Garner*'s aldehyde **(*R*)-98** (9.43 g, 41.1 mmol, 1.0 eq) in dry THF (460 mL), vinylmagnesium bromide (1 M in THF, 90.6 mL, 90.6 mmol, 2.2 eq) was added slowly at -78 °C under an argon atmosphere. The reaction mixture was stirred at -78 °C for 1 h, and then it was allowed to reach room temperature. After the addition of brine (100 mL), the aqueous layer was extracted with Et_2O (2 x 650 mL). The combined organics were washed with brine (2 x 250 mL) and dried over Na_2SO_4. The solvent was removed under reduced pressure, and the resultant crude product was purified by flash chromatography (*n*-hexane/EtOAc 3:1).

Yield: 8.00 g (31.1 mmol, 76%) as a yellowish oil.

1H NMR (500 MHz, DMSO-d_6, 80 °C): δ [ppm] = 1.38 (s, 1 x 3H, C(CH$_3$)$_2$), 1.42 (s, 1 x 9H, *t*-Bu-CH$_3$), 1.42 (s, 1 x 3H, C(CH$_3$)$_2$), 1.43 (s, 1 x 9H, *t*-Bu-CH$_3$), 1.45 (s, 1 x 3H, C(CH$_3$)$_2$), 1.48 (s, 1 x 3H, C(CH$_3$)$_2$), 3.73-3.80 (m, 1 x 1H, H-4), 3.80-3.95 (m, 1 x 2H, 1 x 3H, H-4, H-5), 4.10 (ddd, J = 5.4, 5.4, 5.4 Hz, 1 x 1H, H-1'), 4.37 (ddd, J = 4.9, 4.9, 4.9 Hz, 1 x 1H, H-1'), 4.78 (d, J = 5.4 Hz, 1 x 1H, OH), 4.81 (d, J = 4.9 Hz, 1 x 1H, OH), 5.03 (ddd, J = 10.5, 1.7, 1.7 Hz, 1 x 1H, H-3'a), 5.10 (ddd, J = 10.5, 2.3, 1.3 Hz, 1 x 1H, H-3'a), 5.14-5.21 (m, 2 x 1H, H-3'b), 5.78-5.88 (m, 2 x 1H, H-2'). **13C NMR** (126 MHz, DMSO-d_6, 80 °C): δ [ppm] = 27.09 (C(CH$_3$)$_2$), 27.12 (C(CH$_3$)$_2$), 28.58 (*t*-Bu-CH$_3$), 28.59 (*t*-Bu-CH$_3$, 2 x C(CH$_3$)$_2$), 61.24 (C-4), 61.54 (C-4), 63.34 (C-5), 64.34 (C-5), 70.98 (C-1'), 72.27 (C-1'), 79.57 (C(CH$_3$)$_2$), 79.69 (C(CH$_3$)$_2$), 93.81 (*t*-Bu-C), 93.98 (*t*-Bu-C), 114.98 (C-3'), 116.10 (C-3'), 138.28 (C-2'), 140.12 (C-2'), 152.19 (Boc-C=O), 152.28 (Boc-C=O). **MS** (ESI): m/z = 280.2 [M+Na]$^+$. **HRMS** (ESI): calcd for C$_{13}$H$_{23}$NO$_4$Na 280.1519, found 280.1523 [M+Na]$^+$. **TLC** (isohexane/EtOAc 3:1): R$_f$ = 0.32.

7.2.10.10 Synthesis of (4*R*,1'*RS*)-*N*-Boc-4-(1'-Acetoxy-2'-propenyl)-2,2-dimethyl 1,3-oxazolidine 187

187

To a solution of allyl alcohol **120** (2.00 g, 7.77 mmol, 1.0 eq) in dry CH_2Cl_2 (21 mL), pyridine (1.9 mL, 1.9 g, 24 mmol, 3.1 eq), acetic anhydride (2.2 mL, 2.4 g, 24 mmol, 3.1 eq) and 4-dimethylaminopyridine (93 mg, 0.76 mmol, 0.1 eq) were added sequentially under an argon atmosphere. The reaction mixture was stirred at room temperature for 5 h. The solvent was removed under reduced pressure, and the resultant crude product was purified by flash chromatography (isohexane/EtOAc 3:1).

Yield: 2.18 g (7.28 mmol, 94%) as a white solid.

¹H NMR (500 MHz, DMSO-d$_6$, 80 °C): δ [ppm] = 1.42 (s, 1 x 3H, C(CH$_3$)$_2$), 1.43 (s, 1 x 3H, C(CH$_3$)$_2$), 1.44 (s, 1 x 9H, *t*-Bu-CH$_3$), 1.46 (s, 1 x 9H, *t*-Bu-CH$_3$), 1.47 (s, 1 x 3H, C(CH$_3$)$_2$), 1.48 (s, 1 x 3H, C(CH$_3$)$_2$), 2.03 (s, 1 x 3H, OAc), 2.05 (s, 1 x 3H, OAc), 3.88-3.97 (m, 2 x 2H, H-5), 4.00-4.06 (m, 2 x 1H, H-4), 5.21-5.28 (m, 2 x 2H, H-3'), 5.54-5.56 (m, 1 x 1H, H-1'), 5.60-5.62 (m, 1 x 1H, H-1'), 5.78-5.90 (m, 2 x 1H, H-2'). **¹³C NMR** (126 MHz, DMSO-d$_6$, 80 °C): δ [ppm] = 21.07 (OAc), 21.14 (OAc), 26.55 (C(CH$_3$)$_2$), 26.62 (C(CH$_3$)$_2$), 26.81 (C(CH$_3$)$_2$), 26.85 (C(CH$_3$)$_2$), 28.48 (*t*-Bu-CH$_3$), 28.52 (*t*-Bu-CH$_3$), 58.72 (C-4), 59.32 (C-4), 63.59 (C-5), 63.89 (C-5), 73.25 (C-1'), 73.49 (C-1'), 80.03 (C(CH$_3$)$_2$), 80.07 (C(CH$_3$)$_2$), 94.05 (*t*-Bu-C), 94.18 (*t*-Bu-C), 117.70 (C-3'), 119.08 (C-3'), 133.43 (C-2'), 134.65 (C-2'), 169.39 (Boc-C=O), 169.39 (Boc-C=O), 186.73 (Ac-C=O), 187.12 (Ac-C=O). **MS** (ESI): *m/z* = 322.2 [M+Na]$^+$. **HRMS** (ESI): calcd for C$_{15}$H$_{25}$NO$_5$Na 322.1625, found 322.1627 [M+Na]$^+$. **TLC** (isohexane/EtOAc 3:1): R$_f$ = 0.32.

7.2.10.11 Synthesis of (4*S*)-*N*-Boc-4-(2'-Propenyl)-2,2-dimethyl 1,3-oxazolidine 119

119

To a solution of allyl acetate **187** (1.00 g, 3.34 mmol, 1.0 eq) in dry DMF (40 mL), sodium formate (2.72 g, 40.0 mmol, 12.0 eq), triethylamine (7.90 mL, 5.73 g, 56.6 mmol, 17.0 eq) and tetrakis(triphenylphosphine)palladium (386 mg, 0.334 mmol, 0.1 eq) were added sequentially under an argon atmosphere. The reaction mixture was heated at 65 °C for 48 h. After the addition of Et_2O (40 mL), the mixture was washed with water (2 x 30 mL) and brine (30 mL). The organic layer was dried over Na_2SO_4, and the solvent was removed under reduced pressure. The resultant crude product was purified by flash chromatography (isohexane/EtOAc 5:1).

Yield: 670 mg (2.78 mmol, 83%) as a colorless oil.

^{1}H NMR (500 MHz, DMSO-d_6, 80°C): δ [ppm] = 1.42 (s, 3H, C(CH$_3$)$_2$), 1.44 (s, 9H, t-Bu-CH$_3$), 1.50 (s, 3H, C(CH$_3$)$_2$), 2.22-2.28 (m, 1H, H-1'a), 2.39-2.44 (m, 1H, H-1'b), 3.70 (dd, J = 8.7, 1.8 Hz, 1H, H-5a), 3.84-3.90 (m, 1H, H-4), 3.93 (dd, J = 8.7, 6.0 Hz, 1H, H-5b), 5.04-5.08 (m, 2H, H-3'), 5.74-5.82 (m, 1H, H-2'). **^{13}C NMR** (126 MHz, DMSO-d_6, 80°C): δ [ppm] = 24.41 (C($\underline{C}$H$_3$)$_2$), 28.55 (t-Bu-CH$_3$), 27.72 (C($\underline{C}$H$_3$)$_2$), 37.76 (C-1'), 57.21 (C-4), 67.05 (C-5), 79.44 ($\underline{C}$(CH$_3$)$_2$), 93.44 (t-Bu-C), 117.65 (C-3'), 135.25 (C-2'), 151.82 (Boc-C=O). **MS** (ESI): m/z = 264.2 [M+Na]$^+$. **HRMS** (ESI): calcd for $C_{13}H_{23}NO_3Na$ 264.1570, found 264.1575 [M+Na]$^+$. $[\alpha]_D^{20}$ = 15.5 (c = 1.1, CHCl$_3$). **IR** (ATR): $\tilde{\nu}$ [cm^{-1}] = 2981, 1697, 1389, 1365, 1255, 1174, 1087, 1075, 1040, 1022, 994, 918, 848, 807, 773. **TLC** (isohexane/EtOAc 5:1): R_f = 0.49.

7.2.10.12 Synthesis of (4*S*,2'*S*)-and (4*S*,2'*R*)-*N*-Boc-4-(2',3'-Dihydroxypropyl)-2,2-dimethyl 1,3-oxazolidine (*S*)-91 and (*R*)-91

(2'*S*): (*S*)-**91**, (2'*R*): (*R*)-**91**

Variant 1

A solution of AD-mixα (584 mg) in *tert*-butanol (2.5 mL) and water (2.5 mL) was stirred for 10 min until both phases were clear, and then it was cooled to 0 °C. During this cooling procedure, an orange precipitate was formed. After adding olefin **119** (157 mg, 0.651 mmol, 1.0 eq) at 0 °C, the mixture was stirred for 9 h at 0 °C and 18 h at room temperature. The reaction was quenched by the addition of saturated aqueous $Na_2S_2O_3$ solution (1.5 mL). The mixture was extracted with EtOAc (2 x 5 mL), and the combined organics were dried over Na_2SO_4. After removing the solvent under reduced pressure, the resultant crude product was purified by flash chromatography (isohexane/EtOAc 5:1) to yield the desired compound **91** (140 mg, 0.508 mmol, 78%) as a diastereomeric mixture. A separation of the diastereomers could be achieved by a second flash chromatography (isohexane/EtOAc 1:1→3:1).

Yield (*S*)-91: 71 mg (0.26 mmol, 40%) as a colorless oil.

Yield (*R*)-91: 28 mg (0.10 mmol, 15%) as a colorless oil.

Variant 2

The reaction was carried out following the same procedure as *variant 1*, except with AD-mixβ (585 mg) and olefin **119** (126 mg, 0.522 mmol, 1.0 eq). After the first flash chromatographic purification, the diastereomeric mixture (130 mg, 0.472 mmol) was obtained in 90% yield.

Yield (*S*)-91: 35 mg (0.13 mmol, 25%) as a colorless oil.

Yield (*R*)-91: 47 mg (0.17 mmol, 33%) as a colorless oil.

(*S*)-91: **^{1}H NMR** (500 MHz, DMSO-d_6, 80 °C): δ [ppm] = 1.43 (s, 3H, C(CH$_3$)$_2$), 1.45 (s, 9H, *t*-Bu-CH$_3$), 1.49 (s, 3H, C(CH$_3$)$_2$), 1.42-1.51 (m, 1H, H-1'a), 1.88 (ddd, *J* = 13.4, 3.8, 3.8 Hz, 1H, H-1'b), 3.30 (dd, *J* = 10.9, 5.9 Hz, 1H, H-3'a), 3.33 (dd, *J* = 10.9, 5.9 Hz, 1H, H-3'b), 3.48-3.53 (m, 1H, H-2'), 3.83-3.89 (m, 1H, H-4), 3.89-3.95 (m, 2H, H-5), 4.17 (s$_{br}$, 1H, 2'-OH), 4.20 (s$_{br}$, 1H, 3'-OH). **^{13}C NMR** (500 MHz, DMSO-d_6, 80 °C): δ [ppm] = 23.56 (C(*C*H$_3$)$_2$), 26.60 (C(*C*H$_3$)$_2$), 27.71 (*t*-Bu-CH$_3$), 37.56 (C-1'), 55.66 (C-4), 65.80 (C-3'), 67.38 (C-5), 69.88 (C-2'), 78.52 (*C*(CH$_3$)$_2$), 91.93 (*t*-Bu-C), 150.97 (Boc-C=O). **MS** (ESI): *m/z* = 298.2 [M+Na]$^+$. **HRMS** (ESI): calcd for C$_{13}$H$_{25}$NO$_5$Na 298.1625, found 298.1617 [M+Na]$^+$. **[α]$_D^{20}$** = -13.7 (*c* = 0.82, CHCl$_3$). **IR** (ATR): $\tilde{\nu}$ [cm^{-1}] = 3417, 2981, 2935, 1697, 1667, 1453, 1400, 1365, 1261, 1174, 1151, 1109, 1045, 848, 807, 773. **TLC** (CH$_2$Cl$_2$/MeOH 9:1): R$_f$ = 0.45.

(*R*)-91: **^{1}H NMR** (500 MHz, DMSO-d_6, 80 °C): δ [ppm] = 1.43 (s, 3H, C(CH$_3$)$_2$), 1.45 (s, 9H, *t*-Bu-CH$_3$), 1.50 (s, 3H, C(CH$_3$)$_2$), 1.61 (ddd, *J* = 13.1, 9.3, 3.7 Hz, 1H, H-1'a), 1.73 (ddd, *J* = 13.1, 9.3, 3.2 Hz, 1H, H-1'b), 3.30 (dd, *J* = 10.8, 5.4 Hz, 1H, H-3'a), 3.35 (dd, *J* = 10.8, 5.4 Hz, 1H, H-3'b), 3.51 (dddd, *J* = 9.3, 9.3, 5.4, 5.4 Hz, 1H, H-2'), 3.81 (dd, *J* = 8.8, 2.0 Hz, 1H, H-5a), 3.90 (ddd, *J* = 8.8, 5.4, 1.1, 1H, H-5b), 3.98-4.07 (m, 1H, H-4), 4.11 (s$_{br}$, 1H, 2'-OH), 4.17 (s$_{br}$, 1H, 3'-OH). **^{13}C NMR** (500 MHz, DMSO-d_6, 80 °C): δ [ppm] = 26.61 (C(*C*H$_3$)$_2$), 27.73 (C(*C*H$_3$)$_2$), 30.02 (*t*-Bu-CH$_3$), 36.57 (C-1'), 54.00 (C-4), 65.65 (C-3'), 66.21 (C-5), 68.52 (C-2'), 78.39 (*C*(CH$_3$)$_2$), 92.13 (*t*-Bu-C), 150.78 (Boc-C=O). **MS** (ESI): *m/z* = 298.2 [M+Na]$^+$. **HRMS** (ESI): calcd for C$_{13}$H$_{25}$NO$_5$Na 298.1625, found 298.1633 [M+Na]$^+$. **[α]$_D^{20}$** = +23.2 (*c* = 1.3, CHCl$_3$). **IR** (ATR): $\tilde{\nu}$ [cm^{-1}] = 3237, 2963, 2940, 1697, 1684, 1464, 1395, 1365, 1308, 1261, 1214, 1179, 1092, 1034, 877, 843, 801, 778. **TLC** (CH$_2$Cl$_2$/MeOH 9:1): R$_f$ = 0.39.

7.2.10.13 Attempt to Synthesize (4*S*,2'*S*)- and (4*S*,2'*R*)-*N*-Boc-4-(2',3'-Dimethansulfonylpropyl)-2,2-dimethyl 1,3-oxazolidine (*S*)-188 and (*R*)-188

(2'*S*): (*S*)-188, (2'*R*): (*R*)-188

(*R*)-188: To a solution of the diol **(*R*)-91** (20 mg, 73 µmol, 1.0 eq) in dry CH_2Cl_2 (2 mL), triethylamine (40 µL, 29 mg, 0.29 mmol, 4.0 eq) was added under an argon atmosphere. The mixture was cooled to 0 °C before methanesulfonyl chloride (98%, 15 µL, 22 mg, 0.19 mmol, 2.6 eq) was added dropwise. After stirring the reaction for 1 h at 0 °C and 30 min at room temperature, it was quenched with water (2 mL), and CH_2Cl_2 (20 mL) was added. The mixture was washed with saturated aqueous $NaHCO_3$ solution (5 mL), 1 M aqueous HCl solution (5 mL) and brine (5 mL), and was dried over Na_2SO_4. After removing the solvent, the resultant crude product was purified by flash chromatography (isohexane/EtOAc 1:1).

(*S*)-188: The isomer **(*S*)-188** was prepared in the same way as compound **(*R*)-188** with **(*S*)-91** (23 mg, 84 µmol, 1.0 eq), triethylamine (50 µL, 37 mg, 0.366 mmol, 4.4 eq), methanesulfonyl chloride (98%, 20 µL, 30 mg, 0.26 mmol, 3.1 eq) and dry CH_2Cl_2 (2 mL). The resultant crude product was purified by flash chromatography (isohexane/EtOAc 1:3).

Yield (*R*)-188: 31 mg of a putative unstable product as a white solid (partially decomposed).

Yield (*S*)-188: 42 mg of a putative unstable product as a yellowish solid (partially decomposed).

(*R*)-188: **HRMS** (ESI): calcd for $C_{15}H_{29}NO_9S_2Na$ 454.1176, found 454.1185 $[M+Na]^+$. **TLC** (isohexane/EtOAc 1:1): $R_f = 0.21$.

(*S*)-188: **TLC** (isohexane/EtOAc 1:1): $R_f = 0.21$.

7.3   Fermentation Methodology

7.3.1   General Fermentation Procedures and Preparation of Fermentation Media

All indications of quantities are made for a 200 mL media end-volume, which in all cases was divided into three parts for culture preparation. Erlenmeyer flasks with baffles (250 mL) were used throughout all fermentation processes. All working steps were carried out in a sterile bench.

TSBG: 6 g CASO bouillon (*Sigma Aldrich*) in 150 mL water and 4 g glucose (*AppliChem*) in 50 mL water were sterilized in an autoclave and combined after cooling to room temperature.

BPM21: 7 g soy flour (*Hensel*, low-fat), 1.4 g $CaCO_3$ (*Merck*, for analysis) and 0.5 g L-methionine (*Merck*) in 100 mL water and 16 g maltodextrine (*HBK*, DE 19) and 1 g glucose (*AppliChem*) in 100 mL water were sterilized in an autoclave and combined after cooling to room temperature.

BPM23A: 4 g HY Yest 441, 1.4 g $CaCO_3$ (*Merck*, for analysis), 0.3 g L-methionine (*Merck*), 0.3 g L-leucin (*Merck*) and 0.3 g L-arginine (*Merck*) in 100 mL water and 16 g maltodextrine (*HBK*, DE 19) and 1 g glucose (*AppliChem*) in 100 mL water were sterilized in an autoclave and combined after cooling to room temperature.

7.3.2   Culture Preservation as Glycerol Stocks

Samples of *Streptomyces sp.* NRRL 30471-30477 were obtained from the Agricultural Research Service Culture Collection, Northern Regional Research Laboratory (NRRL). Only the sample NRRL 30471 was used in the following procedures.

The lyophilized cells were suspended in water. This suspension was used to inoculate different TSBG broths (3 x 60 mL). The cultures were incubated at 30 °C and 180 rpm for 5 days. The most densely populated culture was used for preparation of a glycerol stock. The cell culture was divided into 10 x 400 µL aliquots, and 120 µL glycerol (87%) was added to each cell culture and mixed afterwards. The glycerol suspensions were stored at -80 °C for later use.

7.3.3   Fermentation of *Streptomyces* sp. NRRL30471 and Purification of the Fermentation Broth

A TSBG pre-culture (60 mL) was inoculated from a glycerol stock (15 µL) and incubated at 30 °C and 130 rpm for 2 d. This pre-culture (1 mL inoculation volume) was used to inoculate a BPM21 main-culture (60 mL), which was incubated at 26 °C and 180 rpm for 5 d. The resultant fermentation broth was mixed with MeOH (1:1), and the cells/solids were removed by centrifugation (4000 rpm, 2 x 45 min, 20 °C). The methanol-treated broth extracts (20 mL) were applied to a *Bakerbond* spe^{TM} carboxylic acid column (3 mL), which had been equilibrated with MeOH/H_2O (1:1). The column was washed with three column volumes of MeOH/H_2O (1:1) and eluted with acidic MeOH/H_2O (70:30) with 0.05% TFA. The eluent was dried under reduced pressure, and the samples were resuspended in MeOH/H_2O (1:1) and used for LC-MS analysis. Larger amounts of methanolic fermentation broth (150 mL) were purified by using Diaion® WT01S ion exchange resin (50 g).

7.3.4 Verification of Muraymycin Production by LC-MS

Compounds were analytically separated on a *Phenomenex* Synergi column (4 µm, MAX-RP, 150 x 2 mm), which was operated by a *Jasco* 851-AS auto sampler and a *Flux Instruments* Rheos 4000 pump, using a linear gradient from 10% to 100% mobile phase B (MeCN + 0.1% formic acid) in mobile phase A (H_2O + 0.1% formic acid) over 22 min with a flow of 0.8 mL/min. The compounds were detected with a *Finnigan* PDA detector and a *Finnigan* ion trap mass spectrometer LCQ. MS/MS experiments were conducted under the same conditions.

7.3.5 Attempts to Isolate Muraymycin Derivatives by HPLC

Analytical HPLC was conducted using a *Hitachi LaChrom Elite*® system, which was purchased from *VWR* and consists of an organizer, a L-2300 pump, a L-2200 auto sampler, a L-2300 column oven and a L-2455 diode array detector (DAD). The column oven was operated at 24 °C throughout all HPLC runs. For separation, a *VWR* LichroCart®Purospher® RP18e column (5 µm, 4 x 125 mm) with a corresponding precolumn (4 x 4 mm) was used. The following gradient conditions were used with solvent mixtures of water (A) and MeCN (B) and a flow of 1 mL/min:

t [min]	0	10	15	20
B [%]	0	10	50	100

Semi-preparative HPLC was conducted on the same HPLC system using a LichroCart®Purospher® RP18e (10 µm, 10 x 250 mm) column. For injection, the Diaion® WT01S purified fermentation broth residue (628 mg) was dissolved in 1 mL water (injection volume 1 mL). The following gradient conditions were used with solvent mixtures of water (A) and MeCN (B) and a flow of 10 mL/min:

t [min]	0	10	20	25	35	45
B [%]	2	2	10	20	50	100

In the fractions with retention times from 17 to 40 min, different muraymycin derivatives could be detected by LC-MS (method see chapter 7.3.4). No significant separation of muraymycin derivatives could be achieved, but other impurities could be removed. After removing the solvent under reduced pressure, a mixture of muraymycins (72 mg) could be isolated.

7.3.6 Feeding Experiments

The fermentation broth extracts from feeding experiments were prepared as described in chapter 7.3.3 with pulse feeding of uridine, $[5,5\text{-}^2H]$-uridine, glycine and $[1\text{-}^{13}C]$glycine. Prior to use, the compounds were sterilized by filtration. A *Sartorius* membrane filter Minisart RC (diameter 25 mm, 0.2 µm) was used. The following pulse feeding conditions were used.

compound	concentration c [mg/mL]	addition after 1 d	addition after 2 d	addition after 3 d
uridine	7	0.5 mL	0.5 mL	0.5 mL
$[5,5\text{-}^2H]$-uridine	15	1.0 mL	1.0 mL	1.0 mL
	15	0.5 mL	0.5 mL	0.5 mL
glycine	100	1.0 mL	1.0 mL	-
$[1\text{-}^{13}C]$glycine	100	1.0 mL	1.0 mL	-

The fermentation broth extracts were analyzed by LC-MS.

8   References

[1] L. Pasteur, Inaugural lecture as professor and dean of the faculty of science, University of Lille, Douai, France, **1854**, available from *"A Treasury of the World's Great Speeches"*, ed. Houston Peterson, Simon and Schuster inc, New York **1954**, 473.

[2] Helmholtz-Zentrum, *Antibiotika und Antibiotikaresistenzen*, Helmholtz-Zentrum München, FLUGS-Fachinformationsdienst **2007**.

[3] Bundesamt für Verbraucherschutz und Lebensmittelsicherheit, *Antibiotika-Resistenz und -Verbrauch*, GERMAP **2008**.

[4] C. Walsh, Where will new antibiotics come from?, *Nat. Rev. Microbiol.* **2003**, *1*, 65-70.

[5] F. von Nussbaum, M. Brands, B. Hinzen, S. Weigand, D. Häbich, Antibacterial Natural Products in Medicinal Chemistry - Exodus or Revival?, *Angew. Chem. Int. Ed.* **2006**, *45*, 5072-5129; *Angew. Chem.* **2006**, *118*, 5194-5254.

[6] M. S. Butler, A. D. Buss, Natural products - the future scaffolds for novel antibiotics, *Biochem. Pharmacol.* **2006**, *71*, 919-929.

[7] E. Chain, H. W. Florey, A. D. Gardner, N. G. Heatley, M. A. Jennings, J. Orr-Ewing, A. G. Sanders, Penicillin as a chemotherapeutic agent, *Lancet* **1940**, *236*, 226-228.

[8] A. Fleming, On the antibacterial action of a penicillium, with special reference to their use in the isolation of B. influenzae, *Bulletin of the World Health Organization* **2001**, *79*, 780-790 (reprint); *Br. J. Exp. Pathol.* **1929**, *10*, 226-236.

[9] K. Bush, Antibacterial drug discovery in the 21st century, *Clin. Microbiol. Infect.* **2004**, *10*, 10-17.

[10] H. Dellweg, *Mikroben im Kampf gegen Mikroben*, Springer, Berlin/Heidelberg **1994**, 62-69.

[11] J. O. Falkinham, T. E. Wall, J. R. Tanner, K. Tawaha, F. Q. Alali, C. Li, N. H. Oberlies, Proliferation of Antibiotic-Producing Bacteria and Concomitant Antibiotic Production as the Basis for the Antibiotic Activity of Jordan's Red Soils, *Appl. Environ. Microbiol.* **2009**, *75*, 2735-2741.

[12] J. Yanling, L. Xin, L. Zhiyuan, The Antibacterial Drug Discovery, **2013**, available from http://www.intechopen.com/books/drug-discovery/the-antibacterial-drug-discovery, (accessed 05.08.2013).

[13]   J. Brunel, Antibiosis from Pasteur to Fleming, *J. Hist. Med. Allied Sci.* **1951**, *VI*, 287-301.

[14]   P. Vuillemin, Antibiose et symbiose, *Assoc. Franc. Pour l'Avanc. Des Sciences* **1889**, *2*, 525-542.

[15]   S. A. Waksman, History of the Word Antibiotic, *J. Hist. Med.* **1973**, *28*, 284-286.

[16]   K. Lewis, Platforms for antibiotic discovery, *Nat. Rev. Drug Discov.* **2013**, *12*, 371-387.

[17]   E. P. Abraham, E. Chain, An Enzyme From Bacteria Able To Destroy Penicillin (reprinted from *Nature* **1940**, *146*, 837), *Rev. Infect. Dis.* **1988**, *10*, 677-678.

[18]   A. E. Clatworthy, E. Pierson, D. T. Hung, Targeting virulence: a new paradigm for antimicrobial therapy, *Nat. Chem. Biol.* **2007**, *3*, 541-548.

[19]   R. Hakenbeck, T. Grebe, D. Zähner, J. B. Stock, β-Lactam resistance in *Streptococcus pneumoniae*: penicillin-binding proteins and non-penicillin-binding proteins, *Mol. Microbiol.* **1999**, *33*, 673-678.

[20]   X. Z. Li, H. Nikaido, Efflux-Mediated Drug Resistance in Bacteria, *Drugs* **2004**, *64*, 159-204.

[21]   P. A. Lambert, Cellular impermeability and uptake of biocides and antibiotics in Gram-positive bacteria and mycobacteria, *J. App. Microbiol.* **2002**, *92*, 46S-54S.

[22]   M. R. Klevens, M. A. Morrison, J. Nadle, S. Petit, K. Gershman, S. Ray, L. H. Harrison, R. Lynfield, G. Dumyati, J. M. Townes, A. S. Craig, E. R. Zell, G. E. Fosheim, L. K. McDougal, R. B. Care, S. K. Fridkin, Invasive Methicillin-Resistant *Staphylococcus aureus* Infections in the United States, *J. Am. Med. Ass.* **2007**, *298*, 1763-1771.

[23]   B. Wiedemann, Antibiotikaanwendung und Resistenzentwicklung, *Krankenh. hyg. up2date* **2007**, *2*, 21-36.

[24]   B. Spellberg, J. H. Powers, E. P. Brass, L. G. Miller, J. E. Edwards, Trends in Antimicrobial Drug Development: Implications for the Future, *Clin. Infect. Dis.* **2004**, *38*, 1279-1286.

[25]   B. Heym, S. T. Cole, Multidrug resistance in *Mycobacterium tuberculosis*, *Int. J. Antimicrob. Agents* **1997**, *8*, 61-70.

[26]   S. Altschul, Superbugs: A ticking time bomb, **2013**, available from http://www.cbsnews.com/8301-3445_162-57566049/superbugs-a-ticking-time-bomb/, (accessed 25.07.2013).

[27]   M. Roberts, Europe 'losing' superbugs battle, **2011**, available from http://www.bbc.co.uk/news/health-12975693, (accessed 25.07.2013).

[28]   M. Adams, How medicine is killing us all: Antibiotics, superbugs and the next global pandemics, **2013**, available from http://www.naturalnews.com/039425_gram-negative_superbugs_antibiotics.html, (accessed 25.07.2013).

[29]   D. J. Payne, M. N. Gwynn, D. J. Holmes, D. L. Pompliano, Drugs for bad bugs: confronting the challenges of antibacterial discovery, *Nat. Rev. Drug Discov.* **2007**, *6*, 29-40.

[30]   M. G. Kees, Strategien zur Vermeidung von Antibiotikaresistenzen, *Med. Klin. Intensivmed. Notfallmed.* **2013**, *108*, 125-130.

[31]   N. I. Paphitou, Antimicrobial resistance: action to combat the rising microbial challenges, *Int. J. Antimicrob. Agents* **2013**, *42, Supplement 1*, S25-S28.

[32]   Bundesministerium für Bildung und Forschung, Deutsche Antibiotika-Resistenzstrategie - 2. Zwischenbericht, **2011**, available from http://www.bmbf.de/pubRD/Zwischenbericht_DART.pdf (accessed 05.08.2013).

[33]   H. W. Boucher, G. H. Talbot, D. K. Benjamin, J. Bradley, R. J. Guidos, R. N. Jones, B. E. Murray, R. A. Bonomo, D. Gilbert, 10 × '20 Progress-Development of New Drugs Active Against Gram-Negative Bacilli: An Update From the Infectious Diseases Society of America, *Clin. Infect. Dis.* **2013**, *56*, 1685-1694.

[34]   G. Devasahayam, W. M. Scheld, P. S. Hoffman, Newer antibacterial drugs for a new century, *Expert Opin. Investig. Drugs* **2010**, *19*, 215-234.

[35]   N. Y. Yount, M. R. Yeaman, Peptide antimicrobials: cell wall as a bacterial target, *Ann. N. Y. Acad. Sci.* **2013**, *1277*, 127-138.

[36]   K. Lewis, Recover the lost art of drug discovery, *Nature* **2012**, *485*, 439-440.

[37]   J. Clardy, M. A. Fischbach, C. T. Walsh, New antibiotics from bacterial natural products, *Nat. Biotech.* **2006**, *24*, 1541-1550.

[38]   A. A. Brakhage, V. Schroeckh, Fungal secondary metabolites – Strategies to activate silent gene clusters, *Fungal Genet. Biol.* **2011**, *48*, 15-22.

[39]   I. Chopra, L. Hesse, A. J. O. Neill, Exploiting current understanding of antibiotic action for discovery of new drugs, *J. Appl. Microbiol.* **2002**, *92*, 4S-15S.

[40]   A. P. Carter, W. M. Clemons, D. E. Brodersen, R. J. Morgan-Warren, B. T. Wimberly, V. Ramakrishnan, Functional insights from the structure of the 30S ribosomal subunit and its interactions with antibiotics, *Nature* **2000**, *407*, 340-348.

[41]   J. L. Hansen, J. A. Ippolito, N. Ban, P. Nissen, P. B. Moore, T. A. Steitz, The Structures of Four Macrolide Antibiotics Bound to the Large Ribosomal Subunit, *Mol. Cell* **2002**, *10*, 117-128.

[42] F. Schlünzen, R. Zarivach, J. Harms, A. Bashan, A. Tocilj, R. Albrecht, A. Yonath, F. Franceschi, Structural basis for the interaction of antibiotics with the peptidyl transferase centre in eubacteria, *Nature* **2001**, *413*, 814-821.

[43] F. Collin, S. Karkare, A. Maxwell, Exploiting bacterial DNA gyrase as a drug target: current state and perspectives, *Appl. Microbiol. Biotechnol.* **2011**, *92*, 479-497.

[44] W. Vollmer, D. Blanot, M. A. de Pedro, Peptidoglycan structure and architecture, *FEMS Microbiol. Rev.* **2008**, *32*, 149-167.

[45] J. M. Berg, J. L. Tymoczko, L. Stryer, *Biochemie*, 5 ed., Spektrum Akademischer Verlag, Berlin **2003**, 236-237.

[46] T. D. H. Bugg, D. Braddick, C. G. Dowson, D. I. Roper, Bacterial cell wall assembly: still an attractive antibacterial target, *Trends Biotechnol.* **2011**, *29*, 167-173.

[47] J. van Heijenoort, Recent advances in the formation of the bacterial peptidoglycan monomer unit, *Nat. Prod. Rep.* **2001**, *18*, 503-519.

[48] H. Barreteau, A. Kovac, A. Boniface, M. Sova, S. Gobec, D. Blanot, Cytoplasmic steps of peptidoglycan biosynthesis, *FEMS Microbiol. Rev.* **2008**, *32*, 168-207.

[49] T. D. H. Bugg, C. T. Walsh, Intracellular Steps of Bacterial Cell Wall Peptidoglycan Biosynthesis: Enzymology, Antibiotics, and Antibiotic Resistance, *Nat. Prod. Rep.* **1992**, *9*, 199-215.

[50] A. Bouhss, A. E. Trunkfield, T. D. H. Bugg, D. Mengin-Lecreulx, The biosynthesis of peptidoglycan lipid-linked intermediates, *FEMS Microbiol. Rev.* **2008**, *32*, 208-233.

[51] K.-i. Kimura, T. D. H. Bugg, Recent advances in antimicrobial nucleoside antibiotics targeting cell wall biosynthesis, *Nat. Prod. Rep.* **2003**, *20*, 252–273.

[52] T. Tanino, B. Al-Dabbagh, D. Mengin-Lecreulx, A. Bouhss, H. Oyama, S. Ichikawa, A. Matsuda, Mechanistic Analysis of Muraymycin Analogues: A Guide to the Design of MraY Inhibitors, *J. Med. Chem.* **2011**, *54*, 8421-8439.

[53] D. D. Pless, F. C. Neuhaus, Initial Membrane Reaction in Peptidoglycan Synthesis, *J. Biol. Chem.* **1973**, *248*, 1568-1576.

[54] F. C. Neuhaus, Initial translocation reaction in the biosynthesis of peptidoglycan by bacterial membranes, *Acc. Chem. Res.* **1971**, *4*, 297-303.

[55] M. G. Heydanek, W. G. Struve, F. C. Neuhauss, On the Initial Stage in Peptidoglycan Synthesis. 111. Kinetics and Uncoupling of Phospho-*N*-

acetylmuramyl-pentapeptide Translocase (Uridine 5'-Phosphate), *Biochemistry* **1969**, *8*, 1214-1221.

[56]   A. Bouhss, M. Crouvoisier, D. Blanot, D. Mengin-Lecreulx, Purification and Characterization of the Bacterial MraY Translocase Catalyzing the First Membrane Step of Peptidoglycan Biosynthesis, *J. Biol. Chem.* **2004**, *279*, 29974-29980.

[57]   M. Ikeda, M. Wachi, H. K. Jung, F. Ishino, M. Matsuhashi, The Escherichia coli *mraY* Gene Encoding UDP-*N*-Acetylmuramoyl-Pentapeptide:Undecaprenyl-Phosphate Phospho-*N*-Acetylmuramoyl-Pentapeptide Transferase, *J. Bacteriol.* **1991**, *173*, 1021-1026.

[58]   A. Bouhss, D. Mengin-Lecreulx, D. Le Beller, J. van Heijenoort, Topological analysis of the MraY protein catalysing the first membrane step of peptidoglycan synthesis, *Mol. Microbiol.* **1999**, *34*, 576-585.

[59]   A. J. Lloyd, P. E. Brandish, A. M. Gilbey, T. D. H. Bugg, Phospho-*N*-Acetyl-Muramyl-Pentapeptide Translocase from *Escherichia coli*: Catalytic Role of Conserved Aspartic Acid Residues, *J. Bacteriol.* **2004**, *186*, 1747-1757.

[60]   Y. Ma, D. Münch, T. Schneider, H.-G. Sahl, A. Bouhss, U. Ghoshdastider, J. Wang, V. Dötsch, X. Wang, F. Bernhard, Preparative Scale Cell-free Production and Quality Optimization of MraY Homologues in Different Expression Modes, *J. Biol. Chem.* **2011**, *286*, 38844-38853.

[61]   B. C. Chung, J. Zhao, R. A. Gillespie, D.-Y. Kwon, Z. Guan, J. Hong, P. Zhou, S.-Y. Lee, Crystal Structure of MraY, an Essential Membrane Enzyme for Bacterial Cell Wall Synthesis, *Science* **2013**, *341*, 1012-1016.

[62]   M. Winn, R. J. M. Goss, K.-i. Kimura, T. D. H. Bugg, Antimicrobial nucleoside antibiotics targeting cell wall assembly: Recent advances in structure–function studies and nucleoside biosynthesis, *Nat. Prod. Rep.* **2010**, *27*, 279-304.

[63]   J. P. Karwowski, M. Jackson, R. J. Theriault, R. H. Chen, G. J. Barlow, M. L. Maus, Pacidamycins, a novel series of antibiotics with anti-*Pseudomonas* aeruginosa activity, *J. Antibiot.* **1989**, *42*, 506-511.

[64]   M. Inukai, F. Isono, S. Takahashi, R. Enokita, Y. Sakaida, T. Haneishi, Mureidomycins A-D, novel peptidylnucleoside antibiotics with sphereoplast forming activity, *J. Antibiot.* **1989**, *42*, 662-679.

[65]   S. Chatterjee, S. R. Nadakarni, E. K. S. Vijayakumar, M. V. Patel, B. N. Gangul, H.-W. Fehlhaber, L. Vertesy, Napsamycins, new *Pseudomonas* active antibiotics of the mureidomycin family from *Streptomyces* sp. HIL Y-82, 11372, *J. Antibiot.* **1994**, *47*, 595-598.

[66]   Y. Xie, R. Chen, S. Si, C. Sun, H. Xu, A New Nucleosidyl-peptide Antibiotic, Sansanmycin, *J. Antibiot.* **2007**, *60*, 158-161.

[67]   K. Isono, M. Uramoto, H. Kusakabe, K.-I. Kumira, K. Izaki, C. C. Nelson, J. A. McCloskey, Liposidomycin: Novel nucleoside antibiotics with inhibit bacterial peptidoglycan synthesis, *J. Antibiot.* **1985**, *38*, 1617-1621.

[68]   M. Igarashi, N. Nakagawa, N. Doi, S. Hattori, H. Naganawa, M. Hamada, Caprazamycin B, a Novel Anti-tuberculosis Antibiotic, from *Streptomyces* sp., *J. Antibiot.* **2003**, *56*, 580-583.

[69]   Y. Muramatsu, Y. Fujita, A. Aoyagi, M. Kizuka, T. Takatsu, S. Miyakoshi, WO2004/046368 A1, *PCT Int. Appl* **2004**.

[70]   Y. Fujita, M. Kizuka, M. Funabashi, Y. Ogawa, T. Ishikawa, K. Nonaka, T. Takatsu, A-90289 A and B, new inhibitors of bacterial translocase I, produced by *Streptomyces* sp. SANK 60405, *J. Antibiot.* **2011**, *64*, 495-501.

[71]   L. A. McDonald, L. R. Barbieri, G. T. Carter, E. Lenoy, J. Lotvin, P. J. Petersen, M. M. Siegel, G. Singh, R. T. Williamson, Structures of the Muraymycins, Novel Peptidoglycan Biosynthesis Inhibitors, *J. Am. Chem. Soc.* **2002**, *124*, 10260-10261.

[72]   M. Ezaki, M. Iwani, M. Kohsaka, A. Komori, K. Ochi, FR-900493 substance, a process for its production and a pharmaceutical composition containing the same, EP0333177A2, *PCT Int. Appl.* **1989**.

[73]   A. Takatsuki, K. Arima, G. Tamura, Tunicamycin, a new antibiotic. I. Isolation and characterization of tunicamycin, *J. Antibiot.* **1971**, *24*, 215-223.

[74]   K. Eckardt, Tunicamycins, streptovirudins, and corynetoxins, a special subclass of nucleoside antibiotics, *Nat. Prod.* **1983**, *46*, 544-550.

[75]   P. Vogel, D. S. Petterson, P. H. Berry, J. L. Frahn, N. Anderton, P. A. Cockrum, J. A. Edgar, M. V. Jago, G. W. Lanigan, A. L. Payne, C. C. Culvenor, Isolation of a group of glycolipid toxins from seedheads of annual ryegrass Lolium rigidum Gaud infected by Corynebacterium rathayi, *Aust. J. Exp. Biol. Med.* **1981**, *59*, 455-467.

[76]   H. Yamaguch, S. Sato, S. Y. K, M. Itoh, H. Seto, N. Otake, Capuramycin, a new nucleoside antibiotic. Taxonomy, fermentation, isolation and characterization, *J. Antibiot.* **1986**, *39*, 1047-1053.

[77]   S. Suzuki, I. Saburo, K. Isono, N. Kiyoshi, A. Junsaku, T. Akashiba, S. Sasaki, Polyoxin A and B, JP 39015520 B 19640901, *Jpn. Tokkyo Koho* **1964**.

[78]   U. Dähn, H. Hagenmaier, H. Höhne, W. A. König, G. Wolf, H. Zähner, Stoffwechselprodukte von Mikroorganismen, *Archiv. Microbiol.* **1976**, *107*, 143-160.

[79]   C. T. Walsh, W. Zhang, Chemical Logic and Enzymatic Machinery for Biological Assembly of Peptidyl Nucleoside Antibiotics, *Chem. Biol.* **2011**, *6*, 1000-1007.

[80]   Y.-I. Lin, Z. Li, G. D. Francisco, L. A. McDonald, R. A. Davis, G. Singh, Y. Yang, T. S. Mansour, Muraymycins, Novel Peptidoglycan Biosynthesis Inhibitors: Semisynthesis and SAR of Their Derivatives, *Bioorg. Med. Chem. Lett.* **2002**, *12*, 2341-2344.

[81]   A. Yamashita, E. Norton, P. J. Petersen, B. A. Rasmussen, G. Singh, Y. Yang, T. S. Mansour, D. M. Ho, Muraymycins, Novel Peptidoglycan Biosynthesis Inhibitors: Synthesis and SAR of Their Analogues, *Bioorg. Med. Chem. Lett.* **2003**, *13*, 3345-3350.

[82]   C. Bormann, V. Möhrle, C. Bruntner, Cloning and heterologous expression of the entire set of structural genes for nikkomycin synthesis from *Streptomyces tendae* Tü901 in *Streptomyces lividans*, *J. Bacteriol.* **1996**, *178*, 1216-1218.

[83]   L. Kaysser, L. Lutsch, S. Siebenberg, E. Wemakor, B. Kammerer, B. Gust, Identification and Manipulation of the Caprazamycin Gene Cluster Lead to New Simplified Liponucleoside Antibiotics and Give Insights into the Biosynthetic Pathway, *J. Biol. Chem.* **2009**, *284*, 14987-14996.

[84]   M. Funabashi, K. Nonaka, C. Yada, M. Hosobuchi, N. Masuda, T. Shibata, S. G. Van Lanen, Identification of the biosynthetic gene cluster of A-500359s in *Streptomyces griseus* SANK60196, *J. Antibiot.* **2009**, *62*, 325-332.

[85]   W. Zhang, B. Ostash, C. T. Walsh, Identification of the biosynthetic gene cluster for the pacidamycin group of peptidyl nucleoside antibiotics, *Proc. Natl. Acad. Sci. USA* **2010**, *107*, 16828-16833.

[86]   E. J. Rackham, S. Grüschow, A. E. Ragab, S. Dickens, R. J. M. Goss, Pacidamycin Biosynthesis: Identification and Heterologous Expression of the First Uridyl Peptide Antibiotic Gene Cluster, *ChemBioChem* **2010**, *11*, 1700-1709.

[87]   L. Kaysser, S. Siebenberg, B. Kammerer, B. Gust, Analysis of the Liposidomycin Gene Cluster Leads to the Identification of New Caprazamycin Derivatives, *ChemBioChem* **2010**, *11*, 191-196.

[88]   M. Funabashi, S. Baba, K. Nonaka, M. Hosobuchi, Y. Fujita, T. Shibata, S. G. Van Lanen, The Biosynthesis of Liposidomycin-like A-90289 Antibiotics Featuring a New Type of Sulfotransferase, *ChemBioChem* **2010**, *11*, 184-190.

[89]   W. Chen, D. Qu, L. Zhai, M. Tao, Y. Wang, S. Lin, N. J. Price, Z. Deng, Characterization of the tunicamycin gene cluster unveiling unique steps involved in its biosynthesis, *Protein Cell* **2010**, *1*, 1093-1105.

[90]   L. Kaysser, X. Tang, E. Wemakor, K. Sedding, S. Hennig, S. Siebenberg, B. Gust, Identification of a Napsamycin Biosynthesis Gene Cluster by Genome Mining, *ChemBioChem* **2011**, *12*, 477-487.

[91]   L. Cheng, W. Chen, L. Zhai, D. Xu, T. Huang, S. Lin, X. Zhou, Z. Deng, Identification of the gene cluster involved in muraymycin biosynthesis from *Streptomyces* sp. NRRL 30471, *Mol. BioSyst.* **2011**, *7*, 920-927.

[92]   L. Wang, Y. Xie, Q. Li, N. He, E. Yao, H. Xu, Y. Yu, R. Chen, B. Hong, Draft Genome Sequence of *Streptomyces* sp. Strain SS, Which Produces a Series of Uridyl Peptide Antibiotic Sansanmycins, *J. Bacteriol.* **2012**, *194*, 6988-6989.

[93]   X. Chi, S. Baba, N. Tibrewal, M. Funabashi, K. Nonaka, S. G. Van Lanen, The muraminomicin biosynthetic gene cluster and enzymatic formation of the 2-deoxyaminoribosyl appendage, *MedChemComm* **2013**, *4*, 239-243.

[94]   B. C. Tsvetanova, D. J. Kiemle, N. P. J. Price, Biosynthesis of Tunicamycin and Metabolic Origin of the 11-Carbon Dialdose Sugar, Tunicamine, *J. Biol. Chem.* **2002**, *277*, 35289-35296.

[95]   A. E. Ragab, S. Grüschow, D. R. Tromans, R. J. M. Goss, Biogenesis of the Unique 4',5'-Dehydronucleoside of the Uridyl Peptide Antibiotic Pacidamycin, *J. Am. Chem. Soc.* **2011**, *133*, 15288-15291.

[96]   C. Ginj, H. Rüegger, N. Amrhein, P. Macheroux, 3'-Enolpyruvyl-UMP, a Novel and Unexpected Metabolite in Nikkomycin Biosynthesis, *ChemBioChem* **2005**, *6*, 1974-1976.

[97]   G. Oberdorfer, A. Binter, C. Ginj, P. Macheroux, K. Gruber, Structural and Functional Characterization of NikO, an Enolpyruvyl Transferase Essential in Nikkomycin Biosynthesis, *J. Biol. Chem.* **2012**, *287*, 31427-31436.

[98]   X. Chi, P. Pahari, K. Nonaka, S. G. Van Lanen, Biosynthetic Origin and Mechanism of Formation of the Aminoribosyl Moiety of Peptidyl Nucleoside Antibiotics, *J. Am. Chem. Soc.* **2011**, *133*, 14452-14459.

[99]   B. Gust, K. Eitel, X. Tang, The biosynthesis of caprazamycins and related liponucleoside antibiotics: new insights, *Biol. Chem.* **2013**, *394*, 251-259.

[100] S. Barnard-Britson, X. Chi, K. Nonaka, A. P. Spork, N. Tibrewal, A. Goswami, P. Pahari, C. Ducho, J. Rohr, S. G. Van Lanen, Amalgamation of Nucleosides and Amino Acids in Antibiotic Biosynthesis: Discovery of an L-Threonine:Uridine-5'-Aldehyde Transaldolase, *J. Am. Chem. Soc.* **2012**, *134*, 18514-18517.

[101] M. Funabashi, S. Baba, T. Takatsu, M. Kizuka, Y. Ohata, M. Tanaka, K. Nonaka, A. P. Spork, C. Ducho, W.-C. L. Chen, S. G. Van Lanen, Structure-Based Gene

Targeting Discovery of Sphaerimicin, a Bacterial Translocase I Inhibitor, *Angew. Chem. Int. Ed.* **2013**, *52*, 11607-11611; *Angew. Chem.* **2013**, *125*, 11821-11825.

[102] L. Kaysser, K. Eitel, T. Tanino, S. Siebenberg, A. Matsuda, S. Ichikawa, B. Gust, A New Arylsulfate Sulfotransferase Involved in Liponucleoside Antibiotic Biosynthesis in Streptomycetes, *J. Biol. Chem.* **2010**, *285*, 12684-12694.

[103] D. Schwarzer, R. Finking, M. A. Marahiel, Nonribosomal peptides: from genes to products, *Nat. Prod. Rep.* **2003**, *20*, 275-287.

[104] J. F. Martin, α-Aminoadipyl-cysteinyl-valine Synthetases in β-Lactam Producing Organisms, from ABRAHAM'S Discoveries to Novel Concepts of Non-Ribosomal Peptide Synthesis, *J. Antibiot.* **2000**, *53*, 1008-1021.

[105] W. Zhang, I. Ntai, M. L. Bolla, S. J. Malcolmson, D. Kahne, N. L. Kelleher, C. T. Walsh, Nine Enzymes Are Required for Assembly of the Pacidamycin Group of Peptidyl Nucleoside Antibiotics, *J. Am. Chem. Soc.* **2011**, *133*, 5240-5243.

[106] W. Zhang, I. Ntai, N. L. Kelleher, C. T. Walsh, tRNA-dependent peptide bond formation by the transferase PacB in biosynthesis of the pacidamycin group of pentapeptidyl nucleoside antibiotics, *Proc. Natl. Acad. Sci. USA* **2011**, *108*, 12249-12253.

[107] C. Rausch, T. Weber, O. Kohlbacher, W. Wohlleben, D. H. Huson, Specificity prediction of adenylation domains in nonribosomal peptide synthetases (NRPS) using transductive support vector machines (TSVMs), *Nucleic Acids Res.* **2005**, *33*, 5799-5808.

[108] J. Ju, S. G. Ozanick, B. Shen, M. G. Thomas, Conversion of (2*S*)-Arginine to (2*S*,3*R*)-Capreomycidine by VioC and VioD from the Viomycin Biosynthetic Pathway of *Streptomyces* sp. Strain ATCC11861, *ChemBioChem* **2004**, *5*, 1281-1285.

[109] X. Yin, K. L. McPhail, K.-j. Kim, T. M. Zabriskie, Formation of the Nonproteinogenic Amino Acid (2*S*,3*R*)-Capreomycidine by VioD from the Viomycin Biosynthesis Pathway, *ChemBioChem* **2004**, *5*, 1278–1281.

[110] X. Yin, T. M. Zabriskie, VioC is a Non-Heme Iron, α-Ketoglutarate-Dependent Oxygenase that Catalyzes the Formation of (3*S*)-Hydroxy-L-Arginine during Viomycin Biosynthesis, *ChemBioChem* **2004**, *5*, 1274-1277.

[111] H. J. Imker, C. T. Walsh, W. M. Wuest, SylC Catalyzes Ureido-Bond Formation During Biosynthesis of the Proteasome Inhibitor Syringolin A, *J. Am. Chem. Soc.* **2009**, *131*, 18263-18265.

[112] G. H. Hur, C. R. Vickery, M. D. Burkart, Explorations of catalytic domains in non-ribosomal peptide synthetase enzymology, *Nat. Prod. Rep.* **2012**, *29*, 1074-1098.

[113] A. G. Jamieson, N. Boutard, D. Sabatino, W. D. Lubell, Peptide Scanning for Studying Structure-Activity Relationships in Drug Discovery, *Chem. Biol. Drug Des.* **2013**, *81*, 148-165.

[114] T. Kawakami, H. Murakami, Genetically Encoded Libraries of Nonstandard Peptides, *J. Nucleic Acids* **2012**, *2012*, 1-15.

[115] C. T. Walsh, R. V. O'Brien, C. Khosla, Nonproteinogenic Amino Acid Building Blocks for Nonribosomal Peptide and Hybrid Polyketide Scaffolds, *Angew. Chem. Int. Ed.* **2013**, *52*, 7098-7124; *Angew. Chem.* **2013**, *125*, 7238-7265.

[116] W.-H. Lam, K. Rychli, T. D. H. Bugg, Identification of a novel β-replacement reaction in the biosynthesis of 2,3-diaminobutyric acid in peptidylnucleoside mureidomycin A, *Org. Biomol. Chem.* **2008**, *6*, 1912-1917.

[117] K. Tatsuta, N. Mikami, K. Fujimoto, S. Umezawa, H. Umezawa, T. Aoyagi, Structure of Chymostatin, a Chymotrypsin Inhibitor, *J. Antibiot.* **1973**, *26*, 625-646.

[118] A. Okura, H. Miorishima, T. Takita, T. Aoyagi, T. Takeuchi, H. Umezawa, Structure of Elastatinal, an Elastase Inhibitor of Microbial Origin, *J. Antibiot.* **1975**, *28*, 337-339.

[119] H. Yoshioka, T. Aoki, H. Goko, K. Nakatsu, T. Noda, H. Sakakibara, T. Take, A. Nagata, J. Abe, T. Wakamiya, T. Shiba, T. Kaneko, Chemical Studies On Tuberactinomycin. II: Structure Of Tuberactinomycin-O, *Tetrahedron Lett.* **1971**, *12*, 2043-2046.

[120] A. Nagata, T. Ando, R. Izumi, H. Sakakibara, T. Take, Studies on tuberactinomycin (tuberactin), a new antibiotic. I. Taxonomy of producing strain, isolation and characterization, *J. Antibiot.* **1968**, *21*, 681-687.

[121] E. B. Herr, Chemical and Biological Properties of Capreomycin and Other Peptide Antibiotics, *Antimicrob. Agents. Chemother.* **1962**, *1962*, 201-202.

[122] B. W. Bycroft, L. R. Croft, A. W. Johnson, T. Webb, The structure, stereochemistry and reactions of the guanidine moiety of viomycin, *J. Antibiot.* **1969**, *22*, 133-134.

[123] N. Rastogi, V. Labrousse, K. Seng Goh, In Vitro Activities of Fourteen Antimicrobial Agents Against Drug Susceptible and Resistant Clinical Isolates of *Mycobacterium tuberculosis* and Comparative Intracellular Activities Against the Virulent H37Rv Strain in Human Macrophages, *Curr. Microbiol.* **1996**, *33*, 167-175.

[124] H. Wank, J. Rogers, J. Davies, R. Schroeder, Peptide antibiotics of the tuberactinomycin family as inhibitors of group I intron RNA splicing, *J. Mol. Biol.* **1994**, *236*, 1001-1010.

[125] R. E. Stanley, G. Blaha, R. L. Grodzicki, M. D. Strickler, T. A. Steitz, The structures of the anti-tuberculosis antibiotics viomycin and capreomycin bound to the 70S ribosome, *Nat. Struct. Mol. Biol.* **2010**, *17*, 289-293.

[126] M. Yarus, J. Widmann, R. Knight, RNA–Amino Acid Binding: A Stereochemical Era for the Genetic Code, *J. Mol. Evol.* **2009**, *69*, 406-429.

[127] T. Kitagawa, T. Miura, S. Tanaka, H. Taniyama, Relationships between antimicrobial activities and chemical structures of reduced products of viomycin, *J. Antibiot.* **1973**, *26*, 528-531.

[128] S. J. Gould, D. A. Minott, Biosynthesis of capreomycin. 1. Incorporation of arginine, *J. Org. Chem.* **1992**, *57*, 5214-5217.

[129] M. G. Thomas, Y. A. Chan, S. G. Ozanick, Deciphering Tuberactinomycin Biosynthesis: Isolation, Sequencing, and Annotation of the Viomycin Biosynthetic Gene Cluster, *Antimicrob. Agents. Chemother.* **2003**, *47*, 2823-2830.

[130] R. B. Hamed, J. R. Gomez-Castellanos, L. Henry, C. Ducho, M. A. McDonough, C. J. Schofield, The enzymes of β-lactam biosynthesis, *Nat. Prod. Rep.* **2013**, *30*, 21-107.

[131] V. Helmetag, S. A. Samel, M. G. Thomas, M. A. Marahiel, L.-O. Essen, Structural basis for the erythro-stereospecificity of the L-arginine oxygenase VioC in viomycin biosynthesis, *FEBS J.* **2009**, *276*, 3669-3682.

[132] M. Büschleb, M. Granitzka, D. Stalke, C. Ducho, A biomimetic domino reaction for the concise synthesis of capreomycidine and epicapreomycidine, *Amino Acids* **2012**, *43*, 2313-2328.

[133] S. J. Gould, J. Lee, J. Wityak, Biosynthesis of Streptothricin F, 7. The Fate of the Arginine Hydrogens, *Bioorg. Chem.* **1991**, *19*, 333-350.

[134] J. Wityak, S. J. Gould, S. J. Hein, D. A. Keszler, A 1,3-Dipolar Cycloaddition Route to the (3*R*)- and (3*S*)-Hydroxy-(2*S*)-arginines, *J. Org. Chem.* **1987**, *52*, 2179-2183.

[135] A. Lemke, M. Büschleb, C. Ducho, Concise synthesis of both diastereomers of 3-hydroxy-L-arginine, *Tetrahedron* **2010**, *66*, 208-214.

[136] M. D. Jackson, S. J. Gould, T. M. Zabriskie, Studies on the Formation and Incorporation of Streptolidine in the Biosynthesis of the Peptidyl Nucleoside Antibiotic Streptothricin F, *J. Org. Chem.* **2002**, *67*, 2934-2941.

[137] W. Ge, A. Wolf, T. Feng, C.-h. Ho, R. Sekirnik, A. Zayer, N. Granatino, M. E. Cockman, C. Loenarz, N. D. Loik, A. P. Hardy, T. D. W. Claridge, R. B. Hamed, R. Chowdhury, L. Gong, C. V. Robinson, D. C. Trudgian, M. Jiang, M. M.

Mackeen, J. S. McCullagh, Y. Gordiyenko, A. Thalhammer, A. Yamamoto, M. Yang, P. Liu-Yi, Z. Zhang, M. Schmidt-Zachmann, B. M. Kessler, P. J. Ratcliffe, G. M. Preston, M. L. Coleman, C. J. Schofield, Oxygenase-catalyzed ribosome hydroxylation occurs in prokaryotes and humans, *Nat. Chem. Biol.* **2012**, *8*, 960-962.

[138] A. García, M. J. Vázquez, E. Quiñoá, R. Riguera, C. Debitus, New Amino Acid Derivatives from the Marine Ascidian Leptoclinides dubius, *J. Nat. Prod.* **1996**, *59*, 782-785.

[139] E. Higashide, K. Hatano, M. Shibata, K. Nakazawa, Enduracidin, a new antibiotic, *J. Antibiot.* **1968**, *21*, 126-146.

[140] L. Sanière, L. Leman, J.-J. Bourguignon, P. Dauban, R. H. Dodd, Iminoiodane mediated aziridination of α-allylglycine: access to a novel rigid arginine derivative and to the natural amino acid enduracididine, *Tetrahedron* **2004**, *60*, 5889-5897.

[141] H. He, R. T. Williamson, B. Shen, E. I. Graziani, H. Y. Yang, S. M. Sakya, P. J. Petersen, G. T. Carter, Mannopeptimycins, Novel Antibacterial Glycopeptides from *Streptomyces hygroscopicus*, LL-AC98, *J. Am. Chem. Soc.* **2002**, *124*, 9729-9736.

[142] A. Ruzin, G. Singh, A. Severin, Y. Yang, R. G. Dushin, A. G. Sutherland, A. Minnick, M. Greenstein, M. K. May, D. M. Shlaes, P. A. Bradford, Mechanism of Action of the Mannopeptimycins, a Novel Class of Glycopeptide Antibiotics Active against Vancomycin-Resistant Gram-Positive Bacteria, *Antimicrob. Agents Chemother.* **2004**, *48*, 728-738.

[143] D. L. Blanchard, Advanced Studies on Cyclic Amino Acids Involved in GABAergic Neurotransmission and Peptide Antibiotics, Oregon State University, **2008**.

[144] X. Yin, T. M. Zabriskie, The enduracidin biosynthetic gene cluster from *Streptomyces fungicidicus*, *Microbiology* **2006**, *152*, 2969-2983.

[145] N. A. Magarvey, B. Haltli, M. He, M. Greenstein, J. A. Hucul, Biosynthetic Pathway for Mannopeptimycins, Lipoglycopeptide Antibiotics Active against Drug-Resistant Gram-Positive Pathogens, *Antimicrob. Agents Chemother.* **2006**, *50*, 2167-2177.

[146] S. Tsuji, S. Kusumoto, T. Shiba, Synthesis of enduracididine, a component amino acid of antibiotic Enduracucididine, *Chem. Lett.* **1975**, *4*, 1281-1284.

[147] C. J. Schwörer, M. Oberthür, Synthesis of Highly Functionalized Amino Acids: An Expedient Access to L- and D-β-Hydroxyenduracididine Derivatives, *Eur. J. Org. Chem.* **2009**, *2009*, 6129-6139.

[148] F. Sarabia, L. Martín-Ortiz, Synthetic studies on nucleoside-type muraymycins antibiotics based on the use of sulfur ylides. Synthesis of bioactive 5'-epimuraymycin analogues, *Tetrahedron* **2005**, *61*, 11850-11865.

[149] S. Hirano, S. Ichikawa, A. Matsuda, Total Synthesis of Caprazol, a Core Structure of the Caprazamycin Antituberculosis Antibiotics, *Angew. Chem. Int. Ed.* **2005**, *44*, 1854-1856; Angew. Chem. **2005**, *117*, 1888 –1890.

[150] S. Hirano, S. Ichikawa, A. Matsuda, Synthesis of Caprazamycin Analogues and Their Structure-Activity Relationship for Antibacterial Activity, *J. Org. Chem.* **2008**, *73*, 569-577.

[151] T. Tanino, S. Ichikawa, M. Shiro, A. Matsuda, Total Synthesis of (−)-Muraymycin D2 and Its Epimer, *J. Org. Chem.* **2010**, *75*, 1366-1377.

[152] A. P. Spork, Synthetische Untersuchungen zur Nucleosid-Einheit von Muraymycin-Antibiotika, PhD thesis, Georg-August-Universität Göttingen, **2012**.

[153] A. P. Spork, C. Ducho, Novel 5'-deoxy nucleosyl amino acid scaffolds for the synthesis of muraymycin analogues, *Org. Biomol. Chem.* **2010**, *8*, 2323-2326.

[154] A. P. Spork, D. Wiegmann, M. Granitzka, D. Stalke, C. Ducho, Stereoselective Synthesis of Uridine-Derived Nucleosyl Amino Acids, *J. Org. Chem.* **2011**, *76*, 10083-10098.

[155] A. P. Spork, S. Koppermann, C. Ducho, Improved Convergent Synthesis of 5'-epi-Analogues of Muraymycin Nucleoside Antibiotics, *Synlett* **2009**, *2009*, 2503-2507.

[156] A. P. Spork, S. Koppermann, B. Dittrich, R. Herbst-Irmer, C. Ducho, Efficient synthesis of the core structure of muraymycin and caprazamycin nucleoside antibiotics based on a stereochemically revised sulfur ylide reaction, *Tetrahedron Asymmetr.* **2010**, *21*, 763-766.

[157] A. P. Spork, C. Ducho, Stereocontrolled Synthesis of 5'- and 6'-Epimeric Analogues of Muraymycin Nucleoside Antibiotics, *Synlett* **2013**, *24*, 343-346.

[158] G. T. Carter, J. A. Lotvin, L. A. McDonald, Antibiotics AA-896, PCT/US02/13218, *PCT Int. Appl.* **2004**.

[159] A. Dondoni, D. Perrone, Synthesis of 1,1-Dimethyl(S)-4-formyl-2,2-dimethyl-3-oxazolidinecarboxylate by Oxidation of the Alcohol, *Org. Synth.* **2000**, *77*, 64-70.

[160] A. Lemke, Neue Synthesen von 3-Hydroxyarginin für Biosynthese-Studien, diploma thesis, Georg-August-Universität Göttingen, **2009**.

[161] C. Ducho, R. B. Hamed, E. T. Batchelar, J. L. Sorensen, B. Odell, C. J. Schofield, Synthesis of regio- and stereoselectively deuterium-labelled derivatives of L-

glutamate semialdehyde for studies on carbapenem biosynthesis, *Org. Biomol. Chem.* **2009**, *7*, 2770-2779.

[162] T. Masquelin, E. Broger, K. Müller, R. Schmid, D. Obrecht, Synthesis of Enantiomerically Pure D- and L-(Heteroaryl)alanines by asymmetric hydrogenation of (Z)-α-amino-α,β-didehydro esters: eine Veresterungsmethode, *Helv. Chim. Acta* **1994**, *77*, 1395-1411.

[163] J. D. Morrison, R. E. Burnett, A. M. Aguiar, C. J. Morrow, C. Phillips, Asymmetric homogeneous hydrogenation with rhodium(I) complexes of chiral phosphines, *J. Am. Chem. Soc.* **1971**, *93*, 1301-1303.

[164] B. D. Vineyard, W. S. Knowles, M. J. Sabacky, G. L. Bachmann, D. J. Weinkauff, Asymmetric Hydrogenation. Rhodium Chiral Bisphosphine Catalyst, *J. Am. Chem. Soc.* **1977**, *99*, 5946-5952.

[165] M. J. Burk, C2-symetric bis(phospholanes) and their use in highly enantioselective hydrogenation reactions, *J. Am. Chem. Soc.* **1991**, *113*, 8518-8519.

[166] U. Schmidt, A. Lieberknecht, J. Wild, Amino Acids and Peptides; XLIII1. Dehydroamino Acids; XVIII2. Synthesis of Dehydroamino Acids and Amino Acids from *N*-Acyl-2-(dialkyloxyphosphinyl)-glycin Esters; II, *Synthesis* **1984**, 53-60.

[167] U. Zoller, D. Ben-Ishai, Amidoalkylation of mercaptans with glyoxylic acid derivatives, *Tetrahedron* **1975**, *31*, 863-866.

[168] S. D. Debenham, J. Cossrow, E. J. Toone, Synthesis of α- and β-Carbon-Linked Serine Analogues of the Pk Trisaccharide, *J. Org. Chem.* **1999**, *64*, 9153-9163.

[169] M. Büschleb, Synthese von Capreomycidin- und Epicapreomycidin-haltigen Naturstoff-Bausteinen, PhD thesis, Georg-August-Universität Göttingen, **2012**.

[170] C. Douat, A. Heitz, J. Martinez, J.-A. Fehrentz, Stereoselective synthesis of allyl- and homoallylglycines, *Tetrahedron Lett.* **2001**, *42*, 3319-3321.

[171] R. Kaul, S. Surprenant, W. D. Lubell, Systematic Study of the Synthesis of Macrocyclic Dipeptide β-Turn Mimics Possessing 8-, 9-, and 10- Membered Rings by Ring-Closing Metathesis, *J. Org. Chem.* **2005**, *70*, 3838-3844.

[172] H. M. E. Duggan, P. B. Hitchcock, D. W. Young, Synthesis of 5/7-, 5/8- and 5/9-bicyclic lactam templates as constraints for external β-turns, *Org. Biomol. Chem.* **2005**, *3*, 2287-2295.

[173] P. Garner, J. M. Park, Asymmetric synthesis of 5-*O*-carbamoylpolyoxamic acid from D-serine, *J. Org. Chem.* **1988**, *53*, 2979-2984.

[174] N. Micale, A. P. Kozikowski, R. Ettari, S. Grasso, M. Zappalà, J.-J. Jeong, A. Kumar, M. Hanspal, A. H. Chishti, Novel Peptidomimetic Cysteine Protease Inhibitors as Potential Antimalarial Agents, *J. Med. Chem.* **2006**, *49*, 3064-3067.

[175] G. Bold, H. Steiner, L. Moesch, B. Walliser, Herstellung von 'Semialdehyd'-Derivaten von Asparaginsäure- und Glutaminsäure durch Rosenmund-Reduktion, *Helv. Chim. Acta* **1990**, *73*, 405-410.

[176] Y. Jia, J. Zhu, Palladium-Catalyzed, Modular Synthesis of Highly Functionalized Indoles and Tryptophans by Direct Annulation of Substituted *o*-Haloanilines and Aldehydes, *J. Org. Chem.* **2006**, *71*, 7826-7834.

[177] O. Ries, Synthese und Eigenschaften der Lipid-Einheit von Muraymycin-Antibiotika, PhD thesis, Georg-August-Universität Göttingen, **2012**.

[178] G. Papageorgiou, D. Ogden, J. E. T. Corrie, An antenna-sensitised 1-acyl-7-nitroindoline that has good solubility properties in the presence of calcium ions and is suitable for use as a caged L-glutamate in neuroscience, *Photochem. Photobiol. Sci.* **2008**, *7*, 423-432.

[179] T. Kolasa, M. J. Miller, 1-Hydroxy-3-amino-2-piperidone (δ-*N*-hydroxycycloornithine) derivatives: key intermediates for the synthesis of hydroxamate-based siderophores, *J. Org. Chem.* **1990**, *55*, 1711-1721.

[180] R. L. Broadrup, B. Wang, W. P. Malachowski, A general strategy for the synthesis of azapeptidomimetic lactams, *Tetrahedron* **2005**, *61*, 10277-10284.

[181] W. Wang, C. Xiong, J. Yang, V. J. Hruby, An Efficient Synthesis of (2*S*,6*S*)- and *meso*-Diaminopimelic Acids via Asymmetric Hydrogenation, *Synthesis* **2002**, 94-98.

[182] M. Falorni, A. Porcheddu, M. Taddei, Mild reduction of carboxylic acids to alcohols using cyanuric chloride and sodium borohydride, *Tetrahedron Lett.* **1999**, *40*, 4395-4396.

[183] M. Benohoud, L. Leman, S. H. Cardoso, P. Retailleau, P. Dauban, J. Thierry, R. H. Dodd, Total Synthesis and Absolute Configuration of the Natural Amino Acid Tetrahydrolathyrine, *J. Org. Chem.* **2009**, *74*, 5331-5336.

[184] G. S. Nikolova, G. Haufe, Stereoselective synthesis of (2*S*,3*S*,4*Z*)-4-fluoro-1,3-dihydroxy-2-(octadecanoylamino)octadec-4-ene, [(*Z*)-4-fluoroceramide], and its phase behavior at the air/water interface, *Beilstein J. Org. Chem.* **2008**, *4*, 12.

[185] K. Feichtinger, H. L. Sings, T. J. Baker, K. Matthews, M. Goodman, Triurethane-Protected Guanidines and Triflyldiurethane-Protected Guanidines: New Reagents for Guanidinylation Reactions, *J. Org. Chem.* **1998**, *63*, 8432-8439.

[186] D. C. Gowda, K. Abiraj, Heterogeneous catalytic transfer hydrogenation in peptide synthesis, *Lett. Pept. Sci.* **2002**, *9*, 153-165.

[187] G. R. Srinivasa, S. N. N. Babu, C. Lakshmi, D. C. Gowda, Conventional and Microwave Assisted Hydrogenolysis Using Zinc and Ammonium Formate, *Synth. Commun.* **2004**, *34*, 1831-1837.

[188] Z. Gonda, G. L. Tolnai, Z. Novák, Dramatic Impact of ppb Levels of Palladium on the "Copper-Catalyzed" Sonogashira Coupling, *Chem. Eur. J.* **2010**, *16*, 11822-11826.

[189] S. L. Buchwald, C. Bolm, On the Role of Metal Contaminants in Catalyses with FeCl3, *Angew. Chem. Int. Ed.* **2009**, *48*, 5586-5587; *Angew. Chem.* **2009**, *121*, 5694-5695.

[190] S. Hirano, S. Ichikawa, A. Matsuda, Structure-activity relationship of truncated analogs of caprazamycins as potential anti-tuberculosis agents, *Bioorg. Med. Chem.* **2008**, *16*, 5123-5133.

[191] R. Ramapanicker, R. Mishra, S. Chandrasekaran, An improved procedure for the synthesis of dehydroamino acids and dehydropeptides from the carbonate derivatives of serine and threonine using tetrabutylammonium fluoride, *J. Pept. Sci.* **2010**, *16*, 123-125.

[192] T. D. Nelson, R. D. Crouch, Selective Deprotection of Silyl Ethers, *Synthesis* **1996**, *1996*, 1031-1069.

[193] G. Archibald, C.-P. Lin, P. Boyd, D. Barker, V. Caprio, A Divergent Approach to 3-Piperidinols: A Concise Syntheses of (+)-Swainsonine and Access to the 1-Substituted Quinolizidine Skeleton, *J. Org. Chem.* **2012**, *77*, 7968-7980.

[194] U. Jacquemard, V. Bénéteau, M. Lefoix, S. Routier, J.-Y. Mérour, G. Coudert, Mild and selective deprotection of carbamates with Bu4NF, *Tetrahedron* **2004**, *60*, 10039-10047.

[195] D. Fishlock, J. G. Guillemette, G. A. Lajoie, Synthesis of Syn and Anti Isomers of *trans*-Cyclopropyl Arginine, *J. Org. Chem.* **2002**, *67*, 2352-2354.

[196] C. W. Zapf, M. Goodman, Synthesis of 2-Amino-4-pyrimidinones from Resin-Bound Guanidines Prepared Using Bis(allyloxycarbonyl)-Protected Triflylguanidine, *J. Org. Chem.* **2003**, *68*, 10092-10097.

[197] M. Büschleb, Biomimetische Ansätze zur Synthese von Capreomycidin und Epicapreomicidin, diploma thesis, Georg-August-Universität Göttingen, **2008**.

[198] K. Dolbeare, G. F. Pontoriero, S. K. Gupta, R. K. Mishra, R. L. Johnson, Synthesis and Dopamine Receptor Modulating Activity of 3-Substituted γ-Lactam

Peptidomimetics of L-Prolyl-L-leucyl-glycinamide, *J. Med. Chem.* **2003**, *46*, 727-733.

[199] P. Gopinath, T. Watanabe, M. Shibasaki, Studies on Catalytic Enantioselective Total Synthesis of Caprazamycin B: Construction of the Western Zone, *J. Org. Chem.* **2012**, *77*, 9260-9267.

[200] J.-X. Wang, K. Wang, L. Zhao, H. Li, Y. Fu, Y. Hu, Palladium-Catalyzed Stereoselective Synthesis of (*E*)-Stilbenes via Organozinc Reagents and Carbonyl Compounds, *Adv. Synth. Catal.* **2006**, *348*, 1262-1270.

[201] Z.-Y. Peng, F.-F. Ma, L.-F. Zhu, X.-M. Xie, Z. Zhang, Lewis Acid Promoted Carbon−Carbon Double-Bond Formation via Organozinc Reagents and Carbonyl Compounds, *J. Org. Chem.* **2009**, *74*, 6855-6858.

[202] D. A. Evans, P. H. Carter, C. J. Dinsmore, J. C. Barrow, J. L. Katz, D. W. Kung, Mild nitrosation and hydrolysis of polyfunctional amides, *Tetrahedron Lett.* **1997**, *38*, 4535-4538.

[203] F. Saczewski†, Ł. Balewski, Biological activities of guanidine compounds, *Expert Opin. Ther. Pat.* **2009**, *19*, 1417-1448.

[204] I. Rozas, G. Sánchez-Sanz, I. Alkorta, J. Elguero, Solvent effects on guanidinium-anion interactions and the problem of guanidinium Y-aromaticity, *J. Phys. Org. Chem.* **2013**, *26*, 378-385.

[205] T. Ishikawa, Guanidine Chemistry, *Chem. Pharm. Bull.* **2010**, *58*, 1555-1564.

[206] M. P. Coles, Application of neutral amidines and guanidines in coordination chemistry, *Dalton Trans.* **2006**, 985-1001.

[207] M. Reinmuth, C. Neuhäuser, P. Walter, M. Enders, E. Kaifer, H.-J. Himmel, The Flexible Coordination Modes of Guanidine Ligands in Zn Alkyl and Halide Complexes: Chances for Catalysis, *Eur. J. Inorg. Chem.* **2011**, *2011*, 83-90.

[208] S. Li, Y. Lin, H. Xie, S. Zhang, J. Xu, Brønsted Guanidine Acid−Base Ionic Liquids: Novel Reaction Media for the Palladium-Catalyzed Heck Reaction, *Org. Lett.* **2006**, *8*, 391-394.

[209] J. Shao, W. Chen, M. A. Giulianotti, R. A. Houghten, Y. Yu, Palladium-Catalyzed C–H Functionalization Using Guanidine as a Directing Group: Ortho Arylation and Olefination of Arylguanidines, *Org. Lett.* **2012**, *14*, 5452-5455.

[210] Y. Coquerel, J. Rodriguez, Catalytic properties of the Pd/C – triethylamine system, *Arkivoc* **2008**, *xi*, 227-237.

[211] D. Heinrich, U. Diederichsen, M. G. Rudolph, Lys314 is a Nucleophile in Non-Classical Reactions of Orotidine-5′-Monophosphate Decarboxylase, *Chem. Eur. J.* **2009**, *15*, 6619-6625.

[212] D. Wiegmann, Synthetische Untersuchungen zur Nucleosid-Einheit von Muraymycin-Antibiotika und ihren Analoga, master thesis, Georg-August-Universität Göttingen, **2012**.

[213] P. A. Byrne, K. V. Rajendran, J. Muldoon, D. G. Gilheany, A convenient and mild chromatography-free method for the purification of the products of Wittig and Appel reactions, *Org. Biomol. Chem.* **2012**, *10*, 3531-3537.

[214] M. J. Mintz, C. Walling, *t*-Butyl Hypochlorid, *Org. Synth.* **1969**, *49*, 9-12.

[215] S. Hirano, S. Ichikawa, A. Matsuda, Supporting Info: Total Synthesis of Caprazol, a Core Structure of the Caprazamycin Antituberculosis Antibiotics, *Angew. Chem. Int. Ed.* **2005**, *44*, 1854-1856; Angew. Chem. **2005**, *117*, 1888-1890.

[216] B. B. Lohray, V. Bhushan, G. J. Reddy, A. S. Reddy, Mechanistic investigation of asymmetric aminohydroxylation of alkenes, *Indian J. Chem, Sect. B* **2002**, *41*, 161-168.

[217] J. A. Bodkin, M. D. McLeod, The Sharpless asymmteric aminohydroxylation, *J. Chem. Soc., Perkin Trans. 1* **2002**, 2733-2746.

[218] H. Han, C.-W. Cho, K. D. Janda, A Substrate-Based Methodology That Allows the Regioselective Control of the Catalytic Aminohydroxylation Reaction, *Chem. Eur. J.* **1999**, *5*, 1565-1569.

[219] A. J. Morgan, C. E. Masse, J. S. Panek, Reversal of Regioselection in the Sharpless Asymmetric Aminohydroxylation of Aryl Ester Substrates, *Org. Lett.* **1999**, *1*, 1949-1952.

[220] U. Schmidt, H. Griesser, V. Leitenberger, A. Lieberknecht, R. Mangold, R. Meyer, B. Riedl, Diastereoselective Formation of (*Z*)-Didehydroamino acid esters, *Synthesis* **1992**, *1992*, 487-490.

[221] R. Mazurkiewicz, A. Kuznik, M. Grymel, N. Kuznik, ^{1}H-NMR spectroscopic criteria for the configuration of N-acyl-α,β-dehydro-α-amino acid esters, *Magn. Reson. Chem.* **2005**, *43*, 36-40.

[222] R. J. Cox, P. S. H. Wang, Synthesis and in vitro enzyme activity of aza, oxa and thia derivatives of bacterial cell wall biosynthesis intermediates, *J. Chem. Soc., Perkin Trans. 1* **2001**, 2022-2034.

[223] M. E. Swarbrick, F. Gosselin, W. D. Lubell, Alkyl Substituent Effects on Pipecolyl Amide Isomer Equilibrium: Efficient Methodology for Synthesizing Enantiopure

6-Alkylpipecolic Acids and Conformational Analysis of Their *N*-Acetyl *N*ᶜ-Methylamides, *J. Org. Chem.* **1999**, *64*, 1993-2002.

[224] M. Miyano, D. Ito, N. Murai, Twelve-membered cyclomacrolactam derivative, WO2008/126918, *PCT Int. Appl.* **2008**.

[225] H. C. Kolb, M. S. VanNieuwenhze, K. B. Sharpless, Catalytic Asymmetric Dihydroxylation, *Chem. Rev.* **1994**, *94*, 2483-2547.

[226] J. B. Epp, T. S. Widlanski, Facile Preparation of Nucleoside-5'-carboxylic Acids, *J. Org. Chem.* **1999**, *64*, 293-295.

[227] P. S. Baran, R. A. Shenvi, Total Synthesis of (±)-Chartelline C, *J. Am. Chem. Soc.* **2006**, *128*, 14028-14029.

[228] Q. Wang, R. J. Linhardt, Synthesis of a Serine-Based Neuraminic Acid C-Glycoside, *J. Org. Chem.* **2003**, *68*, 2668-2672.

A   Appendix

A.1   Abbreviations

2-OG	2-oxogluterate
A-domain	adenylation domain (NRPS)
Ac	acetyl
AD	asymmetric dihydroxylation
Ala	alanine
Alloc	allyloxycarbonyl
Asp	aspartic acid
ATP	adenosine triphosphate
BAIB	[bis(acetoxy)iodo]benzene
BBC	British Broadcasting Corporation
B.C.	before Christ
Bn	benzyl
Boc	*tert*-butoxycarbonyl
Bu	butyl
CBS	Columbia Broadcasting System
Cbz	benzyloxycarbonyl
CC	complete conversion
C-domain	condensation domain (NRPS)
CL	cytoplasmic loop
CoA	coenzyme A
Cod	cyclooctadiene
COSY	correlation spectroscopy (NMR)
Cp_2ZrHCl	bis(cyclopentadienyl)zirconium(IV) chloride hydride
CR	carbohydrate recognition
CSL	clavaminic acid synthase-like
δ	chemical shift (NMR)
d	doublet (NMR) or day(s)
DABA	2,4-diaminobutyric acid
DAD	diode array detector
DART	Deutsche Antibiotika-Resistenzstrategie
DCC	*N,N'*-dicyclohexylcarbodiimide
DIAD	diisopropyl azodicarboxylate
DIBAL-H	diisobutylaluminium hydride
Diboc	di-*tert*-butoxycarbonyl

DIC	*N,N'*-diisopropylcarbodiimide
diff.	difficult
[DHQD]$_2$AQN	hydroquinidine (anthraquinone-1,4-diyl) diether
[DHQD]$_2$PHAL	hydroquinidine (phthalazine-1,4-diyl) diether
DMAP	4-dimethylaminopyridine
DME	1,2-dimethoxyethane
DMF	*N,N*-dimethyl formamide
DMHH	*N,O*-dimethylhydroxylamine hydrochloride
DMSO	dimethyl sulfoxide
DNA	deoxyribonucleic acid
€	euro(s)
ε	molar extinction coefficient (UV)
EDC	1-ethyl-3-(3-dimethylaminopropyl)carbodiimide hydrochloride
ESI	electrospray ionization (MS)
Et	ethyl
et al.	and others
eq.	equivalents
equilib.	equilibrated
FDA	Food and Drug Administration
FT	Fourier transformation
Glc	glucose
Gln	glutamine
Glu	glutamic acid
h	hour(s)
HOBt	hydroxybenzotriazole
HMBC	heteronuclear multiple bond coherence
HPLC	high performance liquid chromatography
HRMS	high-resolution mass spectrometry
HSQC	heteronuclear single quantum coherence (NMR)
Hz	Hertz
i	*iso*
IBX	2-iodoxybenzoic acid
inoc.	inoculation
IR	infrared
J	scalar coupling constante (NMR)
KHMDS	potassium bis(trimethylsilyl)amide
LC-MS	liquid chromatography – mass spectrometry
λ_{max}	wavelength [nm] (UV)
M	molar

m	multiplet
min	minute(s)
Me	methyl
MF	membrane filter
MIC	minimum inhibitory concentration
mp	melting point
*m*RNA	messenger RNA
MRSA	methicillin-resistant *Staphylococcus aureus*
Ms	methanesulfonyl
MS	mass spectrometry
MS/MS	tandem mass spectrometry
MTA	*S*-methyl-5'-thioadenosine
Mur	muramic acid
m/z	mass to charge ratio (MS)
$\tilde{\nu}$	wavenumber $[\mathrm{cm}]^{-1}$ (IR)
NDP	nucleoside diphosphate
NMM	*N*-methylmorpholine
NMR	nuclear magnetic resonance
No.	number
nOe	nuclear Overhauser effect
NOESY	nuclear Overhauser effect spectroscopy
NP	normal phase
NRPS	nonribosomal peptide-synthetase
NRRL	Northern Regional Research Laboratory
NTP	nucleoside triphosphate
orf	open reading frame
p	pressure
Pbf	2,2,4,6,7-pentamethyl dihydrobenzofuran-5-sulfonyl
PCP	peptidyl carrier protein
Ph	phenyl
PLP	pyridoxal phosphate
Ppant	4'-phosphopantetheine
P_i	phosphate
PP_i	pyrophosphate
ppm	parts per million (NMR)
Pr	propyl
py	pyridine
PyBOB	(benzotriazol-1-yloxy)tripyrrolidinophosphonium hexafluorophosphate

q	quartet (NMR)
quant.	quantitative
quin	quintet (NMR)
rac	racemic
R-domain	reductase domain
R_f	retention factor (TLC)
RNA	ribonucleic acid
RP	reverse phase
(*R,R*)-Me-DUPHOS	(-)-1,2-bis[(2*R*,5*R*)-2,5-dimethylpholano]benzene(1,5-cyclooctadiene)rhodium(I) tetrafluoroborate
rt	room temperature
s	singlet (NMR)
SAM	*S*-adenosyl methionine
SAR	structure activity relationship
s_{br}	broad singlet (NMR)
Ses	2-(trimethylsilyl)ethane sulfonyl
SHMT	serine hydroxymethyltransferase
S_N2	second-order nucleophilic substitution
SPE	solid phase extraction
(*S,S*)-Me-DUPHOS	(+)-1,2-Bis[(2*S*,5*S*)-2,5-dimethylpholano]benzene(1,5-cyclooctadiene)rhodium(I) tetrafluoroborate
t	*tert*
t	triplet (NMR) or time
T	temperature
TBAF	tetra-*n*-butylammonium fluoride
TBAI	tetra-*n*-butylammonium iodide
TBDMS	*tert*-butyldimethylsilyl
TBDPS	*tert*-butyldiphenylsilyl
td	triplet of doublet (NMR)
T-domain	thiolation domain (NRPS)
TE-domain	thioesterase domain (NRPS)
TEMPO	2,2,6,6-tetramethylpiperidine 1-oxyl
Tf	trifluoromethanesulfonate
TFA	trifluoroacetic acid
THF	tetrahydrofuran
TLC	thin layer chromatography
TMP	thymidine monophosphate
TMS	trimethylsilyl
TMSE	2-(trimethylsilyl)ethyl

TOF	time of flight
t_R	retention time (HPLC)
tRNA	transfer RNA
TSBG	tryptic soy broth with additional glucose
tt	triplet of triplet (NMR)
UDP	uridine diphosphate
UMP	uridine monophosphate
UV	ultraviolet
US $	United States Dollar
USA	United States of America
VRE	vancomycin-resistant *Enterococcus faecium*
wt%	weight percent
XDR	extensively drug-resistant

A.2 Acknowledgements

Mein besonderer Dank gilt Prof. Dr. Christian Ducho für die interessante Aufgabenstellung, seine stete Diskussions- und Hilfsbereitschaft sowie die sehr angenehme Atmosphäre in seinem Arbeitskreis.

Zudem danke ich Prof. Dr. Ulf Diederichsen für die Übernahme des Zweitgutachtens dieser Arbeit.

Für die Teilnahme in der Prüfungskomission gilt mein Dank zusätzlich Prof. Dr. Daniel B. Werz, Prof. Dr. Dietmar Stalke, Prof. Dr. Claudia Steinem und Dr. Inke Siewert.

Beim CaSuS-Programm möchte ich mich nicht nur für die finanzielle Unterstützung in Form eines Lichtenberg-Stipendiums bedanken, sondern auch für die Möglichkeit, an vielen interessanten, interdisziplinären Veranstaltungen und Workshops teilnehmen zu können. Hanna Steininger danke ich dabei für die mühevolle Koordination und Hilfe bei allen verwaltungstechnischen Fragen und den anderen CaSuS-Doktoranden für die immer gute Atmosphäre, vielen Diskussionen und netten Abende.

Meinen Bachelor-Studenten Kim, Florian und Marta danke ich für ihre engagierte Arbeit zur Deuterierung von Uridin und zur Synthese von Enduracididin und die gute Arbeitsatmosphäre.

Für das gründliche und schnelle Korrekturlesen, auch über die Ländergrenzen hinaus, danke ich Daniel, Olli und insbesondere meiner Schwester Maike.

Der NMR-Abteilung um Herrn Machinek in Göttingen und der NMR-Abteilung um Herrn Egold in Paderborn danke ich für die stets schnellen und zuverlässigen Messungen unzähliger Spektren. Bei der Massenabteilung um Herrn Dr. Frauendorf in Göttingen und bei der Massenabteilung um Herrn Dr. Weber in Paderborn möchte ich mich für die Messung der Massenspektren bedanken und die Möglichkeit eigenständig LC-MS-Messungen durchführen zu dürfen. Frau Pfeil danke ich für die Messung einiger IR- und UV-Spektren. Frau Lissy danke ich für ihre Hilfe bei jeglichen Fragen zur Durchführung von Fermentationsprozessen.

Bei den Mitarbeitern meiner Arbeitsgruppe Stephi, Olli, Ole, Daniel, Martin, Philipp, Marius, Kristin, Anatol und Boris möchte ich mich für die gute Atmosphäre, ihre Hilfsbereitschaft bei allen auftretenden Problemen und Fragen sowie natürlich auch für die netten Kaffee-, Eis- oder Frisbeepausen und unvergesslichen AK-Fahrten bedanken. Mein besonderer Dank gilt dabei Stephi, die nicht nur den Umzug nach Paderborn zu etwas Besonderem gemacht hat, sondern mir vor allem auch in der Endphase der Promotion immer aufmunternd und mit einem offenem Ohr für alle Probleme zur Seite stand. Ole danke ich außerdem für die stets zuverlässige und schnelle Hilfe bei allen großen und kleinen Computernotfällen und Daniel, Stephi und Kristin trotz unterschiedlicher Musikgeschmäcker für den großen Spaß im Labor.

Meinen Mädels Julia, Nora G., Toni, Nora H. und Henrike danke ich für die immer lustigen und intensiven Mädels-Wochenenden, die auch in Zukunft einen festen Platz in meiner jährlichen Urlaubsplanung haben werden. Dem Rest der chaotischen Göttinger-Reisegruppe danke ich für eine unvergessliche Studienzeit und dafür, dass ich immer wieder mit offenen Armen und einen Schlafplätzchen in Göttingen willkommen war. Den Paderbornern danke ich dafür, dass sie mir in kürzester Zeit das Gefühl gegeben haben, hier zu Hause zu sein. Es waren zwei wundervolle Jahre hier, wozu nicht zuletzt auch meine fantastischen WG-Mädels Tabea, Stephi, Laura und Marie beigetragen haben.

Zuletzt möchte ich mich ganz besonders bei meiner Familie bedanken. Danke, dass ihr mich in allem unterstützt habt und immer für mich da seid.